UNIVERSITY OF STRATHCLYDE
30125 004901
KV-119-858

ANDERSONIAN LIBRARY
WITHDRAWN
FROM
LIBRARY
STOCK
UNIVERSITY OF STRATHCLYDE

Understanding Natural Flavors

Understanding Natural Flavors

Edited by

J.R. PIGGOTT and A. PATERSON
Department of Bioscience and Biotechnology
University of Strathclyde
Glasgow

BLACKIE ACADEMIC & PROFESSIONAL
An Imprint of Chapman & Hall
London · Glasgow · Weinheim · New York · Tokyo · Melbourne · Madras

Published by
Blackie Academic and Professional, an imprint of Chapman & Hall, Wester Cleddens Road, Bishopbriggs, Glasgow G64 2NZ

Chapman & Hall, 2–6 Boundary Row, London SE1 8HN, UK

Blackie Academic & Professional, Wester Cleddens Road, Bishopbriggs, Glasgow G64 2NZ, UK

Chapman & Hall GmbH, Pappelallee 3, 69469 Weinheim, Germany

Chapman & Hall Inc., One Penn Plaza, 41st Floor, New York NY 10119, USA

Chapman & Hall Japan, Thomson Publishing Japan, Hirakawacho Nemoto Building, 6F, 1–7–11 Hirakawa-cho, Chiyoda-ku, Tokyo 102, Japan

DA Book (Aust.) Pty Ltd, 648 Whitehorse Road, Mitcham 3132, Victoria, Australia

Chapman & Hall India, R. Seshadri, 32 Second Main Road, CIT East, Madras 600 035, India

First edition 1994

Typeset in 10/12 pt Times by Photoprint, Torquay, Devon
Printed in Great Britain by T.J. Press (Padstow) Ltd., Padstow, Cornwall

ISBN 0 7514 0180 3

A catalogue record for this book is available from the British Library
Library of Congress Catalog Card Number: 94–70037

Printed on acid-free text paper, manufactured in accordance with ANSI/NISO Z39.48–1992 (Permanence of Paper)

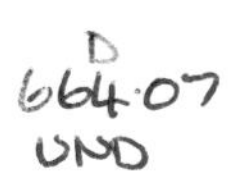

Preface

There has been increasing interest in recent years in the concept and production of natural foods. Advertising claims that food is natural, without additives or artificial ingredients, have taken on great importance in marketing. Consumption of food that can be considered natural is currently central to the sophisticated lifestyle. However, there is only a limited published literature on what constitutes natural food flavours. Much of the flavour and fragrance industry has worked on development of synthetic or 'nature-identical' flavours which represent a chemist's simulation of the natural character. As marketing claims become more strident it is necessary to gain a better understanding of natural food flavours in order to safeguard food quality and for prevention of fraud. There have been great advances recently in analytical chemistry, and partly as a result of this progress there seems to be a never-ending increase in the number of volatile compounds identified in foods. Unfortunately, this has not always been matched by an equal increase in the understanding of how these volatile compounds arise, or how they contribute to the sensation which we call flavour. Throughout the development of Western society, quality of food, particularly flavour, has been highly regarded. The amateur or professional cook with the skills to optimize and maintain standards in flavour has been held in the highest respect. As food supply exceeds the level of demand, it becomes important for producers to generate foods that have natural, or indeed processed, flavours that match consumer expectations. This volume has been compiled with the aim of reviewing the present state of knowledge of the origins and perception of natural and processed food flavours, omitting the lists of volatiles that do not assist understanding.

Flavour is thus not the volatiles in a food; it is an interaction of the components in the aroma headspace above the food and the consumer. Any study of flavour must therefore consider the observer, or consumer, and his or her perception. The first section of the book addresses this and related issues – the importance of flavour and its effects on food acceptability, how to measure flavour, and how to ensure that the flavour presented in a food is the optimum for the consumer. Such factors are of the greatest importance to the producer since they relate to his or her success in the marketplace. The core chapters discuss the origins of flavours in a range of foods. Rather than provide a catalogue of food types and volatiles, we asked contributors to help us to build a framework, describing the sources of the important components of flavour and how and why flavour varies in a particular food. Some foods are traditionally subjected to substantial processing, and so examples of these have been included. It was considered that the balance between the human assessment of flavour, and food origins of flavour, is likely to provide a major field for researchers in the coming years. The methods of biotechnology provide an exciting opportunity for future development of improved flavours in natural foods, and so progress in this area is also recorded. Although traditional plant breeding has made major contributions to increasing yield and production in crops, it is often considered to be the central reason why our food fails to meet expectations of flavour laid down in our memory in early years. By combining an understanding of mechanisms of human flavour assessment and more sophisticated strategies in applied molecular biology and biotechnology, it should become possible to dissect complex flavours. A foundation of knowledge of fundamental science will enable crop specialists to gain greater insights into flavour and quality factors. This should enhance quality of life for all and restore the reputation of science in food.

To the extent that we have been successful in achieving our aims, we must thank the contributors for the excellent job they have done. Where we have been unsuccessful, we must bear the responsibility. We also wish to acknowledge editorial assistance from Allan James, and the publishers for their continuing encouragement and patience.

JRP
AP

Contributors

B.D. Baigrie Reading Scientific Services Ltd., Whiteknights, PO Box 234, Reading RG6 2LA, UK

J. Bakker Institute of Food Research, Earley Gate, Whiteknights Road, Reading, Berkshire RG6 2EF, UK

F. van den Berg TNO Biotechnology and Chemistry Institute, PO Box 360, 3700 AJ Zeist, The Netherlands

R.G. Berger Institut für Lebensmittelchemie, Universität Hannover, Wunstorfer Strasse 14, D-30453 Hannover, Germany

M. Bertuccioli Dipartimento di biologia difesa e biotecnologie agro forestali, Università degli studi della Basilicata, Via N. Sauro 85, 85100 Potenza, Italy

J. Bricout Centre de Recherche Pernod Ricard, 120 av. du Maréchal Foch, 94015 Créteil Cedex, France

P. Brunerie Centre de Recherche Pernod Ricard, 120 av. du Maréchal Foch, 94015 Créteil Cedex, France

K. Drew School of Leisure and Food Management, Sheffield Hallam University, Pond Street, Sheffield S1 1WB, UK

P. Dürr Swiss Federal Research Station for Fruit-Growing, Viticulture and Horticulture, CH-8820 Wädenswil, Switzerland

C. Eriksson Royal Veterinary and Agricultural University of Denmark, Food Science and Technology Center, Thorvaldsensvej 40, DK-1871 Frederiksberg C, Denmark

F. Gmitter University of Florida, IFAS, Citrus Research and Education Center, Lake Alfred, FL 33850, USA

J. Grosser University of Florida, IFAS, Citrus Research and Education Center, Lake Alfred, FL 33850, USA

C. Javelot Centre de Recherche Pernod Ricard, 120 av. du Maréchal Foch, 94015 Créteil Cedex, France

J. Koziet Centre de Recherche Pernod Ricard, 120 av. du Maréchal Foch, 94015 Créteil Cedex, France

D.G. Land Taint Analysis and Sensory Quality Services, 8 High Bungay Road, Loddon, Norwich NR14 6JT, UK

B.A. Law Institute of Food Research, Earley Gate, Whiteknights Road, Reading, Berkshire RG6 2EF, UK

H. Maarse TNO Biotechnology and Chemistry Institute, PO Box 360, 3700 AJ Zeist, The Netherlands

J. du Manoir Centre de Recherche Pernod Ricard, 120 av. du Maréchal Foch, 94015 Créteil Cedex, France

H. Martens Consensus Analysis A/S, Ski Business Park, N-1400 Ski, Norway

M. Martens Techné as Gamlevn. 13, N-1430 Ås, Norway

D.S. Mottram University of Reading, Department of Food Science and Technology, Whiteknights, Reading RG6 2AP, UK

A.C. Noble Department of Viticulture and Enology, University of California, Davis, CA 95616, USA

A. Paterson Department of Bioscience and Biotechnology, University of Strathclyde, Glasgow G1 1SD, UK

J.R. Piggott Department of Bioscience and Biotechnology, University of Strathclyde, Glasgow G1 1SD, UK

P.H. Punter Oliemans Punter & Partners, PO Box 14167, NL-3508 SG Utrecht, The Netherlands

E. Risvik MATFORSK, Norwegian Food Research Institute, Osloveien 1, N-1430 Ås, Norway

I. Rosi Dipartimento di biologia difesa e biotecnologie agro forestali, Università degli studi della Basilicata, Via N. Sauro 85, 85100 Potenza, Italy

R. Rouseff University of Florida, IFAS, Citrus Research and Education Center, 700 Experiment Station Road, Lake Alfred, FL 33850–2299, USA

H.G. Schutz Professor Emeritus, Food Science and Technology, University of California, Davis, CA 95616, USA

S. van Toller Department of Psychology, University of Warwick, Coventry CV4 7AL, UK

L.C. Verhagen Heineken Technisch Beheer B.V., PO Box 510, 2380 BB Zoeterwoude, The Netherlands

A.A. Williams Sensory Research Laboratories Ltd, 4 High Street, Nailsea, Bristol BS19 1BW, UK

Contents

1 Predicting acceptability from flavour data

H.G. SCHUTZ

Abstract

The paper presents the design, procedural and analysis considerations in the conduct of studies that predict acceptability from flavour data. The primary applications of such information are discussed. Examples are given of the multiple regression procedure of prediction for carrot texture, rice flavour and strawberry jam flavour, which illustrate the factors to be considered in such studies.

1.1 Introduction

The prediction of acceptability from flavour data, whether in the form of sensory and/or analytical information, is a very appealing and at the same time challenging task. Historically the methodology for the three aspects – sensory, analytical and acceptance – have developed independently from one another. In the sensory description of food products there has been an evolvement of methods for describing the flavour of food products from relatively unreliable qualitative descriptions to highly reliable profiling and quantitative descriptive techniques. The same can be said for analytical measures where simple techniques of measurement have been supplemented by more complex and sophisticated methods. The area of acceptance measurement as well has progressed from rather rudimentary indications of like and dislike to techniques that take into account the contextual aspects of acceptance, as well as specific target populations. Although progress in the development of methodology in these three areas has produced a greater understanding of food and flavour, the critical issue of relating sensory and/or analytical data to acceptability has only become an area of dedicated study in the last 15 years.

What has been missing from the independent approaches is essential information that allows one to determine the relative importance of the sensory and/or analytical flavour characteristics to food acceptability. It is quite possible that those characteristics, which are of low intensity, make important contributions to the liking of a product, whereas those attributes that may be quite evident are of little importance. Unless this type of information is available, one cannot be expected to make intelligent decisions with regard to a variety of flavour and acceptance issues.

The objectives for defining this relationship fall into four general areas. The first is theoretical, that is, an interest in determining a fundamental relationship between some underlying sensory or analytical property of flavour and the liking for a particular product. Typically in theoretical relationships there is a concentration on the sensory or analytical characteristic itself rather than an interest in the product *per se*. For example, one might be interested in studying the relationship of sweetness as represented by sucrose, or bitterness by quinine, and acceptance for various intensities of these materials in a medium such as distilled water. Included in this type of objective is research relating individual sensory-preference determinations to nutritional objectives, such as reducing salt in foods [1,2]. A second is for agricultural decision-making. Here the interest is in relating various flavour properties that are the result of differing agricultural practices, including pre- and post-harvest activities, to the acceptance of the resulting food products. This information can help in deciding which of these practices are likely to lead to greater acceptance of the food involved. Third, is in the development of new product-improvement guidelines. By understanding the relationship between sensory and/or analytical flavour data and acceptance, one can offer food-product development scientists guidelines as to which property should be emphasized when making product-development decisions. This decision process includes processing, ingredient and economic considerations. Related to this objective is the growing interest in the area of product optimization in which the various sensory and/or analytical characteristics of a product can be combined in such a manner so as to yield products that have higher acceptability than those presently available in the marketplace. A fourth goal is in quality assurance and control. Data obtained on the relative importance of various sensory and/or analytical flavour data to acceptance can provide guidelines in the development of quality control targets and tolerance bands. Depending on which of these objectives or combination of objectives are desired, different design characteristics would be selected.

1.1.1 Procedure considerations

Before discussing the specific factors considered in the design of studies, which relate sensory and/or analytical flavour properties to acceptance, it is appropriate to discuss in more detail some of the measurement problems and characteristics associated with each measure. When we consider sensory characteristics it is fairly easy to obtain agreement about the basic tastes of sweet, salt, sour, bitter. However, we are less comfortable with aspects of odour, texture and appearance, where there are no universally accepted classification systems that can be used in the development of quantitative descriptions of the flavour of food products. Specific terms have been developed for particular products in order to more accurately describe their flavour by using procedures involving trained judges. These terms can inclusively and reliably reflect all aspects of the flavour of a product.

As mentioned earlier, analytical variables can be very simple in construct, such as pH or viscosity measures, or very sophisticated as with chromatographic or mass spectrometric data. Reliability of such methods and sensitivity to differences in products has improved immensely over time. However, this has not always added clarity to the understanding of the role of flavour in acceptability, due in part to the overwhelming number of measures that the various instrumental techniques are now capable of producing. Relating these measures to sensory data is not the purpose of this discussion as this will be discussed in Chapter 5.

When it comes to acceptance methodology, one enters a much less clear area [3]. Simple measures of degree of liking or hedonic value have obvious value, but it is well known that they vary in validity because of both measurement procedures as well as the type of respondent used. When one examines the literature, the terms of preference, hedonic liking, acceptance and quality have been used in many cases to describe what is the same phenomenon. Except for quality, which surely must be relegated to judgements that involve some type of standard, the other terms all relate to one type or another of effective judgement. Preference had an earlier specific meaning of 'prefering one item over another', which is now practically synonymous with hedonic value, the absolute degree of liking or disliking of a product. Whereas preference and hedonic value have been measured relatively context free, measures of acceptance have in many cases involved determinations that include aspects of the context of consumption, such as the time of day, meal occasion, etc., and even actual consumption of a food product. Naturally the choice of a particular measure of effect depends on the nature of the problem and objectives of a particular study. In addition to the considerations just described, to make matters even more complex, the acceptance of an item in the marketplace is influenced not only by the sensory and/or analytical characteristics of flavour but by such factors as packaging, nutrition, price, brand and advertising.

1.1.2 Design considerations

When examining research study designs involving the sensory/analytical prediction of acceptance, one finds a wide range of methods, with differing numbers of products, respondents, attributes, methods of measuring acceptance, as well as the statistical procedure used in the determination of relationships. In the development of designs one must consider aspects of reliability, internal and external validity in relation to the objectives of the study.

Let us examine some of the relevant concepts so as to be able to understand the advantages, disadvantages and appropriateness of various procedures used in this research area. The respondents used for acceptability testing could either be trained panellists or a consumer population. A trained panel could have high internal validity but not have the external validity that would be obtained with a representative sample of consumers from a selected population. The product items could range from model systems [4], which are based on a combination of attributes in food-like materials, to actual products obtained in the marketplace. The measurement of sensory characteristics can involve various degrees of sophistication with highly trained judges using standard term-development and testing procedures. If term-development is not thorough, the data produced can be unreliable and not include all relevant characteristics. For analytical measures the number and type chosen depends in part on resources available, the typical measurements made for that product class and the instrumentation available.

An overarching characteristic in guiding the selection of products, sensory measurement techniques and analytical variables is the statistical approach chosen. The two major approaches are response surface methodology (RSM) [5] and multiple regression (MR) [6–8]. The RSM methodology approach is usually used when the variables important to acceptability are few, known and can be systematically varied. The MR approach is more appropriate when there are large numbers of variables, where relative importance is not known and where systematic variation is not feasible. Both these approaches usually use trained judges to evaluate the sensory characteristics and untrained consumers to evaluate the acceptability dimension. However, a variant of both these approaches uses consumers alone in generating the relationships between underlying sensory attributes and preference [9,10]. The major advantage to this procedure is simplicity of design, the major disadvantage is the limitations of the consumer in making skilled descriptive judgement and the possible difficulty in communicating 'consumer words' to food technologists.

1.1.3 Analysis considerations

For bivariate relationships the preferred method of analysis is a product moment correlation coefficient. However, if a scatterplot between the sensory or analytical variable and preference indicates any sign of curvilinearity, a suitable curvilinear coefficient should be computed. For the types of designs involving more than one predictor variable, some type of multivariate statistical technique is necessary.

One of the characteristics of the MR approach is enigmatic to the researcher, in that the large number of variables, both sensory and/or analytical, leads to high levels of redundancy and possibly to multicollinearity. This can produce aberrations in the interpretation of the results of the MR procedures used to analyse such studies. The RSM, since it uses just a few variables, usually does not yield this problem. In order to be able to handle efficiently the data produced in the MR approach, one can reduce the redundancy through principle component analysis (PCA) and thus yield a smaller orthogonally related set of sensory or analytical variables.

Statistically the MR analysis used to relate the predictor variables with preference in the regression approach is a linear additive model, whereas in RSM a complete polynomial is determined. However, if appropriate, terms involving curvilinear and interaction characteristics can be included in the MR analysis.

Another elegant approach to the analysis of multivariate data, which combines principal components with regression analysis, is partial least squares, which can produce a model relating sensory/analytical and preference data [11].

1.1.4 Practical considerations

Some considerations in the application of these techniques, that can be crucial in developing meaningful results, concern themselves with procedures of testing and selection of products. In the area of the measurement of sensory attributes, it is clear that in the RSM technique one must be confident of the importance of the variables to preference and in their ability to be manipulated in order to have a successful RSM analysis. In the MR approach it is essential that in developing the attribute terminology every attempt is made to cover all of the possible attributes that might be used to describe a particular product class. Certainly the training of individuals in the use of the descriptive technique is an essential component in the successful application of such a method.

One especially difficult practical problem is the comparability of samples used in the various types of measurement stages. It is obvious, but not always easy, to ensure that the products tested in the sensory and/or analytical phases of the study are identical to the ones that the consumers evaluate in the preference phase of the study.

Another source of external validity problems, not so much in the area of theoretical relationships but in the product development area, is the way in which products are evaluated. Simple model systems or aqueous solutions are relatively easy to control and the use conditions are typically quite simple. However, when one uses real foods, the simplicity rapidly disappears. One realizes that in order to represent the sensory properties and preferences in a way that would realistically predict consumers' liking for a product, it becomes quite difficult. For example, products in which there are multiple-use possibilities (e.g. vegetable oil) could either have sensory characteristics that are not present under other conditions and/or change the relative importance of these characteristics. This could lead to a relationship that has high internal validity but would not predict preference in a real-life consumption situation. This challenges the ingenuity of the researcher to evaluate the products under conditions that are representative of varied consumption conditions.

In all of these types of studies one must keep in mind the cultural *milieu* in which they are embedded. The sensory attributes and analytical variables themselves may be invariant. This does not mean new attributes may not appear in products that are not presently available. More importantly, the preference order for a particular product class may change with the advent of new products in the market or through heavy advertising of certain sensory characteristics. For that matter, preferences for simple aqueous solutions can vary over time owing to changes in food habits of the consumer. For example, it is not difficult to assume a condition in which degree of liking for salt, sweetness and fattiness in many types of foods could change, based on nutrition-related motivations.

In addition, a new array of products could develop based on food processors' desire to capitalize on changing consumer desires. This can lead to the exposure of the consumer to a different set of stimuli, which in turn could result in preference changes. Thus any relationship determined in real foods is of a cross-sectional nature, and longitudinal studies are essential in order to be able to ascertain the invariance of such relationships.

1.2 Examples of the multiple regression approach

In order to better understand the details of design, procedure, analysis and interpretation, three examples using the MR approach will be presented. The first involves raw carrot texture, the second cooked rice and the third strawberry jam.

1.2.1 Predicting hedonic values of raw carrot texture from sensory descriptive attributes

This first example represents an early attempt to predict hedonic value from descriptive ratings of only a texture nature using the multiple regression approach [6]. Five trained panellists evaluated 24 raw carrot samples on four texture attributes. Forty-one consumers (convenience sample of students) judged the same samples for hedonic value of texture. The treatments, ratings of characteristics and hedonic average scores are given in Table 1.1.

Since there was only one variety of carrot, variations in texture were produced by different treatments, also shown in Table 1.1. Note that the

Table 1.1 Average texture judgements and hedonic ratings for 24 carrot samples

Carrot samples and treatment[a]	Flexibility	Hardness	Chewiness	Juiciness	Hedonic rating
Outside (1)[b]	4.0[c]	4.8	5.2	3.2	1.5
Outside (2)	3.0	4.4	5.6	2.8	2.1
Outside (3)	1.4	5.4	4.6	4.8	5.5
Fresh carrot	1.4	4.4	4.8	4.2	5.4
Open, no sun (1)	2.2	5.6	5.6	3.4	5.3
Open, no sun (2)	2.0	5.4	5.0	4.4	5.4
Open, no sun (3)	2.2	4.8	5.6	3.8	5.4
Open, no sun (5)	3.8	4.8	5.0	4.0	4.1
Freezer (1)	5.6	2.6	2.6	5.4	1.7
Freezer (2)	5.0	2.2	4.0	6.4	1.2
Freezer (3)	6.6	1.6	2.4	6.6	1.3
Freezer (5)	6.4	2.2	4.2	6.6	1.5
Plastic bag (1)	1.4	5.8	5.0	4.2	4.4
Plastic bag (2)	1.0	6.4	5.4	5.0	5.1
Plastic bag (3)	2.8	5.2	4.4	4.2	5.2
Plastic bag (5)	1.2	6.2	4.8	4.8	5.7
Cold water (1)	1.2	6.2	5.2	3.8	5.3
Cold water (2)	1.2	5.2	4.4	4.4	4.6
Cold water (3)	1.4	5.6	4.6	5.0	5.1
Cold water (5)	1.0	4.8	3.4	4.4	5.1
Cupboard (1)	1.6	6.0	4.8	3.8	5.1
Cupboard (2)	2.0	4.6	4.8	3.8	5.6
Cupboard (3)	1.4	5.4	4.6	4.8	5.5
Cupboard (5)	2.8	5.2	5.2	4.2	4.2

[a] Treatments took place under the following storage conditions: outside, exposed to the sun; frozen in the refrigerator; unwrapped in the cupboard; in a plastic bag in the refrigerator; in refrigerator, packed in air-tight water-filled container; in the open air, protected from the sun; and fresh, no treatment.
[b] Numbers in parentheses refer to total days in storage.
[c] 1 = low degree of characteristic; 7 = high degree of characteristic.

requirement is met of a range of scores for the attributes and hedonic ratings, which allows for the possibility of obtaining higher correlation coefficients between the sensory attributes and hedonic value.

As an illustration of how the data may be viewed in their simplest form Figures 1.1 and 1.2 present a diagram of the pattern of ratings. The average ratings for the four sensory characteristics are given for the three highest hedonic rated carrots in Figure 1.1 and for the three lowest hedonic rated carrots in Figure 1.2. The differences between a well liked and disliked carrot texture are clear from the examination of the figures.

A stepwise regression was conducted with the four sensory characteristics as independent variables and the hedonic ratings as the dependent variable. Using a significance criterion of $p = 0.05$ or less, only three sensory characteristics appear in the final equation. The final equation is:

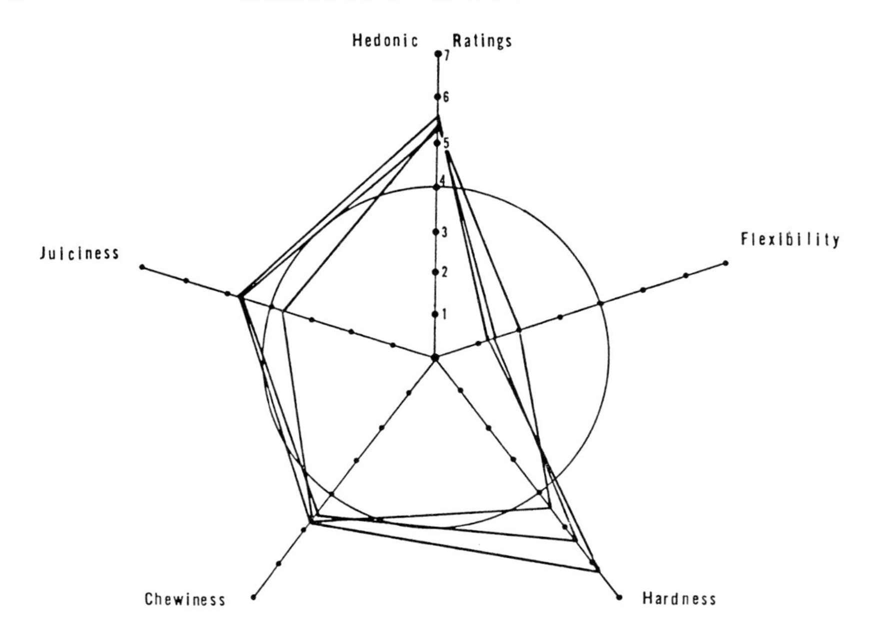

Figure 1.1 Sensory texture patterns of carrot samples with three highest average hedonic ratings.

$$\text{Predicted mean hedonic value} = 4.18 - 0.54 \text{ flexibility} + 0.42 \text{ hardness} - 0.10 \text{ chewiness.}$$

The multiple correlation, R was 0.91 and R^2, which indicates the amount of variance accounted for by the three independent variables, was 0.83 (83% of variance). Another way of looking at the role of the three variables is in terms of relative importance. Using the beta values that were obtained the relative importance can be expressed in the following manner: flexibility accounts for 61.4%, hardness for 33.6% and chewiness for 5% of the variation in hedonic judgements.

The R of 0.91 is unusually high. The magnitude obtained is not common in predicting food preferences, or for that matter in predicting any complex human response. However, one must be cautious when generalizing the results to the general population (older consumers might find hardness less desirable), to flavour as a whole, or to other carrot varieties.

1.2.2 Predicting hedonic values of cooked rice from sensory descriptive attributes

In this example the method is applied to a cooked product with many sensory variables, including taste, aroma, texture and appearance, as well as a more representative consumer population [7].

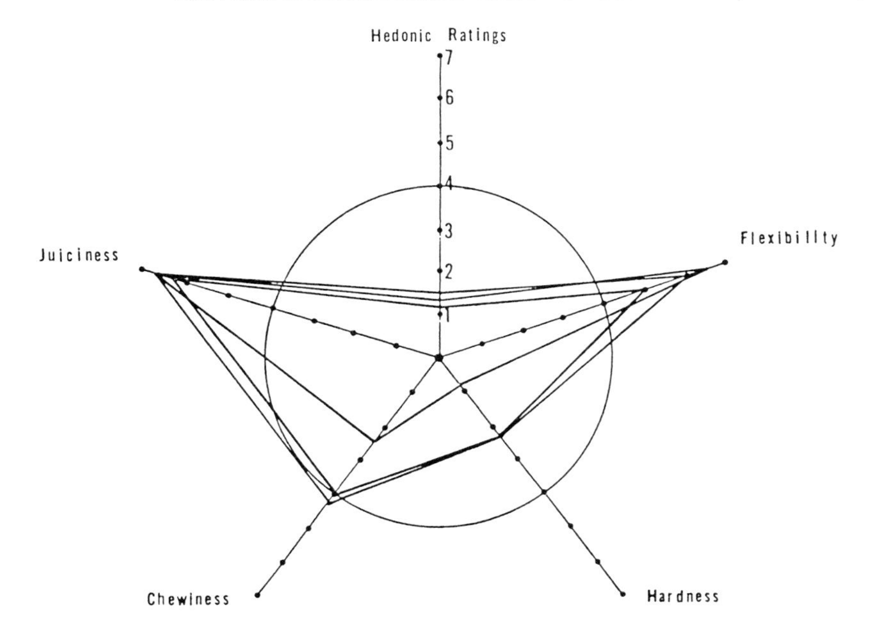

Figure 1.2 Sensory texture patterns of carrot samples with three lowest average hedonic ratings.

Six trained panellists evaluated 20 cooked-rice samples on 15 sensory characteristics using a seven-point line scale. One hundred volunteer females evaluated the 20 samples in a central location setting on a 7 point hedonic scale. The description of the rice samples along with the one on hedonic values is shown in Table 1.2. The range of hedonic values from a mean of 1.4 to 5.6 indicates the spanning of a wide preference space, a requirement for a good prediction model.

One of the practical problems in this study was control over rice temperature and thus flavour characteristics, which could vary with time and temperature. By using appropriate storage containers, by checking temperature regularly, and by limiting the length of time before testing, temperature at serving time was still adequate for tasting. The 15 sensory characteristics were size of grain, yellowness, compactness, moistness, dryness, stickiness, fluffiness, chalkiness, starchiness, rubberiness, firmness, tenderness, doneness, rice flavour and other flavour. The mean ratings for each product on each of the 15 attributes are not presented here but they had a large enough range to ensure the possibility of high r values with hedonic value.

The number of sensory variables was not large enough to require any reduction for a stepwise regression analysis, and therefore all 15 were entered into such an analysis. The results of a five variable equation are

Table 1.2 Description of rice samples

Rice type	P of rice/water (g/cm^2)	Cook time (min)	Hedonic value
1. Short grain	0.30	19	2.8
2. Short grain	0.44	32[a]	3.3
3. Short grain	0.30	30[a]	3.1
4. Short grain	0.34	19	3.7
5. Medium grain	0.34	14[a]	2.5
6. Medium grain	0.55	15[a]	3.9
7. Medium grain	0.55	15	3.6
8. Medium grain	0.30	25	2.4
9. Medium grain	0.21	15	1.4
10. Parboiled medium grain	0.48	25	4.3
11. Parboiled medium grain	0.34	20[a]	2.1
12. Parboiled medium grain	0.50	20	3.8
13. Parboiled medium grain	0.37	25	3.1
14. Parboiled medium grain	0.75	27	3.3
15. Long grain	0.47	20	2.7
16. Long grain	0.34	25	3.9
17. Long grain	0.28	20	3.5
18. Precooked long grain	0.43	10	5.6
19. Parboiled long grain	0.30	27	5.2
29. Precooked long grain	0.10	15	4.9

[a] Washed before preparation.

shown in Table 1.3 (a 13 variable equation actually had a slightly superior R, but the ratio of sensory scales to products, 13 to 20, made the smaller model a better one to use). The R of 0.90 is again quite high and, in addition, an adjusted R^2 is presented, which is a more conservative estimate of the likely results found if the study were to be repeated with other samples and another set of respondents drawn from the same population.

Table 1.3 Multiple regression analysis for predicting hedonic ratings of rice using five sensory scales[a]

Variables	Regression coefficient
Yellowness	0.157
Moistness	−0.199
Starchiness	−0.057
Doneness	0.520
Rice flavour	−0.696
Constant	3.139

[a] Multiple R = 0.896; adjusted R = 0.870; R^2 = 0.803; adjusted R^2 = 0.757; SE of estimate = 0.511.

Table 1.4 Prediction of optimum rice preference[a]

Variable	A Regression weight	B Target value	A × B
Yellowness	0.157	6.7	1.052
Moistness	−0.199	1.2	−0.239
Starchiness	−0.057	1.3	−0.074
Doneness	0.520	6.3	3.276
Rice flavour	−0.696	1.7	−1.183
Constant			+3.139

[a] Predicted optimum = 5.971

Again one should be cautious in generalizing the results to new rice products, for example, brown rice, wild rice and perhaps to men. Also particular ethnic groups may weight the importance of the various sensory characteristics in a different manner than the respondents in this study.

An analysis not conducted at the time of the original analysis of the data was performed to illustrate how the regression equation may be used to describe an 'optimum' product. Table 1.4 gives the information for this calculation. For each sensory characteristic the data for each rice sample are examined to find the highest (or lowest if the regression weight is negative) value. This is substituted in the equation and the equation solved. In this case, the resulting predicted mean value represents the most likely practical best rice product in terms of the five sensory variables. The value of 5.97 is higher than the highest rated rice product in the test array (5.6 for precooked long grain).

The reasons the target values are chosen the way they were, rather than taking the extremes of the scale (1 or 7) are: (i) the values may represent practical or technical limits in achieving these levels of the attributes; and (ii) there is less stress on the equation due to joint extrapolation of the data.

It is also of value to recognize that there are actually a number of 'optimum' products that could be predicted with target values close to those in the example, which might predict a preference score higher than the best in this test array.

In practice the flavour technologist can use the relative important information (not reported here for rice) in guiding their product-development decisions. One can also use the equation to help decide which products are appropriate for larger more-expensive market-research tests by performing trained-panel evaluations on the candidate products, calculating the attribute means and solving the equation with the obtained attribute values. Those products with the highest predicted hedonic score (or any other agreed upon limiting hedonic value) would be identified for further consumer evaluation with considerable cost savings.

1.2.3 Predicting strawberry jam preference from sensory and analytical variables

The last example represents a more complex and sophisticated application of the MR approach to predicting acceptance. In this instance, both sensory and analytical variables are used, the redundancy among these variables is dealt with, the segmentation of the respondent population by user groups and preference groups is looked at, and the relationship between selected sensory and analytical variables is examined (H.G. Schutz, unpublished data).

Twelve trained panellists using standard QDA (quantitative descriptive analysis) evaluated 24 strawberry jam products obtained in the marketplace, on 41 sensory attributes, including taste, aroma, appearance and texture. In addition to the sensory variables, 22 analytical measures typically used to measure 'quality' of jam were obtained for the same 24 products. Preference judgements on a 9-point hedonic scale were collected from 156 consumers of strawberry jam for the 24 products. For both trained panellists and consumers the products were evaluated on an unsalted cracker.

The question of the appropriate testing conditions, both the vehicle and the use instructions, as mentioned earlier, can have a major impact on external validity of the results. For example, in the present study, if the product were simply eaten off a spoon, the evaluations of both sensory attributes and preference could differ not only in magnitude but also in order. Spreading the product may also significantly influence the role of various attributes. On the other hand, if the vehicle were of a different texture, i.e. a soft bread, evaluations of appearance and texture could vary. The choice of testing conditions should be based on practical limitations and such factors as typical or sensitive use conditions.

Owing to the proprietary nature of the data, the results in Table 1.5 only indicate the 'number' for the various strawberry jam products, but this in no way diminishes the general value of information presented in relation to the objectives of this chapter. The products covered a wide range of characteristics of piece size and appearance, colour, sweetness, viscosity and amount of strawberries, so as to ensure a variation of attribute and analytical values. It could be said that the product attributes spanned the attribute space of products in the marketplace. Table 1.5 also gives the mean hedonic values for the total population, three user groups and three preference groups. The three user groups were identified by background questions asked of each respondent. The preference groups were obtained by conducting a PCA in which individuals were the variables and their 24 hedonic ratings for the jam were the cases. Thus a 156 $\times$ 156 matrix with 24 cases was analysed. Although there are more than three factors that could be extracted, there were not sufficient people with high enough weighting

Table 1.5 Mean preference values for seven strawberry jam groups

Product	Total population	Light users	Medium users	Heavy users	Preference Group I	Group II	Group III
10.	7.13	7.04	6.85	7.49	7.16	7.81	7.52
11.	7.04	6.96	7.08	7.09	7.32	7.08	7.04
1.	6.96	7.12	6.75	7.00	7.39	7.81	6.88
24.	6.97	6.86	7.02	7.02	7.39	6.88	7.64
7.	6.67	6.39	7.06	6.57	7.39	5.96	6.64
8.	6.69	6.53	6.58	6.96	6.88	6.34	7.16
4.	6.56	6.50	6.63	6.57	7.21	6.88	6.20
9.	6.58	6.94	6.33	6.49	6.54	7.04	7.40
6.	6.44	6.82	6.04	6.45	6.71	6.42	6.80
5.	6.41	6.22	6.35	6.66	6.46	6.77	5.36
12.	6.18	6.04	6.46	6.04	6.55	6.15	6.52
2.	5.99	5.27	6.29	6.39	5.66	7.27	5.96
3.	5.94	5.84	5.92	6.04	6.00	7.38	4.84
21.	5.91	5.53	6.13	6.06	6.82	5.04	5.56
23.	5.61	5.74	5.81	5.28	6.30	4.88	5.68
19.	5.38	5.57	5.35	5.24	6.23	4.88	4.36
20.	5.27	5.18	5.21	5.41	6.71	3.61	4.28
18.	5.00	5.49	4.75	4.77	4.18	5.65	5.92
22.	4.26	4.24	4.52	4.02	4.80	4.35	3.68
16.	3.94	3.90	4.02	3.89	3.70	2.61	5.00
17.	3.83	3.67	3.92	3.89	3.20	2.31	6.00
15.	3.31	3.59	3.19	3.16	3.07	4.04	2.12
13.	3.04	2.86	2.85	3.41	2.73	2.65	3.40
14.	2.99	2.56	3.17	3.19	2.66	3.35	2.84
N	156	51	52	53	56	26	25

on the factors beyond three. Those individuals who had a factor weight of 0.50 or higher made up each factor. Their ratings on the 24 samples were averaged to come up with the values in Table 1.5. Examination of this table reveals a wide range of preference scores for each group, but it is not clear how much the groups differ.

Table 1.6 gives the correlation coefficients among all groups. It is clear from examination of this table that the only groups which vary from one other are the three preference groups. Thus use level would not lead to meaningful different preference model than the total population and need not be used as dependent variables. On the other hand preference groups may yield preference models with some meaningful differences from one another.

Since there are 41 sensory variables, this is an instance where redundancy among characteristics is quite likely and the use of PCA as a simplifying tool appropriate. Table 1.7 presents the results of the PCA with seven sensory factors accounting for 93% of the variance. Only those factors weighing above 0.50 are printed for ease of interpretation and use. The communality represents the amount of variance in each sensory

Table 1.6 Pearson correlation coefficients among all preference groups

	Total population	Heavy users	Medium users	Light users	Preference group 1	2	3
Total population	1.00						
Other brand	0.99						
Heavy users	0.99	1.00					
Medium users	0.99	0.97	1.00				
Light users	0.98	0.95	0.95	1.00			
Preference group 1	0.94	0.91	0.94	0.92	1.00		
2	0.92	0.91	0.90	0.90	0.79	1.00	
3	0.87	0.86	0.85	0.86	0.73	0.73	1.00

attribute accounted for by the seven factors. Examination of the sensory characteristics heavily weighted on each factor yields a nominal description of the factor. Factor I is primarily a taste and aroma dimension; II, a fruit piece; III, viscosity; IV, sourness; V, seediness; VI, overall flavour intensity; VII, overall aroma intensity.

The same need for redundancy reduction is present for the 22 analytical measures and the PCA results accomplishing this objective are shown in Table 1.8. There are six factors accounting for 83% of the variance. One measure 'total organics' is not represented well by these six factors as shown by its communality of 0.40. If the heavily weighted measures on each factor are examined, the dimensions are as follows: Factor I, moisture; II, weight; III, flavanol; IV, sugars; V, colour; VI, viscosity.

For both the sensory and analytical PCA results it would be possible to conduct a PCA multiple regression in which the factor dimensions are used as independent variables. However, if this procedure is used, it is difficult to provide practical guidelines for product development or quality assurance since the factor dimensions do not define a simple attribute to modify. Another, more practical way to use the PCA results is to select a single attribute from each factor to represent that factor in a stepwise multiple regression.

This selection process in its simplest form is to choose the variable that weights the highest on each factor, since this would minimize the multicollinearity in the subsequent regression analysis. However, in practice, other considerations should be taken into account, such as evidence of nonlinearity, reliability of the attribute, cost and ease of measurement, and the zero-order correlation with preference measures. For the present analysis, the variables selected to represent the sensory dimensions were: Factor I, old fruit aroma; II, mashed appearance; III, gelatinous appearance; IV, sour aftertaste; V, seedy texture; VI, cloudy appearance; VII, overall aroma intensity.

Table 1.7 Summary of principal component analysis for 41 QDA variables by 24 products

	Factor weights							
Sensory attribute	I	II	III	IV	V	VI	VII	Communality
Old fruit aroma	0.96							0.96
Other fruit flavour	0.96							0.98
Brown colour	0.93							0.90
Chemical aroma	0.92							0.94
Strawberry aroma	−0.92							0.95
Old fruit flavour	0.91							0.88
Brightness	−0.89							0.88
Red colour	−0.87							0.81
Other fruit aroma	0.87							0.90
Chemical flavour	0.87							0.90
Strawberry flavour	−0.86							0.96
Cooked flavour	0.85							0.94
Sweet aroma	−0.81							0.91
Bitter aftertaste	0.78							0.92
Sour aroma	0.77							0.88
Bitter flavour	0.61			(0.61)				0.97
Mealy texture	0.58							0.81
Chunky texture		0.93						0.95
Amount of fruit		0.93						0.95
Size of fruit pieces		0.92						0.93
Firm fruit pieces		0.91						0.96
Mashed fruit appearance		−0.89						0.98
Firm appearance		0.89						0.98
Spreads easily		−0.81						0.94
Fruit uniformity		0.67						0.79
Thickness appearance			0.91					0.95
Thickness texture			0.91					0.94
Gelatinous texture			0.87					0.96
Gelatinous appearance			0.83					0.87
Sticky appearance			0.75					0.92
Drying texture	(0.60)			0.72				0.96
Sour flavour	(0.63)			0.67				0.94
Sour aftertaste	(0.65)			0.66				0.94
Seedy texture					0.96			0.96
Seedy fruit appearance					0.88			0.94
Overall flavour intensity						0.69		0.83
Cloudy appearance	(0.53)					−0.63		0.90
Sweet aftertaste	(−0.52)					0.58		0.99
Stringy fruit texture						−0.54		0.93
Sweet flavour				(−0.51)		0.54		0.98
Overall aroma intensity							0.89	0.95
Percentage variance accounted for	47.3	15.0	10.9	6.7	6.1	4.0	2.9	92.7

Table 1.8 Principal component analysis for 22 analytical variables by 24 products

	Factor						
Variable	I	II	III	IV	V	VI	Communality
Percentage moisture	−0.96						0.95
Percentage soluble solids	0.94						0.93
Water activity	−0.92						0.89
pH	−0.77						0.78
Percentage drained weight		0.93					0.92
Piece weight		0.91					0.91
Piece count		0.90					0.93
Percentage insoluble solids		0.86					0.85
Anthocyanin content		0.63					0.75
Flavanol content			0.83				0.79
Percentage fructose			0.80				0.84
Percentage total acid			0.73				0.65
Hunter *a*			−0.71				0.74
Total organics			−0.61				0.40
Percentage maltose				0.87			0.91
Percentage sucrose				−0.85			0.91
Percentage dextrose	(0.52)			0.67			0.91
Hunter *b*					0.93		0.93
Hunter *L*					0.93		0.93
Bostwick						0.80	0.80
Gel						−0.78	0.83
Spread						0.65	0.77
Percentage variance accounted for	22.4	18.4	16.1	10.9	8.8	6.8	83.2

An example of the use of these variables in predicting preference is shown in Table 1.9, where the 'best' equation for predicting the total population preference scores is given. The term 'best' is used since the actual analytical process consists of examining various combinations of the independent variables and not just accepting those that appear in a straight-

Table 1.9 Summary of multiple regression 'best' equation predicting total population preference scores from selected QDA trained-panel variables[a]

Sensory attribute	*B*	Beta	*r*	(Beta) × (*r*)	% Relative contribution
Cloudy appearance	−0.057	−0.263	−0.499	0.132	13.5
Seedy texture	0.038	0.157	0.306	0.048	4.9
Gelatinous appearance	−0.033	−0.154	−0.378	0.058	5.9
Sour aftertaste	−0.074	−0.439	−0.812	0.357	36.5
Mashed appearance	0.034	0.250	−0.219	0.055	5.6
Old fruit aroma	0.080	−0.396	−0.825	0.327	33.5

[a] Constant = 8.94; $R = 0.93$; $R^2 = 0.87$, Adjusted $R^2 = 0.82$; SE = 0.57.

Table 1.10 Summary of multiple regression 'best' equation predicting total population preference scores from selected analytical variables[a]

Analytical variable	*B*	Beta	*r*	(Beta) × (*r*)	% Relative contribution
Percentage moisture	−0.038	−0.226	−0.452	0.102	13.7
Percentage drained weight	0.030	0.220	0.084	0.018	2.4
Flavanol content	−0.298	−0.556	−0.647	0.360	48.5
Percentage maltose	0.102	0.413	0.576	0.238	32.1
Hunter *L*	−0.188	−0.322	−0.073	0.024	3.2

[a] Constant = 10.64; R = 0.86; R^2 = 0.74; Adjusted R^2 = 0.67; SE = 0.77.

forward stepwise regression. In essence the 'best' equation is one in which the R^2 is highest, the SE is lowest, and the residual analysis indicates the least nonlinearity and smallest error of prediction for all the products. From examination of Table 1.9 it can be seen that the ability to account for the preference of strawberry jam products is quite high with 87% of the variance accounted for with six sensory variables. The relative contribution data are calculated in a slightly different fashion than presented earlier in that the beta value is multiplied by the zero order correlation before summing and dividing by the total and multiplying by 100 to obtain a percentage. This statistic is sensitive to both the joint and individual importance of the variable. From these importance data it is clear that the absence of 'sour aftertaste' and 'old fruit aroma' are the critical factors in preference.

The variables chosen to represent the six analytical measures in predicting preference were as follows: Factor I, % moisture; II, % drained weight; III, flavanol content; IV, % maltose; V, Hunter *L*; VI, Bostwick; and a unique variable not well accounted for by the six factors, total organics. The results of the regression analysis predicting preference for the total population are shown in Table 1.10. Again the prediction is quite good, but with a lower level of variance (74%) is accounted for in preference than with the sensory variables. Examination of the relative importance data reveals that absence of flavanol content and higher levels of maltose are of most importance in accounting for preference.

A valuable analysis is to combine the selected sensory and analytical variables in one stepwise regression predicting preference. This allows for whatever independent contributions of the two variable sets to operate is the possible improvement of preference prediction. For the present data set the results of this type of analysis are shown in Table 1.11. In fact, the obtained 'best' equation results in a higher amount of preference variance accounted for (93%) than either the sensory or analytical analysis by themselves.

Table 1.11 Summary of multiple regression 'best' equation predicting total population preference scores from selected QDA trained panel and analytical variables[a]

Sensory attribute or analytical variable	*B*	Beta	*r*	(Beta) × (*r*)	% Relative contribution
Old fruit aroma	−0.059	−0.295	−0.825	0.244	26.1
Sour aftertaste	−0.065	−0.386	−0.812	0.313	33.5
Percentage maltose	0.071	0.286	0.576	0.165	17.6
Percentage moisture	−0.027	−0.162	−0.452	0.073	7.8
Hunter *L*	−0.116	−0.199	−0.073	0.015	1.6
Flavanol content	−0.103	−0.192	−0.647	0.125	13.4

[a] Constant = 10.68; $R = 0.97$; $R^2 = 0.93$; Adjusted $R^2 = 0.91$; SE = 0.40.

Similar regressions can be conducted for other segments of the population, which have been identified as important and with low correlations with each other, such as the preference groups given in Table 1.5.

As with the rice example given earlier, an 'optimum' product can be calculated in terms of the variables in the 'best' prediction equation. An example of this for the combination of sensory and analytical variables is shown in Table 1.12. The procedure for selecting target values for each variable is the same as described earlier. The predicted optimum of 8.48 is significantly higher ($p < 0.01$) than the highest rated product in the test array at 7.13. At the bottom of this table are given guidelines for the factor representative variables not in the equation. These allow those involved in product development to have checkpoints with which to ensure that changes made in the variables in the equation do not adversely affect other dimensions of the product. These guidelines are based on one to two standard error distances from the mean of each variable depending on the size and deviation of the zero-order correlation with preference. Guidelines for factor representative variables not in equation are as follows:

Table 1.12 QDA and analytical optimization for total population[a]

Variable	Regression weight	Target value	Regression weight × target
Old fruit aroma	−0.0592	5.69	−0.337
Sour aftertaste	−0.0648	13.21	−0.856
Percentage maltose	0.0706	15.1	−1.066
Percentage moisture	−0.0271	28.9	−0.783
Hunter *L*	−0.1165	7.2	−0.839
Flavanol content	−0.1031	4.33	−0.446

[a] Constant = 10.677; predicted optimum = 8.48 ± 0.318.

Cloudy appearance	23.70–26.21
Seedy texture	21.08–22.20
Gelatinous appearance	33.26–34.55
Mashed appearance	30.88–35.92
Overall aroma intensity	28.53–29.89
Percentage drained weight	13.30–17.28
Total organics	29 298–32 733
Bostwick	4.35–4.68

One of the byproducts of the collection of both sensory and analytical data is the ability to determine relationships between these two sets of data. If one knew what analytical measures predict each sensory attribute, it would be easier to understand how to change some underlying ingredient or process to bring about a desired change in a sensory attribute. An illustration of this for the present data set is given in Table 1.13. Here all but one of the sensory variables can be predicted with an R, which is significant with a p of 0.05 or lower.

1.3 Conclusion

When considering the discussion and examples given in the paper on predicting acceptability from flavour data, one can come to some reasonable conclusions. In spite of the very real problems in design, data collection, analysis and their influence on validity, it can be seen that it is

Table 1.13 Multiple regression results for seven QDA attributes predicted by seven analytical measures[a]

	Cloudy appearance (B)	Seedy texture (B)	Gelatinous appearance (B)	Sour aftertaste (B)	Mashed appearance (B)	Old fruit aroma (B)
Percentage moisture	0.510	−0.171	−0.410	–	–	–
Percentage drained weight	−0.239	–	–	−0.202	−0.794	−0.196
Flavanol content	0.580	–	–	1.785	–	1.222
Percentage maltose	–	–	–	–	–	−0.409
Hunter L	–	−0.700	–	–	–	–
Total organics	–	−2.462	–	–	–	–
Bostwick	–	–	−1.322	–	–	–
Constant	5.708	43.992	42.12	10.441	32.37	9.69
R	0.82	0.67	0.52	0.61	0.83	0.70
p	<0.00001	0.006	0.03	0.007	<0.00001	0.0031
R^2	0.68	0.45	0.27	0.37	0.69	0.49
Adjusted R^2	0.63	0.37	0.20	0.31	0.66	0.42
SE	3.7	4.35	5.63	6.60	5.78	5.10

[a] There were no analytical variables that predicted 'overall aroma intensity' at $p < 0.05$.

possible to predict acceptance successfully from flavour data and that these results are necessary and critical in the decision-making process in agriculture, product development and quality assurance.

References

1. Shepherd, R., Farleigh, C.A. and Land, D.G., The relation between salt intake and preference for salt levels in soup. *Appetite*, 1984, **5**, 281–290.
2. Tourila-Ollikainen, H., Lahteenmeki, L. and Salounaara, H., Attitudes, norms, intentions and hedonic responses in the selection of low-salt bread in a longitudinal choice experiment. *Appetite*, 1986, **7**, 127–139.
3. Tourila, H., Hedonic responses to colour, sweetness, saltiness and fattiness in selected foods as related to corresponding attitudes and other behavioural measures. Academic Dissertation, University of Helsinki, 1986.
4. Drewnowski, A., Bronzell, J.D., Sande, T., Iverius, P.H. and Greenwood, M.R., Sweet tooth reconsidered: Taste responsiveness in human obesity. *Physiol. Behav.*, 1985, **35**, 617–622.
5. Giovanni, M., Response surface methodology and product optimization. *J. Food Technol.*, 1983, **37**, 41–45.
6. Schutz, H.G., Damrell, J.D. and Locke, B.H., Predicting hedonic ratings of raw carrot texture. *J. Text. Stud.*, 1972, **3**, 227–232.
7. Schutz, H.G. and Damrell, J.D., Prediction of hedonic ratings of rice by sensory analysis. *J. Food Sci.*, 1974, **39**, 203–206.
8. Schutz, H.G., Multiple regression approach to optimization. *Food Technol.*, 1983, **37**, 46–49.
9. Moskowitz, H., Subjective ideals and sensory optimization in evaluating perceptual dimensions in food. *J. Appl. Psychol.*, 1972, **56**, 60–66.
10. Fishken, J., Consumer-orientated product optimization. *J. Food Technol.*, 1983, **37**, 49–52.
11. Martens, M. and Martens, H., Partial least square regression, in *Statistical Procedures in Food Research*, (ed. J.R. Piggott), Elsevier, London, 1986, pp. 293–360.

2 Sensory analysis of flavours

P.DÜRR

Abstract

Sensory analysis may be defined as the measurement of both the flavour and the assessor's characteristics using human senses. The quality of flavours and other items depends on two sources of variables, namely the flavour and its judge. Flavours vary in composition, whereas judges vary in sensitivity, expectations, experiences and actual disposition. A major factor in sensory analysis is hedonism. Hedonic response affects to a great degree the evaluation of parameters of luxury and stimulant foods. Visual, tactile, audible and chemical stimuli are perceived with highly specific receptor cells of different organs, although independent use of human senses is difficult. A complex stimulation of all chemosensory systems is called flavour, because the brain blends all the information to a single perceptual gestalt.

Sensory analysis starts with a clear analytical question. The information content and significance level of the results may be limited by the cost of the analyses. Practicable sensory measurements may be classified as yes/no decisions, ranking, classifying, scoring, intensity measurement and verbal analysis. Verbal analysis is the most complex, controversal, demanding and rewarding sensory work. Sensory analysis uses scales based on human characteristics. Humans respond more exponentially than linearly to sensory stimuli. Sensory analysis requires a skilled analyst, available assessors, scales adapted to human abilities, a suitable method and a controlled testing environment. The assessors can be divided into three types: the small expert group, the research panel and the consumer group.

2.1 Introduction

2.1.1 Definition

Sensory analysis may be defined as the measurement of both the flavour and the assessor's characteristics using human senses. To 'measure' means to produce data that can be scaled and are reproducible. Human senses are part of the human instrument, with typical stimulus/response relations.

2.1.2 Quality: two sources of variables

The term 'quality' is widely used in the field of sensory analysis. One aspect is important: quality of flavours and other items depends on two sources of variables, namely the flavour and its judge. Flavours vary in composition, whereas judges vary in sensitivity, expectations, experiences and actual disposition.

2.2 The sensory instruments

The human instruments can be divided into three types: the small expert group, the research panel and the consumer group.

The expert is characterized by excellent product knowledge and training in sensory techniques. The response variation is small to medium, therefore the group must consist of at least three experts with a strict control and feedback of individual performance. Experts have an extended vocabulary to describe product characteristics but also severe conceptual problems, such as the use of different words for one stimulus or the same word for distinctly different stimuli.

Owing to statistical requirements, a research panel consists of 15–25 assessors, highly trained in methods and techniques. Familiarity with particular products is unnecessary but is acquired through experience. Such a group shows only a small response variation. The individual performance and the work in the sensory research laboratory are both easy to control. However, a research panel is too small to measure consumer response.

Consumer groups are often used to measure product characteristics but are also themselves objects of investigation, for example when measuring consumer preference or attitude toward a product. Since consumer groups often show a wide variation in response, relatively large groups are used. The final number can be determined in a stepwise procedure. It is difficult to control the performance of the individual. The working place of a consumer group should be close to consumer reality.

2.3 Psychology

Humans perceive their world within the limits of their senses. Our reality is not the real world. Our senses react qualitatively differently to certain physical and chemical stimuli and quantitatively between threshold and saturation. Our perception is more than the sum of all stimulus sensations. Incoming information is screened by our attention and combined with our acquired memory and integrated to a final complex perception. Humans are able to selectively attend to one message but unable to attend two messages at once.

A major factor in sensory analysis is hedonism. People tend to describe what they like and dislike, rather than describing directly the characteristics of the sample. Hedonic responses are not monotonic with increasing magnitude of the stimulus but show a peak above and below which the ratings decline. Hedonic response affects to a great extent the evaluation of parameters of luxury and stimulant foods. Acceptance includes both hedonic response (preference) and attitude. The latter two factors often contradict one another [1].

2.4 Psychophysics

Psychophysics deals with the quantitative relation between stimulus and perceived intensity at and above threshold values.

Odour and taste thresholds, unless properly determined, have no practical value. A look at published data [2] shows the extreme variation of thresholds of a given compound. Fechner [3] described the three classical psychophysical methods of determining thresholds. With the method of constant stimuli, different concentrations of the compound are presented in random order. Presenting the stimuli in ascending or descending order is called the 'method of limits'. In the 'method of adjustment' the concentrations of the stimulus is slowly changed either by the subject or the experimenter until the subject can just detect the stimulus. The method of adjustment is the least accurate but the fastest and is well established in olfactometry. Factors other than the subject's sensitivity may influence the results of a threshold determination. Therefore new procedures were created on the basis of the signal detection theory [4]. A short-cut signal detection measure of the size of small differences has been proposed by O'Mahony [5].

Above threshold, up to saturation, the relation between stimulus concentration and perceived intensity is described by Stevens' power law [6]. The underlying technique is called magnitude estimation, which has found many applications as well as critics in practical sensory analysis [7,8].

2.5 Physiology: oral, nasal, retronasal and trigeminal perception

Visual, tactile, audible and chemical stimuli are perceived with highly specific receptor cells of different organs, although independent use of human senses is difficult. It may be necessary to blindfold the eyes or to close the nostrils. Important to flavours are our chemical senses, the odour mucosa in the upper nose, the taste buds on the tongue and the trigeminal nerve ending in the skin of the mouth and nose.

A pure odour sensation is possible through the nose, whereas a pure taste sensation from complex samples is almost impossible because of retronasal and trigeminal effects. Some important stimuli like ethanol or pepper evoke nasal, retronasal, oral and trigeminal sensations. Trigeminal sensation reacts more slowly to a stimulus and tends to increase with time (e.g. burning of red pepper), whereas olfactory sensation reacts fast but soon adapts (e.g. environmental odours). A complex stimulation of all chemosensory systems is commonly called taste, professionally called flavour, because the brain blends all the information to a single perceptual gestalt [9].

2.6 Sensory analysis

2.6.1 The possibilities

Sensory analysis starts with a clear analytical question. More often than not, the sensory analyst is faced with unspecific or complex or even impossible demands and uses a considerable amount of his or her time in attempting to clarify matters. Furthermore, the information content and significance level of the results are limited by the cost of the analyses.

2.6.2 The tool box

Practicable sensory measurements may be classified as follows:

1. 'Yes/no decisions'. For example: is there a perceptible difference between two samples?; is a sample suitable for a specific purpose?; is a product free of taints? etc. The human instrument can also be used as a specific detector on specified compounds like chloroanisoles in poultry meat, ethyl acetate in wine, etc.
2. Ranking. Samples (2–6), presented together, are ranked according to a defined criterion, which can be preference or the intensity of a sample character.
3. Classifying, scoring. Samples, presented in random order, are put into defined classes. The definition includes a classification criterion and a magnitude (class). Popular as a measure of magnitude are limited point scales. The criterion can be preference, competition medals or a sample character.
4. Intensity measurement. A crude estimate of odour and taste intensity is achieved by classifying. Another approach is based on the ratio of the concentration of a compound to its threshold concentration. These

values are called odour units. This concept neglects the fact, that odour intensities above the threshold increase at different rates according to Stevens' power law. However, it may give an idea as to the importance of a compound in the odour of a mixture [10]. Where data with ratio character are required, intensity is measured by magnitude estimation [7].

5. Verbal analysis. This is the most complex, controversial, demanding and rewarding sensory work and worth consideration. Language is used as an information carrier, words become descriptors. People may have different concepts on what a descriptor means. Therefore many descriptors used are ambiguous (e.g. floral) or even meaningless (e.g. fruity) and some are synonyms (e.g. thin, watery, short). One approach is to define product characteristics and their descriptors. Although difficult, concepts of words can be modified by training. Problems arise especially when product experts are testing products of high hedonic value like alcoholic beverages. It has been shown, that even experts can fail to understand their own descriptions [11].

To overcome the problems with such strongly fixed personal vocabularies, Williams [12] proposed the free choice profiling. The assessor works with his or her personal vocabulary and a short point scale to measure the intensity of attributes of his or her choice. This information is used to give each sample a position in a multidimensional space. By using rotational, stretching and shrinking techniques, it is possible to match each assessor's spaces to others and yet maintain the relationship between samples. The difficulty is to relate the axes of such multidimensional spaces to verbal attributes, which find general acceptance.

An alternative that is closer to the language may be called the semantic approach. Free choice descriptors given from 10–20 product experts, who must be good describers, are collected. Hedonic as well as obviously opposite descriptors are omitted from the collection. The critical part is to group descriptors with similar meanings, which requires some interpretation of the experts word concepts by the sensory analyst. The result is a brief description formed by the most mentioned descriptors on a nominal scale base.

2.6.3 Quality of judgement

Sensory judgements may have different information levels and different degrees of accuracy. The information level depends on the scale used, the accuracy on the number of judgements, the performance of the judges, the data treatment and on the chosen level of significance. The quality of the judgement is part of the analytical question to be asked at the beginning of the sensory work. Is a hint or idea sufficient? Is a magnitude or an exact numerical value required? At which confidence level should the result be?

2.6.4 Scales

Any measurement needs a scale. Sensory analysis uses scales based on human characteristics. Humans respond more exponentially than linearly to sensory stimuli. Therefore, three types of scales with their own set of statistical methods are appropriate:

The nominal scale is related to the 'equal or different' question. Is there a perceptual difference between two samples? Do you smell acetic acid in the wine? The answer is yes or no, it is very precise but not very informative.

The ordinal scale is related to the 'more or less' question. It is more informative but does not provide quantitative data. It is nonlinear and therefore connected to nonparametric statistics. The ordinal scale is the most popular scale in sensory analysis and used in ranking or classifying samples along a given product attribute or along a hedonic dimension (preference).

The various category, rating, scoring and point scales are often misused as the apparently equal intervals on these scales are disregarded, which gives data on the ordinal level. In practice, many of these scales are used and varied in design. The results should always be treated as only ordinal, unless the data produced can be shown to have nominal, interval or ratio properties. Land [8] recently discussed the properties of sensory rating scales. Furthermore, with popular point scales, the assessors often confuse the hedonic dimension with the requested magnitude of a sample characteristic.

The ratio scale is related to the 'how many times more or less' question and has a true zero. The appropriate measure is the geometric mean and parametric statistic techniques may be used. It provides quantitative data but is less precise. An example is the sensory measurement of odour intensity of beverages [13]. The way the assessors are instructed is critical. The ratio scale is commonly used in instrumental or chemical measurements.

2.6.5 Methods

The sensory methods are described in almost any textbook on sensory analysis. Piggott [14] compiled a concise review of modern sensory analysis. Some of the methods were developed pragmatically, others derive from psychophysics. O'Mahony [15] presented the psychologist's approach to sensory statistics, a recommended reading for the sensory analyst. A detailed laboratory manual of sensory analysis has been written by Jellinek [16]. The methods can be grouped into difference tests, ranking procedures, scoring and classification tests, descriptions, profiling methods and intensity measurements.

2.6.6 Sensory technique

Sensory analysis requires a skilled analyst, available assessors as already described, scales adapted to human abilities, a suitable method, a controlled testing environment, and much time. Sampling techniques are used to make sure that an assessor's sample of only 10–20 g represents truly the total item. The sample coding system should be transparent to the analyst but not to the assessors. A double-blind procedure is used in critical cases. To minimize neighbour effects from sample to sample, a balanced or random one-by-one presentation is necessary.

To get independent data, communication between the assessors is avoided by sample coding, spatial and temporal separation. It is important that assessors work with the same information on the samples and the test, and give the results in written or electronic form. The performance of assessors is routinely controlled by double samples and replications of whole tests. Assessors only provide useful data if they are fully motivated for sensory work. This is achieved by professional organization of the work, fast data feedback and orientation on the test problem, training sessions, some free gifts or pay. Under- and overcharge of assessors must be avoided. The test place should be friendly, with fresh air and without noise, whether this is a sensory experimental laboratory, a common room, a supermarket, a restaurant or the assessor's home.

References

1. Tuorila, H., Hedonic responses and attitudes in the acceptance of sweetness, saltiness and fattiness of foods, in *Food Acceptance and Nutrition*, (eds. J. Solms, D.A. Booth, R.M. Pangborn and O. Raunhardt), Academic Press, London, 1987, pp. 337–351.
2. Stahl, W.H., Compilation of odor and taste threshold values, American Society for Testing and Materials, Committee E-18, Philadelphia, 1973.
3. Fechner, G.T., *Elemente der Psychophysik*, Vol. 1, Breitkopf und Harterl, Leipzig, 1860.
4. Swets, J.A., Signal detection and recognition by human observers, Wiley, New York, 1964, p. 435.
5. O'Mahony, M., Short-cut signal detection measure for sensory analysis, *J. Food Sci.*, 1979, **44**(1), 302–303.
6. Stevens, S.S., On the psychophysical law, *Psychol. Review*, 1957, **64**, 153–181.
7. Moskowitz, H.R., Application of sensory measurement to food evaluations: II. Methods of ratio scaling, *Lebensm. Wiss. Technology*, 1975, **8**(6), 249–254.
8. Land, D.G., Scaling and ranking methods, in *Sensory Analysis of Foods*, (ed. J.R. Piggott), Elsevier Applied Science, London, 1988, pp. 1–15.
9. Mariunak, J.A., The sense of smell, in *Sensory Analysis of Foods*, (ed. J.R. Piggott), Elsevier Applied Science, London, 1988, pp. 25–68.
10. Maarse, H. and van den Berg, F., Current issues in flavour research, in *Distilled Beverage Flavour* (eds. J.R. Piggott and A. Paterson), Ellis Horwood, Chichester, 1989, p. 1–15.
11. Dürr, P., Wine description by experts and consumers, Proceedings of the Second International Cool Climate Viticulture and Oenology Symposium, Auckland, New Zealand, January 1988.
12. Williams, A.A. and Langron, S.P., A new approach to sensory profile analysis, in

Flavour of Distilled Beverages, (ed. J.R. Piggott), Ellis Horwood, Chichester, England, 1983, pp. 219–224.
13. Dürr, P. Sensorische Beurteilung der Intensität von Fruchtsaftaromen, *Ber. Intern. Fruchtsaftunion*, 1978, **15**, 155–161.
14. Piggott, J.R., (ed.), *Sensory Analysis of Foods*, 2nd edn., Elsevier, London, 1988.
15. O'Mahony, M., *Sensory Evaluation of Foods: Statistical Methods and Procedures*, Marcel Dekker, New York, 1986.
16. Jellinek, G., *Sensory Analysis of Food*, Ellis Horwood, Chichester, 1985.

3 Food acceptability

A.A. WILLIAMS

Abstract

Food acceptance is a complex field influenced by many factors requiring both acceptance, perceptual and physical and chemical information if it is to be understood. In this chapter, some of the problems and approaches available for acquiring acceptance and perceptual information, with examples, are reviewed. Also the various procedures and underlying psychological/behavioural models that enable links to be developed between the two and consequently lead to a deeper understanding of how consumers select food are explored.

3.1 Introduction

The acceptability of foods and beverages is determined by how they are perceived in sensory, utilitarian, imagery and attitudinal terms, coupled with the consumers reaction to and trading-off of these various perceived characteristics [1,2].

Within this scenario, compositional and, hence, functional and sensory factors are those that the technologist can influence. It is important, however, to remember that the consumer also pays attention to less tenuous factors such as perceived nutritional value, health risks and associated social and personal acceptability. The only way to manipulate these latter aspects is by promotion, packaging, advertising and pricing.

All consumers look at the world differently. Everyone has their own individual window on the world – filtering out and processing information based on personal aspirations, past personal experiences and pressures imposed by their peers and environment. This window allows individuals to create their own internal world, permitting them to handle the magnitude of information being thrust at them by reducing it to manageable proportions. It is this internal world that determines how a person reacts to a product.

If manufacturers are to produce successful products it is important that they understand the factors influencing this personal world, they must know how their products are being perceived, how the consumer relates to them, what other products are viewed in a similar light in the marketplace and how these compare sensorially, functionally and in an imagary sense with their own products.

Without such information and without understanding how it relates to product acceptance and choice on the one hand and to technical, chemical and physical information on the other, developing new markets and new products, designing advertising and promotion and even establishing quality control criteria becomes a hit-and-miss operation.

To understand acceptance it is essential to be able to measure product acceptability and preference and provide some evaluation of the underlying factors that influence them.

3.2 Tools for measuring and understanding product acceptance and perception

How a product is perceived and evaluated is a consequence of the reaction between its chemical and physical properties and the person evaluating it. To help unravel the complex array of interacting factors influencing acceptance the scientist must explore the problem from both sides.

From the human side, the consumer of a product, or group of products is the only person who can truly provide information on acceptance. Unfortunately the responses of such people, as shall be discussed later, are subject to all sorts of variations and provide very little information of direct value to the product developer.

Human beings acting as either lay or trained panels together with various market research and sensory techniques provide the scientist with an insight into some of the factors influencing acceptance as perceived. The more analytical such assessments, the more closely they relate to the real world but the less representative are the views expressed of what is influencing acceptance and choice [3–5].

On the product side there is a vast array of chemical and physical techniques available for providing detailed information on the underlying composition of foods and beverages from which all sensory and many other stimuli must derive ultimately. It is this information (the real world) to which many aspects of acceptance must be related if it is ever to be properly understood and controlled by the manufacturer.

The objective of the food acceptance scientist is to measure and understand the consumers' reactions to products and relate them to factors upon which manufacturers can act. In this context, it is often tempting, particularly so far as sensory attributes are concerned, to try to relate acceptance information directly to production and analytical factors. Attempting to short-cut the underlying steps by which stimuli influence acceptance is fraught with potential pitfalls and often leads to purely mathematical relationships being derived. Such relationships only hold

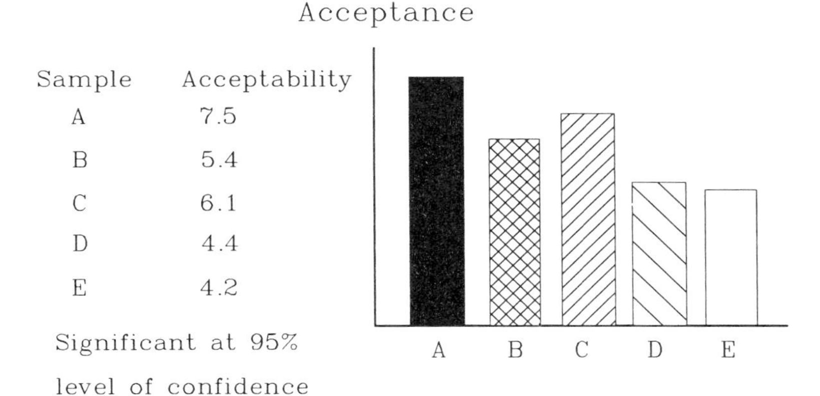

Figure 3.1 Typical acceptance information on a set of five products obtained using a nine-point acceptance scale.

within the context of the information being examined and are potentially useful in a predictive sense; however, they may have no underlying causative implications and are of little value in truly understanding perception or in deciding how products may be improved.

Consumer and analytical descriptive information and product image analysis provides a valuable bridge, ensuring that only causative models are postulant and false links are not made.

3.3 Measuring acceptance

Product acceptance measurements, as already stated, must come from the consumer. They may be obtained either directly by asking consumers what they think of products or by observing behaviour [5,6]. Although more time-consuming, the latter approaches probably provide a more realistic measure of ultimate acceptance in the marketplace.

Direct methods of measuring acceptance usually involve some form of ranking, scaling, paired comparison or product grouping. Typical of the scales used are: (i) the nine-point hedonic scale of Peryam [7]; (ii) the standard graphic line scale; (iii) the relative to ideal scale [8]; and (iv) the percentage likelihood to select scales in a choice situation. Typical results obtained in such research are shown in Figure 3.1.

Unlike indirect behavioural methods, direct questioning can also be used to elicit acceptability information on imaginary products, brand names and concepts.

Indirect methods involve the monitoring of rates of consumption or usage of comparative products in a free-choice situation [9], completion of

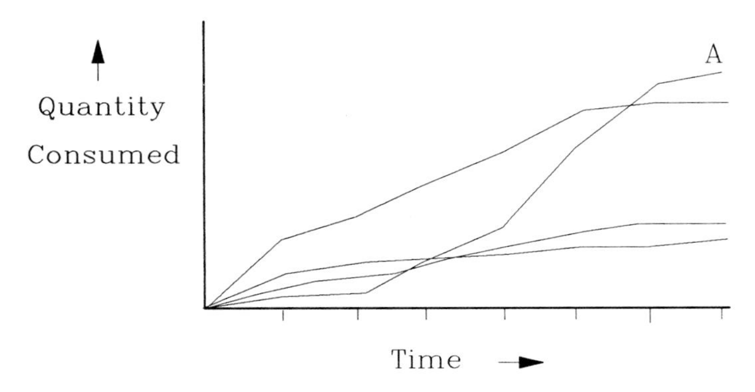

Figure 3.2 Rates of consumption of four ciders during an evening's drinking.

diaries, or even examining sample purchasing patterns. Typical information obtained as a result of the session testing of a set of ciders is presented in Figure 3.2.

In this example the lines relate to the amount of different ciders drunk throughout an evening's drinking by a group of 100 cider drinkers following exposure to all products. More of cider A is consumed in total despite not being considered the most acceptable product on sip evaluation.

Consumers, because of biological variation, do not show common sensitivities to the complex array of stimuli presented to them in a food; neither do they, as already stated, act passively to these stimuli. They take information and distort it to fit in with their own aspirations and experiences. As a consequence, any observation about likes and dislikes is bound to be subject to variability.

Before attempting to understand the underlying reasons why people rate products in the way they do, it is important to ensure at least that one is dealing with a homogeneous segment of the population with common likes and dislikes. Examining data from non-homogeneous populations will not only result in identifying the highest commonly acceptable product, which strictly speaking may not truly satisfy anyone, as the best product, but, when linked to descriptive data, provide false clues as to the underlying reasons for preference.

Various spacial [3–5,9,10] and hierarchical [11] linkage clustering techniques can be used to explore consumer acceptance data for homogeneity and potential segmentation. Figures 3.3 and 3.4 show the results of an internal preference mapping exercise on colas and of a furthest neighbour clustering of consumers of a confectionery product. In the map the position of samples reflects relative overall acceptances, samples having similar acceptances being grouped together. Vectors in these maps show the direction of maximum acceptance for individual assessors, in this instance

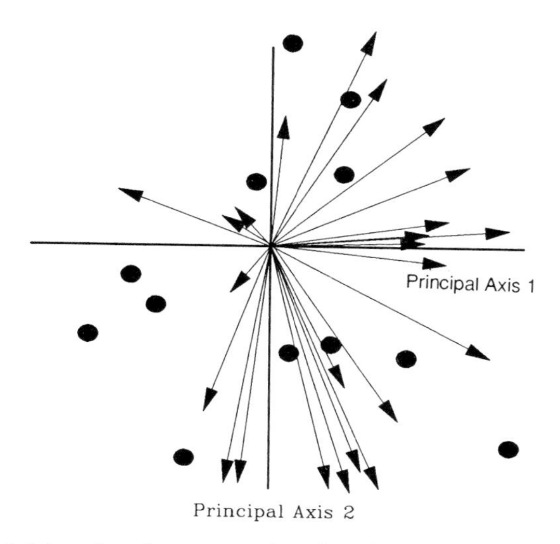

Figure 3.3 Internal preference map based on the acceptance of colas.

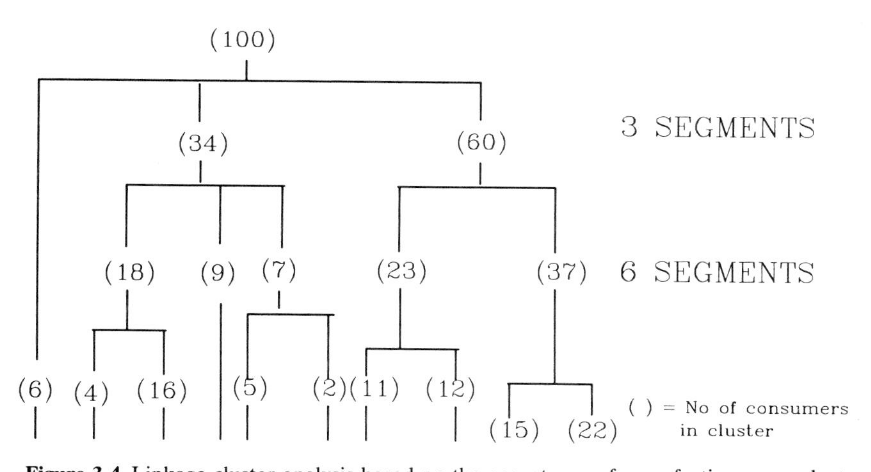

Figure 3.4 Linkage cluster analysis based on the acceptance of a confectionery product.

showing a wide variation in preferences. The hierarchical clustering shows one definite odd group of assessors with the remainder falling into two or five groups, depending on the criteria used.

Although many of the factors influencing acceptance are outside the control of an operator [12], from a practical viewpoint, by designing questions carefully and ensuring that consumers examine products in the same light, it is often possible to reduce some of the sources of variation and hence provide the operator with more consistent answers. Some of the more obvious factors to be borne in mind are given below.

3.3.1 Environment and context

The acceptability of a food often depends on the meal it is being considered for, other items it is to be eaten with, the actual eating environment and the importance of the occasion at which it is to be consumed.

3.3.2 Products being considered

When examining general commodities, declared acceptability will probably depend on the specific item the consumer is thinking of at the time of being questioned.

3.3.3 Product associations

If one is dealing with branded products or products that are easily recognizable or are immediately associated with recognizable products, factors such as brand images, and associated price structures may well influence how a product is rated.

3.3.4 Availability and familiarity

If people are unfamiliar with a product, or product type, they will have no firm views as to its acceptability. Minor factors may then sway judgements, giving rise to unreliable responses.

3.3.5 Variety

In many instances consumers like variety in what they eat. The instant acceptability of a product therefore may well be influenced by what that person has been consuming over the last week or so. Preference information is not static.

3.3.6 Interviewer influence

People are often accused of giving responses that they believe the interviewer wants to hear. Responses may therefore vary depending on the nature of the interview and interviewer.

3.4 Methods for exploring underlying causes of acceptance

Informal interviewing of consumers of a product is one obvious way of discovering underlying causes of acceptance. Over recent years, however, such approaches have been augmented by a number of more formal structured interviewing, evaluation and statistical modelling techniques [9,10,13]. Although there are obvious overlaps and links, these may be divided broadly as follows:

1. Regression modelling between potential causative characteristics and acceptance
   (a) Simple modelling
   (b) Formal descriptive assessment/objective mapping coupled with fixed and free surface response modelling of acceptance
2. Trade-off modelling and conjoint analysis.
3. Behavioural modelling.
4. Information process network modelling

All of these in their first stages involve some form of discussion and eliciting of factors that potentially influence the consumer. Group discussions and various repertory grid interviewing techniques [14] are often used in such explorations. In the more analytical of these approaches, terms are defined and perceptual maps produced showing how various samples relate to one another and the significant aspects giving rise to variation. These techniques also allow the scientist to determine how individuals compare in the way they view samples. In the case of approaches using freely elicited terminology, such maps and comparisons are made independently of the terms used, the mathematics permitting the operator to discover not only how products relate to one another on an individual basis but what terms individuals are using to describe common dimensions.

Similarity/dissimilarity, grouping and individual mapping approaches can also be used to explore underlying dimensions pertaining to acceptance but require external information if dimensions are to be characterized.

Free-choice profiling and Procrustes analysis [15,16] is particularly useful in the context of understanding acceptance as it operates with the minimum of training, providing information closer to how consumers view products and the factors they consider important. When dealing with sensory data, relating such evaluations to those obtained by trained panellists, determines, in so far as the latter evaluations reflect norms and reality, how such individual perceptions relate to the real world. As a consequence, it forms an ideal link between acceptance and trained-panel product evaluation.

3.4.1 Regression modelling

Modelling of such consumer perceptions using a variety of techniques, (e.g. least-squares regression, partial least-squares regression, Procrustes analyses, OVERALS and allied techniques [17]) either using raw data, or

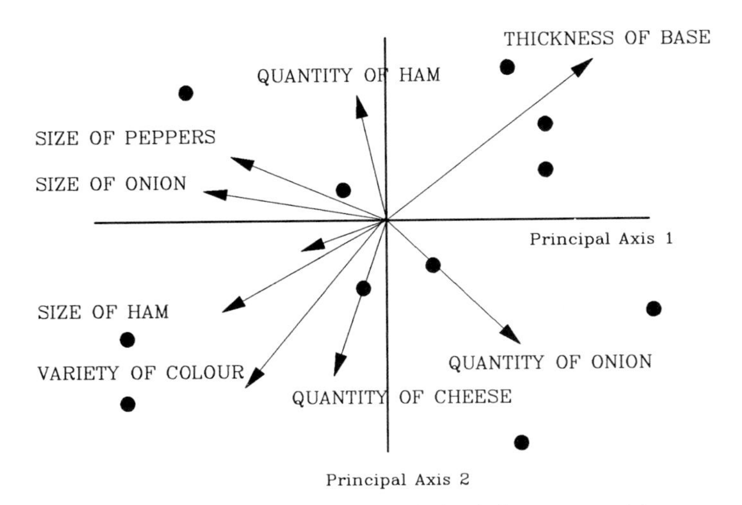

Figure 3.5 Principal component analysis of pizza sensory data.

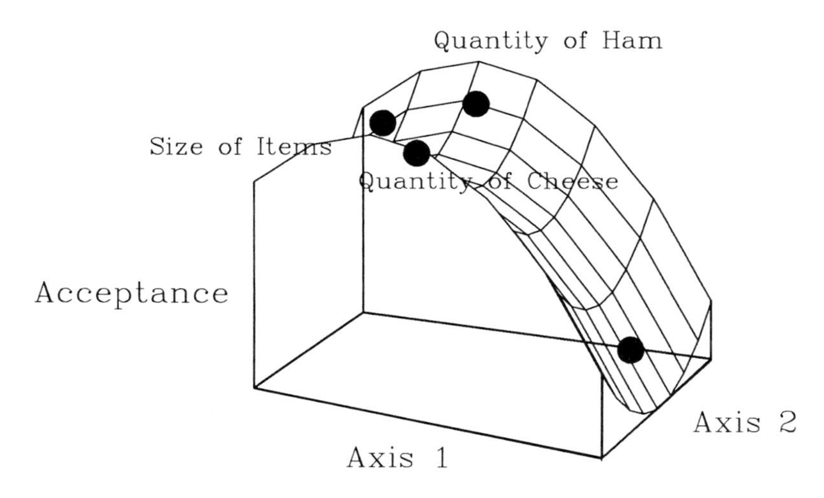

Figure 3.6 Surface response modelling of the acceptance of pizzas using predefined models.

data rationalized by some form of perceptual mapping provides the link between acceptance and perception.

Regression modelling in various guises using raw data has been widely used by several researchers [8,10,13,18,19]. Our own approaches have tended to be based on the modelling of acceptance information against objective descriptive data rationalized by principal component or other multivariate descriptive techniques [3,9]. Figures 3.5 and 3.6 provide an

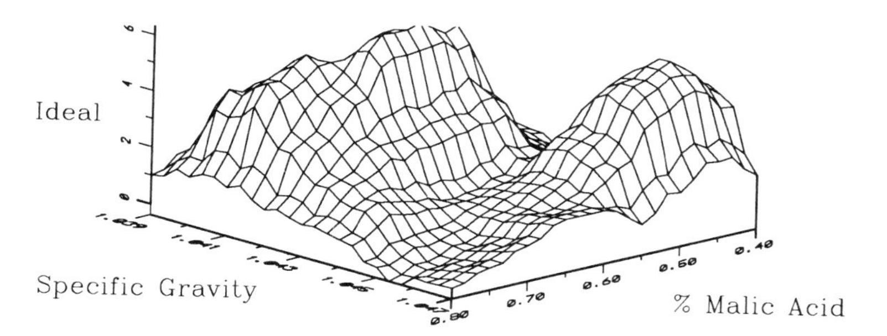

Figure 3.7 Surface response modelling of the acceptance of apple juices using free modelling.

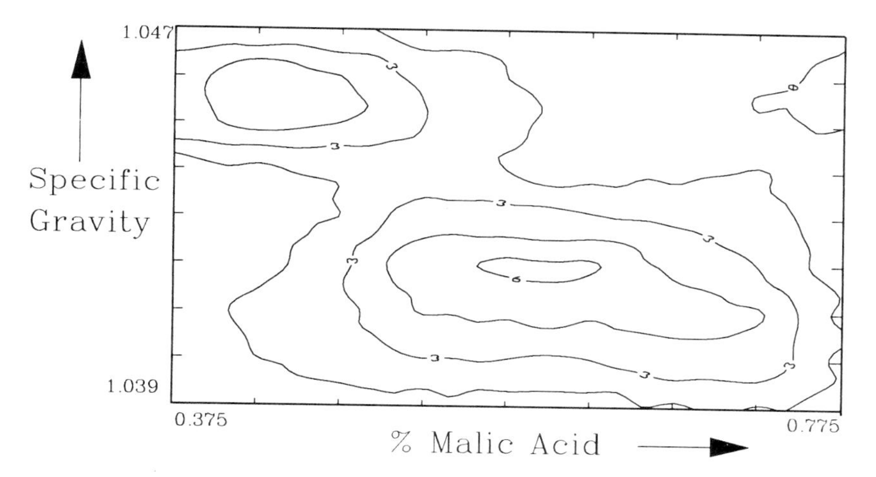

Figure 3.8 Contour plot of acceptance of apple juice using free modelling.

example of such modelling based on predefined models, clearly showing what the consumer wanted in a pizza. In this instance, the requirements were distinct ingredients of a reasonable size and not too thick a base.

A second example (Figures 3.7 and 3.8) shows the free modelling of acceptance information against sweetness and acidity in apple juices [20]. It shows clearly that two populations exist, each requiring different levels of acidity in their products.

From investigations of usage, demography and lifestyle information, it was shown that such groups emerged because of the different usages. One group was using apple juice as a breakfast drink, requiring something fresh and crisp, while the others were rating products as a drink to be drunk with meals any time of the day.

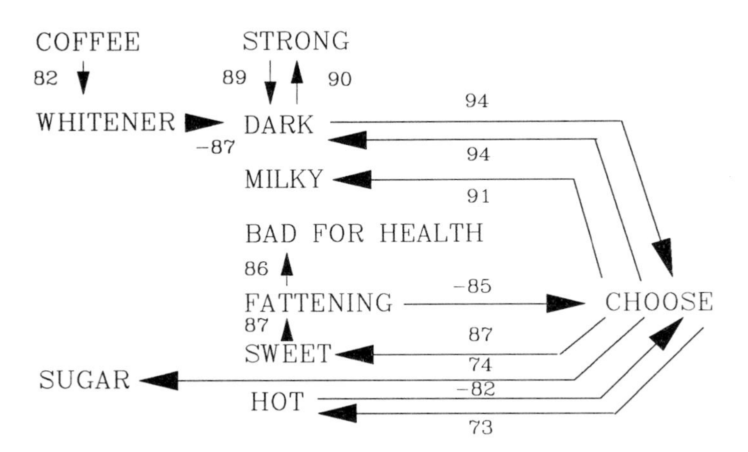

Figure 3.9 Significant linear relationships between ingredient attributes and choice in a series of coffee formulations (after Booth and Blair [13]).

Free-modelling approaches are obviously less restrictive than fixed modelling and recently have provided more reliable pictures of the factors influencing acceptance in several commercial products.

Booth and Blair [13], using ideal relative generated data, used simple regression modelling to determine the association between descriptors, ingredients and choice in a series of coffee/sugar/whitener formulations. Their results are illustrated in Figure 3.9 (only significant relationships are given).

By exploring inter-relationships between potential causative factors [21] one is moving towards a more structured understanding of food acceptance.

All the examples so far have concentrated on exploring the use of modelling approaches for optimizing sensory characteristics. The techniques can, however, be used equally well to explore non-sensory, imagery and functional aspects and how they are traded-off one against the other.

3.4.2 Trade-off modelling and conjoint analysis

Such models [22,23] have not yet found wide application in the sensory area but have proved extremely valuable in evaluating the importance of factors in more material goods and services.

By presenting assessors with a series of pairs of combinations of characteristics real and/or imaginary and asking them to make trade-offs and state which combination they would choose, the procedure tells the operator a great deal about respondents' values, thus building up a picture of the combination of attributes a product must have to be successful.

MOST LIKELY TO CHOOSE PRODUCT 1	**PRODUCT 1**
1	Flavour 2%
2	Sugar 1%
3	Acid 0.75%
4	Colour 1%
5	
6	**PRODUCT 2**
7	
8	Flavour 0.5%
9	Sugar 1.5%
	Acid 0.5%
MOST LIKELY TO CHOOSE PRODUCT 2	Colour 1%

Figure 3.10 Typical trade-off modelling questionnaire.

Preliminary questioning ensures the most appropriate combination of factors are initially selected for evaluation. The procedure can also make use of results already acquired to help select the most appropriate combinations of attributes for subsequent comparisons, hence arriving at important factors with the minimum of questioning and sampling.

A typical example of the sort of question asked in such trade-off modelling is given in Figure 3.10.

3.4.3 Behavioural modelling

In an attempt to understand the factors underlying people's choice of foods and beverages, several behavioural models have been postulated [24–27]. Those of Fishbein and Ajzen [24] and also Triandis [26] seem to be the most popular and productive in the food area.

All models assume some reasoned behaviour. The Fishbein model states that behavioural intent is determined by a person's attitude (A_{att}) towards the product (what he or she thinks about it) and his or her subjective norm (SN), i.e. what he or she believes people who have influence over them believe they should do. Behavioural intent can therefore be represented by the following equation.

$$I = w_1 \times A_{att} + w_2 \times \text{SN} \tag{3.1}$$

The Triandis approach is based on two models for predicting human behaviour. The first of these relates to behavioural intention, which is determined by three factors:

1. Behaviour (A). The individual's appreciation of the product.
2. The perceived consequences of the behaviour (C).
3. Social factors (S). The pressures imposed by the outside world.

The second model states that the probability to act depends on two factors:

1. The strength of habit.
2. Facilitating conditions.

In the *locus of control model* developed by Rotter [28] behaviour is controlled by internal and external factors.

Typical questions explored to give information on the various aspects of such models are shown in Table 3.1 [29].

Both the Fishbein and Triandis models have been used successfully with foods [29,30]; that of Triandis because it takes into account the development of habits, possibly proving the more appropriate to the food situation. Both models, however, fall down on several issues:

1. They are all linear.
2. They do not allow for interactions.
3. They do not allow for correlations between factors.

Perceptual mapping of potential contributing factors and the estimation of salient dimensions using such techniques as principal component analysis [29] would enable orthogonal factors to be abstracted. The use of various nonlinear modelling techniques could also enable more interactive models to be created and trade-offs to be established.

3.4.4 Information-processing-network approaches

The way in which consumers process the information they receive is obviously closely related to the way their minds work and how they see and react to a product. Understanding the framework within which an individual processes the information will enable the food scientist to get a deeper understanding of the way people are influenced when they are presented with foods and beverages.

Human minds, unlike a computer, cannot use all information freely, neither do they always act rationally; however, they select information and process it according to predefined pathways.

Networking, means-end or schemata approaches [31–35], which attempt to understand these pathways, have been used primarily to assist in the design of advertising. However, understanding the cognitive structure activated at the time a stimulus whether from advertising or foods themselves is received, how that information is processed, the various factors influencing that process and at what stage, will also help one to

Table 3.1 Items scored in the behavioural modelling of ice cream

Component	Operation
Attitude/effect	Rate your degree of liking of ice cream. Base your rating on the types, flavours or brands you use most frequently. Dislike extremely 1 2 3 4 5 6 7 Like extremely
Subjective norm	Most people who are important to me think that I should eat ice cream. Strongly disagree 1 2 3 4 5 6 7 Strongly agree
Consequences	Rate ice cream according to how good or bad it is for you. Extremely bad 1 2 3 4 5 6 7 Extremely good
Social norm	According to nutritionists, I should eat ice cream. Strongly disagree 1 2 3 4 5 6 7 Strongly agree
Facilitating conditions	Rate the ease of difficulty in obtaining ice cream when you feel like eating. Extremely difficult 1 2 3 4 5 6 7 Extremely easy.
Habit	I eat ice cream out of habit. Not at all 1 2 3 4 5 6 Very much so
Intention	I intend to have ice cream whenever I get the next opportunity, either as part of a meal or as a snack. Extremely unlikely 1 2 3 4 5 6 7 Extremely likely.
Frequency of consumption	How often do you eat ice cream? Never or almost never = 1; 1–2 times a month = 2; once a week = 3; 2–3 times a week = 4; at least once a day = 5.
Consumption compared with others	Estimate your consumption of ice cream, when compared with the consumption of other people of your age. Very low = 1; low = 2; the same = 3; high = 4; very high = 5.
Nutritional locus of control	My family's food choices often determine what I eat. Strongly disagree 1 2 3 4 5 6 7 Strongly agree
Nutritional locus of control	My choice of foods is greatly influenced by unpredictable circumstances. Strongly disagree 1 2 3 4 5 6 7 Strongly agree

Source: Tourila and Pangborn [29].

understand more fully the relationship between relevant factors and acceptance in its entirety.

Means-end models or schemata can be considered as structured images of a product based on all the characteristics of that product and their

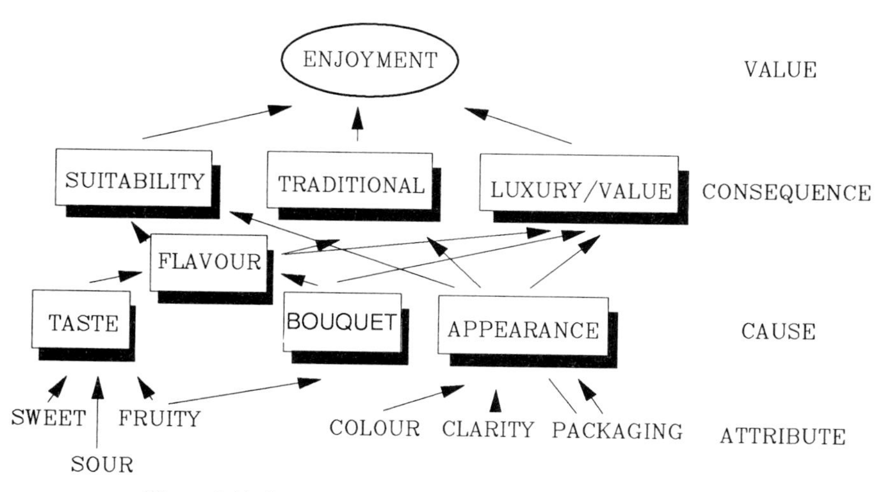

Figure 3.11 Suggested pathway for the enjoyment of a wine.

evaluations developed within a person's mind as a result of personal experiences.

All consumers have several pathways in their heads. In total they form an interconnecting network enabling the respondent to draw on all past knowledge and experience. Interpreting a stimulus is not a passive process, as might be implied from earlier models but a structured/organized step-by-step active process. The more frequent a pathway is used the more habitual the process becomes. From the point of view of understanding product acceptance two basic pathway types play a role:

1. Self schemata. These enable the consumer to put products into a personal world and compare them with aspirations – what sort of person we believe we project of ourselves when purchasing and consuming certain products. Self schemata enable people to behave consistently. Deviations from norms cause tension that a person is motivated to reduce. This need to react distorts the person's view of reality.
2. Brand products schemata. These play an important role in acceptance. Experience with brands as well as product attributes, risks and functionality are important cognitive factors, contributing to consumers' thought processes in the area of product acceptance.

Most pathways follow an attribute consequence value hierarchy typical of that shown in Figure 3.11.

Determining such pathways depends much, at the moment, on using interviewing techniques and the use of repertory grid, laddering [36] and pyramidal approaches for determining increasingly subordinate constructs. However, linking such ideas with other analytical techniques (e.g. conventional and free-choice profiling and various step-by-step regression modelling between potential interdependent factors using both linear and non-linear models) could provide some insight into the way in which information is processed and the interdependence of factors in a chronological sense. Once pathways have been established, more complex and useful models than those derived from either the Triandis or Ajzen and Fishbein approaches for understanding food choice could be established.

3.5 Conclusions

The various problems encountered in measuring and understanding acceptance in foods and beverages have been described in this chapter. Acquiring acceptance data, although itself possessing its own problems, is nothing to the unravelling of the causes of food choice. Various descriptive approaches and associated mapping techniques – particularly those that permit information to be obtained from consumers with the minimum of training, are helping to bridge the understanding gap between acceptance and objectivity, providing a picture of how individuals perceive the real world. Linking such information to acceptance data using the advanced modelling methods now available, as well as the various behavioural concepts of people like Fishbein, gives some insight into causes of choice in foods but fails to take into account the true chronological thought processes going on within an individual when making acceptance decisions. The concepts of internal network information processing already finding applications in the advertising world, when associated with various modern mathematical modelling approaches, could begin to throw new light onto this complex problem.

References

1. Thomson, D.M.H., (ed), *Food Acceptability*. Elsevier Applied Science, London, 1988.
2. Williams, A.A. and Atkin, R.K., *Sensory Quality in Foods and Beverages Definition, Measurement and Control*, Ellis Horwood, Chichester, 1983.
3. Williams, A.A., Optimising flavours and fragrances for international markets, in *Seminar on Research for Flavours and Fragrances*, Esomar, Amsterdam, 1989, pp. 109–120.
4. Williams, A.A. and Martin, D.C., The use of modern sensory analysis in market research, in *Proceedings of the Market Research Society 31st Annual Conference*, The Market Research Society, London, 1988, pp. 289–305
5. Williams, A.A., Optimising formulations in cosmetics, toiletries and associated products, *Parfumerie und Kosmetik*, 1990, **71**, 766–774.
6. Meiselman, H.L., Consumer studies of food habits, in *Sensory Analysis of Foods*, (ed. J.R. Piggott), Elsevier Applied Science, London, 1988, pp. 267–334.
7. Peryam, D.R. and Pilgrim, F.J. Hedonic scale for measuring food preferences. *Food Technol.*, 1957, **11**, 9–14.
8. Conner, M.T., Haddon, A.V. and Booth, D.A., Very rapid, precise assessment of

effects of constituent variation on product acceptability: Consumer sweetness preferences in a lime drink. *Lebensm-Wiss. u Technol.*, 1986, **19**, 486–490.

9. Williams, A.A., Procedures and problems in optimising sensory and attitudinal characteristics in foods and beverages, in *Food Acceptability*, (ed. D.M.H. Thomson), Elsevier Applied Science, London, 1987, pp. 297–310.
10. Moskowitz, H.R., Sensory segmentation and simultaneous optimisation of products and concepts for developing and marketing of foods, in *Food Acceptability*, (ed. D.M.H. Thomson), Elsevier Applied Science, London, 1987, pp. 311–328.
11. Green, P.E. and Schaffer, C.M., A reduced-space approach to the clustering of categorical data in market research. *J. Market Res.*, 1989, **30**, 267–288.
12. Khan, M.H., Evaluation of food selection patterns and preferences. *CRC Critical Reviews in Food Science and Nutrition*, 1981, 129–53.
13. Booth, D.A. and Blair, A.J., Objective factors in the appeal of a brand during use by the individual consumer, in *Food Acceptability*, (ed. D.M.H. Thomson), Elsevier Applied Science, London, 1987, pp. 329–346.
14. Fransella, F. and Banister, D., *A Manual for Repertory Grid Technique*, Academic Press, London, 1977.
15. Williams, A.A. and Langron, S.P., The use of free choice profiling for the evaluation of commercial ports. *Journal of Science of Food and Agriculture*, 1984, **35**, 558–568.
16. Arnold, G.M. and Williams, A.A., The use of generalised Procrustes techniques in sensory analysis, in *Statistical Procedures in Food Research*, (ed. J.R. Piggott), Elsevier Applied Science, London, 1986, pp. 223–252.
17. Rogers, C.A., Williams, A.A. and Collins, A.J., Relating chemical/physical data in food acceptance studies. *Food Quality and Preference*, 1988, **1**, 25–31.
18. Conner, M.T., Booth, D.A., Clifton, V.J. and Griffths, R.P., Individualised optimisation of the salt content of white bread for acceptability. *Journal of Food Science*, 1988, **53**, 549–554.
19. Booth, D., Individualised objective measurements of sensory and image factors in product acceptance. *Chemistry and Industry*, 1987, 441–446.
20. Williams, A.A., Langron, S.P. and Arnold, G.M., Objective and hedonic sensory assessment of ciders and apple juices, in *Sensory Quality* in *Foods and Beverages Definition, Measurement and Control*, (eds A.A. Williams and R.K. Atkin), Ellis Horwood, Chichester, 1983, 310–323.
21. Lewis, V.J. and Booth, D.A., Causal influences within an individual's dieting thoughts, feelings and behaviour. Report of an EC Workshop, Giessen, West Germany, 1–4 May, (eds J.M. Diehl and C. Leitzmann), Euronut, 1985.
22. Cattin, P. and Wittink, D.R., Commercial use of conjoint analysis, a survey. *J. Marketing*, 1981, **46**, 44–53.
23. Johnson, R.M., Adaptive Conjoint Analysis. Private publication, Sawtooth Software Inc., Idaho, USA, 1987.
24. Ajzen, I. and Fishbein, M., *Understanding Attitudes and Predicting Social Behaviour*, Prentice Hall, Englewood Cliffs, New Jersey, 1980.
25. Becker, H., Maiman, L.A., Kirscht, J.P., Haefner, D.P. and Drachman, R.H., The health belief model and prediction of dietary compliance: A field experiment. *Journal of Health and Social Behaviour*, 1977, **18**, 348–366.
26. Triandis, H., Values attitudes and interpersonal behaviour, in *Nebraska Symposium on Motivation*, (eds H.E. Howe and M.M. Page), University of Nebraska Press, Lincoln/London, 1980, pp. 195–259.
27. Wallston, B.S. and Wallston, K.A., Locus of control and health: A review of the literature. *Health Education Monographs*, 1978, **6**, 107–117.
28. Rotter, J.B., Generalised expectancies for internal versus external control of reinforcement. *Psychol. Monographs*, 1966, **80**, all page.
29. Tuorila, H. and Pangborn, R.M., Behavioural models in the prediction of consumption of selected sweet, salty and fatty food, in *Food Acceptability*, (ed. D.M.H. Thomson), Elsevier Applied Science, London, 1987, pp. 267–282.
30. Shepherd, R., Consumer attitudes and food acceptance, in *Food Acceptability*, (ed. D.M.H. Thomson), Elsevier Applied Science, London 1987, pp. 253–266.

31. Gutman, J. A means-end chain model on consumer categorization process. *Journal of Marketing*, 1982, **46**, 60–72.
32. Gunert, K.G., Research in consumer behaviour: beyond attitudes and decision-making. *European Research*, 1988, **16**, 172–83.
33. Zeithman, V.A., Consumer perceptions of price quality and value: A means-end model and synthesis of evidence. *Journal of Marketing*, 1988, **52**, 2–22.
34. Walker, B., Celsi, R. and Olson, J.C., Exploring the structural characteristics of consumer knowledge, in *Advances in Consumer Research 16*, (eds. Wallendorf *et al.*, Association of Consumer Research, Provo (UT), 1987, pp. 17–21. Also in *Proceedings of the Annual Conference of the Association of Consumer Research (ACR) 16–18 Oct Toronto*, 1987.
35. Beijk, J. en van Raaij, Schemata: Informatieverwerking Beinvloedingsprocessen en Reclame; Pre-Advies aan de VEA (Nederlandse Vereniging van Erkende reclame-adviesbureaus). Amsterdam: s.n. 1989.
36. Reynolds, T.J. and Gutman, J., Laddering theory methods analysis and interpretation. *J. Advertising Research*, 1988, **28**, 11–31.

4 Psychology and psychophysiological measurements of flavour

S. VAN TOLLER

Abstract

Psychologists have traditionally analysed smell flavour by using oral and psychometric methods, which are often contaminated by subjective interpretations. One way of overcoming these types of problem has been to use the methods of psychophysiology in which bodily signals are used as indices of a measure of a subject's reaction. The chapter presents a general review of the olfactory research being carried out by the Olfactory Research Unit in the Department of Psychology.

Over the last 5 years, techniques involving the use of brain electrical activity mapping (BEAM) have been explored in a variety of olfactory investigations. The main value of BEAM is that it allows brain cortical activity to be analysed in real time. Studies denote that statistically reliable effects have been obtained showing the effects of fragrances and food smells in subjects from 12 weeks of age to mature adults.

4.1 Introduction

Intriguingly, taste is extremely limited both in terms of actual number of basic receptor types and also in the total number of taste cells located on the tongue and in the buccal cavity. They number in the order of 10 000 which is few in terms of the usual expected number of receptor cells. When people speak of a food 'tasting nice' they are usually referring to its flavour, which is derived from a complex interaction of smell, texture, temperature and colour, as well as a host of individual variables relating to learning, cultural and social attitudes. However, one of the most important variables is the smell. Attention has been drawn to the importance of sensory, attitudinal and psychosocial factors in food choice and eating behaviour [1]. It has been stated that there is a need to create a scientific understanding of food acceptance and nutrition based on the quantification of these factors [1]. Boakes *et al.* [2] have edited an important book that draws attention to the need to understand how food habits are learned. Although in this chapter comments will be restricted to *Homo sapiens*, much research has been carried out on that other omnivore *Rattus rattus*.

In the introduction to their book when talking about differences in the eating behaviour of rat colonies sharing identical ecological niches, Boakes *et al.* have highlighted neatly the interrelationship between human and animal food behaviour studies in their remark, '*one rat's meat can become another rat's poison*'. In particular, the study of animal behaviour can shed light on the specific brain mechanisms of motivated behaviour concerned with eating and drinking. An example is the suggestion that there are specific patterns of motivated behaviour in the brain that contain models of relevant aspects of both the internal physiological and the external world [3]. The interaction between these two aspects enables animals to direct their behaviour appropriately and purposefully. For example, in the laboratory situation, animals have been shown to be better at locating food that is stored in several places in a maze but multiple location is a handicap so far as water is concerned. In real life this reflects the fact that whereas the water-hole is in a fixed location, food may be found in many places. This suggests the presence of an evolutionary brain mechanism predisposing the animals to this fact. Is it likely that this type of consideration has implications for humans? Perhaps the seemingly universal liking for sweetness has a genetic basis.

4.2 The psychological study of eating behaviours

Psychologists have concentrated traditionally on the abnormal and the cognitive abilities of eating behaviour; however, the 'how' and 'why' of food preference has been largely ignored. Similarly, there has been very little interest in how emotional and feeling states effect and result from eating. One area that has been investigated concerns abnormal food behaviours. Abnormality can often be revealing; however, abnormal and normal are not the opposites but rather two ends of a continuum. Bulimia is an example of an eating disorder involving overeating or binging, while anorexia nervosa involves massive undereating and both of these conditions can reveal interesting facts relating to human dietary intake. The classical notion about Billy Bunter being a fat schoolboy who would make any effort to gain food is in reality far from the truth. It has been shown that fat people actually make less effort to gain food. Similarly, overweight rats are easily dissuaded from feeding if they are required to make an effort to get their food. In one study, obese subjects drank less milkshake that had been adulterated with quinine than normal-weight subjects [4]. It seems that overweight people are less dominated by their internal physiological eating cues and are easily influenced by external sensory cues. Thus, people within the 'normal-for-their-body-size' weight range tend to label their state of hunger according to their bodily needs, while overweight people tend to regulate their food intake according to external stimuli. For a general review of this area, Logue [5] has written an excellent textbook on the psychology of eating and drinking.

Psychologists have shown that humans and animals have dietary boundaries rather than internal physiological set-points and these boundaries can be modified by cognitive factors. For example, it has been shown that colouring an odorous solution results in the solution being rated as a stronger solution [6]. The authors concluded that this arose because the colour induced a weak odour percept combined with the odorant-induced percept. Sensitivity to a particular odour note can be critical. The problem that parents often encounter in getting their children to eat green vegetables may, in some cases, relate to differences in sensitivity. The child with a lower threshold for bitterness is more sensitive than the parents whose threshold for bitterness is higher. The parents, using their own bitterness standards, deny that there is anything wrong with the taste and try to get the child to eat the food. This raises issues about eating and food appreciation in the elderly [7]. Such problems are clearly critical in food and flavour appreciation and they await research clarification.

How are food-related perceptions acquired? Is the brain genetically 'hardwired' for this type of sensory information, or is it 'softwired', i.e. the result of learning? The possibility that there may be brain mechanisms predisposing us to respond in certain ways has already been mentioned. Cain [8] points out that taste is often held to be an analytical sense, while smell is held to be a synthetic sense. Using this argument, taste is hardwired into the brain, while smell is a software package. Cain argues that taste and smell categorizations, like other perceptual categorizations, are not inborn but unfold from our experiences and are therefore idiosyncratic and highly personal. McBurney [9] has questioned the breaking down of food perception into component parts. He draws attention to the fact that, in a Gibsonian sense, food percepts are holistic and to break down flavour into its separate parts is a highly artificial thing to do. In the perception of food, attention is focused on the object. This is also the case for the sense of smell where the name given to an odour is nearly always that of the object producing the smell. One speaks of a fishy smell rather than giving the chemical name trimethylamine, or the smell of a rose rather than phenylethyl alcohol, which is the chemical name of the rose smell. However, one must remember that it is the component sensations that produce the hedonic tone, making an object desirable or undesirable. For example, odour memory is largely episodic and a food odour associated with food poisoning can have powerful aversive effects in a single trial.

Despite the paucity of humans in decribing smells, chefs, wine tasters and perfumers in particular, can use sensory descriptors and categorize smells in meaningful ways. However, 'smell experts' can often and fairly easily be fooled by smells, and it is a rare perfumer who receives the accolade of being called a 'nose'. Training panellists for food evaluation can be a very difficult and time-consuming business. O'Mahoney [10] has published a review outlining the difficulties in getting concept alignment between panellists using descriptive analytical techniques.

4.3 Psychophysiological techniques

One way psychologists have used to get around the difficulty subjects have in labelling has been to use techniques of psychophysiology. These involve using subvert signals from the body as indices of a subject's reaction to experimental manipulation. Characteristically the measures used include heart rate, skin conductance, movement of the pupillary muscles of the eye, blood pressure, the constriction and dilation of peripheral blood vessels, muscle tone, and electrical activity of the brain. The advantages of psychophysiological techniques are that they are recorded in real time and a subject usually has low awareness of the particular physiological system being recorded.

Of course, the use of such techniques brings problems. Probably the worst is the need to attach subjects to the measuring instruments being used and also for the subjects to sit still during recordings. Muscle activity from movement can easily obliterate delicate internal electrical signals from a single physiological system. This usually means that it is not a particularly natural environment for the subjects. However, in certain situations, this may not be too much of a problem and with increasingly refined and miniaturized measuring techniques the processes are gradually becoming less intrusive. Even given these contraints, psychophysiological methods can reveal information unavailable by other techniques. Their greatest value is that they overcome subjective responding. The great 'bugbear' in most psychological investigations arises from subjects attempting to guess the purpose of the experiment. In the author's experience the instruction to subjects, 'play it off the top of your head', is the hardest one for a subject to obey.

4.4 R.W. Moncrieff

It is appropriate and a privilege to be able to mention here the work of the great Scottish olfactologist, R.W. Moncrieff, who was so influential in the mid decades of this century for his attempts to understand the role of smell in human behaviour. Among his many olfactory interests were his pioneering attempts to use psychophysiological techniques to examine the influence of odour. Credit is due to him for recognizing the value of techniques that did not rely on subjective reports. At various times he examined techniques for comparing threshold concentrations of ocular and olfactory trigeminal irritants [11] and a subject's psychogalvanic response to smells [12]. In addition, he also examined electroencephalographic (EEG) activity in response to odours [13,14].

Moncrieff's EEG technique involved placing a line of four electrodes along either side of his subject's head. He used a range of pleasant smells that included night-scented stock, French lavender, synthetic ambrette musk, patchouli as well as the herb chive and the spice cinnamon. The unpleasant smells he used were: pyridine, carbon disulphide and ammonium sulphide. He reported that the alpha (8–14 Hz) band of EEG activity decreased during the presentation of the odours; sometimes this was for a short, initial period during an odour presentation. Moncrieff mentioned that with some odours he obtained localization effects in that certain of his electrodes showed changes not found in the others. From the suggestion by Grey Walter that alpha-band EEG was a scanning response, Moncrieff proposed that the alpha frequency was a scanning response for smell. Thus the waxing and waning of the EEG alpha potential stopped when a smell was perceived. He candidly reported that his proposal received no support from two consulted EEG authorities. He concluded by saying, 'If regularities of EEG response were found to come from certain groups of odorants, the groupings might prove to be of practical value to the compounder of perfumes.' [13]. The recording techniques at that time were limited and his studies lacked appropriate controls. From a drawing of his equipment one infers that the subjects might well have been aware when an odour was about to be introduced into the airflow blowing into their faces. Moncrieff tended to generalize from the particular and he used few subjects in his studies. This is well illustrated in his later book dealing with olfactory preferences where he states 124 laws relating to odour preferences [15]. Many of these laws would not evoke criticism but Moncrieff's biases and lack of proof for his generalizations are very obvious. However, his laws have a value in that they are the undoubted opinions and distilled wisdom of an expert in the area of olfaction, but they lack vital requirements for consideration as valid research studies.

Subsequently, other workers have related EEG to stimulation of the olfactory sense. In an early paper dealing with evoked potentials and odour administration the authors quote three papers published during the early 1960s that report the recording of EEG during odour administration [16]. A review by Lorig discusses certain aspects of EEG and odour administration [17].

4.5 Averaged evoked potentials

Over the last decade the techniques and apparatus to measure EEG have improved enormously. Basically there are now two main research techniques for measuring brain electrical activity in research investigations. The first involves averaging repeated stimulations in a technique called evoked potentials (EPs). The reason for the averaging is that one is dealing with a tiny electrical signal, perhaps 12 μV in amplitude, in the very noisy electrical environment of the brain where overall, the total electrical activity will be in the order of 60 μV. Thus, there is a need to repeat the stimulation many times. As the number of trials increase the EP signal will slowly become apparent as a nonrandom activity. The value of this technique is that it allows us to study the electrical activity to a standard sensory stimulation. Kobal and his colleagues [18] have carried out an extensive and very careful set of studies looking at the effects of stimulation of olfactory evoked potentials (OEPs) and chemosensory evoked potentials (CEPs) on brain cortical activity.

An early report on taste-evoked potentials was made by Funakoshi and Kawamura [19]. They examined summated cerebral evoked potentials (CEPs) to a series of dilutions of sucrose, sodium chloride, tartaric acid and quinine hydrochloride; the control was demineralized water. The average number of trials was 40 and, at the moment of recording, 1 ml of the stimulus solution was poured onto the tongue. An early activity peak was elicited by the tactile stimulation of the liquid touching the tongue but the authors reported a later gustatory activity peak. The early activity peak occured at 150 ms, while the later activity occurred at 500–1500 ms. Gustatory EP activity was found to acid and salt but not to the sweetness of the sucrose solution nor the bitterness of the quinine solution. The late EP activity was not produced by the demineralized water and could be enhanced by increasing concentrations of the acid solutions. The reason for the lack of response to the sweet and bitter stimulation was said to be due to the fact that the sweet receptors are found on the tip of the tongue while bitter receptors are found at the base of the tongue. However, the solutions were reported to be kept in the mouth for at least 5 s so this should have resulted in stimulation of all the tongue during this period. A method of contingent negative variation (CNV) effects from various odours has been reported [20]. The CNV is a DC current measure that is found when a subject is anticipating an expected signal. In this case the subject had been given the odour and was then given a sound signal that indicated a light would come on after 2.3 s. Subjects were required to press a button as soon as the light was shown.

4.6 Psychophysiology, EEG and brain electrical activity mapping

The second method has already been outlined in the description of Moncrieff's EEG study. Basically it is the recording of EEG in real time so that the moment the cortex registers reception of the olfactory stimulus is recorded. One of this author's major interests in psychology concerns the study of emotion [21] and it has been a research ambition to attempt to uncover the emotional components of olfaction [22,23].

This author's original research involved measuring the skin conductance of subjects who were receiving smell. From the psychophysiological theories of the time about the decay time of the electrodermal reponse (EDR) that looked to see if pleasant smells resulted in a longer decay time for the EDR and if unpleasant smells resulted in a shorter decay time [24]. The findings did not support this type of interpretation and it was at about this time that it was shown that the decay time of the EDR was a function of the length of the interstimulus period. We then turned our attention to use of the pupillary movement of the eye as a hedonic measure. Initially pleasant smells were thought to produce pupillary dilation and unpleasant smells pupillary constriction; however, it was discovered that all smells produce dilation regardless of hedonic tone.

From the early pilot studies we had shown that subjects placed in olfactory experiments were easily distracted by extraneous visual and auditory stimuli. Moreover, they would actively seek such information to help them with any task they were set. Better olfactory processing from the subjects was achieved when they were perceptually isolated. Accordingly, a series of studies was set up in which the subjects wore visually blacked out goggles and given headphones to reduce extraneous noises. The headphones also allowed the researcher to communicate with the subjects during the experiments. The result was that the subjects, in a condition of partial perceptual isolation and relaxation in a comfortable armchair, found it easy to concentrate on the olfactory tasks. This basic experimental paradigm has been used in all our subsequent studies.

This author realized that a technique was needed to allow the study of EEG brain activity using single trials. It would only be on a single trial that we could be confident of observing effects of novelty and emotion. The brain is very good at ignoring repeated incoming information and multiple presentations tend to be ignored cognitively. This is because our brains 'pigeon-hole' any incoming stimuli. From many experimental studies psychologists have shown that the brain 'sees' what it 'knows' rather than the converse. In our earlier skin-conductance studies, we had found that males took longer than females to process perfume-like smells. We argued that females would be more likely to have a conceptual 'pigeon-hole' for perfumes and could quickly place any incoming fragrant odour into this brain 'slot'. Males, lacking a convenient cognitive brain label, needed to devote more processing time to any incoming perfume-like smell.

Another example was that male subjects who rated the smell of 5-alpha-adrostanone as 'pleasant' produced skin conductance responses that were larger than females with similar ratings. We presumed that females were cognitively classifying the odour as 'perfume-like'. Overall our conclusions from the pilot psychophysiological experiments confirmed the view that they could be effective but in terms of examining brain activity we had a very small time-window in which to observe any effects.

The earlier EEG technique used by Moncrieff [13] used ink pens producing analogue traces on long lengths of folded paper. This method made interpretation very difficult if not impossible. However the introduction of the technique of fast Fourier analysis (FFA) has enabled computers to generate topologies of EEG activity and produce coloured visual maps [25]. Over the last 5 years at Warwick, we have been using brain electrical activity mapping (BEAM) techniques to try to understand what happens on the surface of the brain when the olfactory sensory system is stimulated. Before discussing this, some studies that have involved stimulating and recording from the cerebral cortex will be discussed.

4.7 Neuropsychology and the cerebral cortex

The original cortical-stimulation discoveries were made by the Canadian surgeon, Penfield, and his co-workers who defined the well known cortical homunculus, which is situated either side of the central sulcus. Movement is found on the precentral part of the central sulcus and sensation is represented by postcentral areas of the central sulcus. The areas of the cortex are disproportionate because the cortical area is larger for more critical parts of the body. For example, the face and the finger cortical regions are larger than those concerned with the foot or knee.

Since the original observations, both projection and function maps have been constructed for the neocortical covering of the brain. Functional maps are produced by stimulating areas of the cortex to see what behaviour is elicited, and projection maps are constructed by tracing axons of sensory nerves cells [26]. The value of cortical maps has also been discussed [27]. As mentioned previously, electrical stimulation of the precentral areas immediately in front of the central sulcus produce movement, while stimulation of postcentral areas immediately behind the central sulcus produce somatic sensation. It has also been shown that the visual and the auditory systems have precise neocortical representation with distributions that are as precise as those found for the motor systems. Taste has a tongue-shaped area in the postcentral gyrus region of the neocortex. It has recently been stated that it is not known if the olfactory system has such a representation on the cortex [25]. It has been one of the purposes of the research carried out in Warwick University to discovery if such a cortical representation exists for the sense of smell.

4.8 The Warwick olfaction research methodology

Subjects are brought into the laboratory and prepared for EEG recordings. This involves fitting them with an appropriately sized fabric headcap containing 28 electrodes; the design of the cap positions the electrodes at required points on the scalp. Ground or reference electrodes are clipped onto each ear. The headcap is secured onto the head by a chestband attachment and each of the 28 electrodes is then filled with a saline gel to ensure a good contact between the electrode and the scalp. The normal recording impedance level between the electrode and the skin is better than 4 kΩ.

Initially, subjects were tested in a large well ventilated room but more recently subjects are tested in a specially constructed ventilated low-odour room. As mentioned earlier, subjects are always tested under conditions of perceptual isolation and seated in a relaxed manner in a comfortable chair. Before the experimental trials begin, subjects are instructed to breathe in through their nose and out through their mouth. This process ensures that any residual odour in the mouth or lungs does not stimulate the olfactory receptors via retronasal stimulation. The experimenter checks during each trial that subjects are obeying this instruction.

The experimental procedure is then explained to the subjects and they are reminded that they must breathe in the required manner during the subsequent trials. They are also informed that during a trial they may or may not be presented with an odour and that in any event they are to relax and not consciously think about any smells they receive. Subjects are then fitted with the blindfold and the headphones. All communications during the actual tests are made via a microphone worn by the experimenter and the subject's headphones.

The Neuroscience Imager produces an averaged map for each 2.56 s time-period or frame. During each odour trial, an initial baseline sequence of four frames is recorded (4 × 2.56 s), and this is followed by a further sequence of four frames during which time an odour (or a blank strip in the case of the control trials) is presented to the subject. The odour is removed at the end of the eighth frame and a final four postodour frames are recorded. During the intertrial periods subjects are told that they can move. At the end of the intertrial period (average length 2 min) subjects are told that another trial is about to begin and they are reminded of the need to sit still and to breathe in the required manner.

For each odour trial, twelve frames, each of 2.56 s duration, are recorded. The odours are presented to subjects using perfumers' smelling strips held just beneath both nostrils. In any series of experiments, odours are diluted to isointense odour levels and are presented to a subject in a shuffled sequence. After all the trials have been given, the headphones, blindfold and electrode cap are removed from the subjects. They are then given the previously presented odours to smell again and asked to rate them on three psychometric bipolar scales for pleasantness/unpleasantness; familiar/unfamiliar; strong/weak. They are invited to name the odours and, in certain studies, are asked to provide adjectives or descriptors for the odours. This latter task is always performed poorly by the subjects.

Our EEG recording procedure provides us with real brain time EEG activity recorded from a subject's cortex during the period they are receiving an olfactory stimulation. We also record psychometric ratings of the odours from subjects. Thus, we are in a position to analyse the cortical patterns evoked by the odours and also look for correlations between the psychometric ratings and cortical activity.

4.8.1 The quantification of BEAM maps

The standard topographic maps were useful in the early periods of the research because they gave clues about what sort of cortical activity one could expect. For example, we found that subjects with certain psychometric rating would, after removal of the odour, repeat the cortical patterning they had shown when the odour was first presented [28]. It is important to note that this phenomenon, which has been called 'the reprise', fell in the second frame after removal of the odour. Thus, the odour was presented and produced a certain cortical pattern, upon removal of the odour the original cortical pattern was repeated in a 'reprise'. This phenomenon occurs with odours that subjects feel ambivalence or interest towards. Odours such as the putative male pheromone 5-alpha-androstanone or musks tend to produce the 'reprise' effect. We have also found the same effect with trigeminal stimulants. The point at which subjects rate an odour at the extremes of a psychometric scale has not been found; that is, where they positively like or dislike an odour.

The coloured topographic maps can be very seductive and we decided to download the raw data being collected at the electrodes and to subject these numbers to further analyses. This was achieved in two main ways. The first was to use other graphical techniques to represent the electrode output. For example, subjects will react to 5-alpha-androstanone in one of three ways. They may report that they like the smell; alternatively that they dislike the smell; finally they may show a specific anosmia to the compound and not be able detect it. If these three groups are taken and the individual electrode values are plotted over time, clear differences in the three plots are found. Subjects reporting that they like the smell show a high level of integrated activity. If they report that they dislike the smell they show less integration. Subjects who are unable to detect it show little or no integrated electrode activity. As you might expect, cortical activity in the latter case looks like the control or blank presentation. Incidentally, the control blank presentations have never shown significant differences from pre-odour baseline data.

The other main method was to subject the raw data to statistical analysis using techniques such as multidimensional scaling (MDS). MDS is a descriptive method producing two-dimensional plots where items (in our case electrodes) showing similarities are shown close together on the plot and items showing dissimilarities are shown further apart. This enables us to see which electrodes are reacting to the odours and to see if the electrodes showing reactions are in cortical positions that one would expect. In addition, the psychometric ratings can be studied to see if they correlate with electrode activity. Detailed analysis, using the alpha band (8–14 Hz) of EEG activity, has shown that three statistically significant bands of electrodes are active on the cerbral cortex during odour presentation [29]. Psychometric scores have also been used to construct an MDS configuration that indicates a more frontal statistically significant evaluative area of cortical activity for these measures.

4.9 A BEAM investigation using infants and a longitudinal adult study

Kendal-Reed [30,31] has extended the above experiments by examining EEG responses to food related odours in 12-week-old babies. In his investigations he found that EEG activity in babies of this age was confined mainly to the delta band (0–3.5 Hz) of EEG activity. Young babies do not display the higher EEG frequencies that are characteristic of the adult patterns and, during their first year, mainly show delta EEG activity. Kendal-Reed has concluded that human infants at the age of 3 months show statistically significant patterns of cortical activity in response to the range of food smells used. Moreover, he found active electrodes in a limited area of the cortex. It is interesting to note that the active electrodes he found are a subset of the active electrodes found in the study using adults [29]. Both the adult and the baby EEG studies conclude that the phenomenon was not due solely to an odour producing EEG arousal. In the adult study, some of the responses fell below the level of the blank control stimulus; that is, were both negative and positive. In the infant EEG study, the evoked odour cortical activity was different from cortical activity evoked by a tone used to produce an orienting response. In addition, Hotson and Van Toller [32] have presented preliminary results from a longitudinal study examining EEG responses to olfactory stimulation in a group of young adults over a 3-year period.

4.10 Event-related potentials and food odours

More recently we have begun to use an evoked-potential technique to allow us to examine cognitive factors produced by placing a subject in an ambient odour environment. The technique involves event-related potentials (ERP) [33]. It is often called the 'oddball' paradigm because subjects are required to listen to a series of tones occurring at two sound levels and to decide how many occur at the lower and less frequent sound level. The tones are arranged so that the lower tone level occurs about 10–20% of the time.

In a recent pilot study, we examined changes in the hedonic tone of food smells that occur during a meal. For example, a dessert smell can be aversive if smelt at the beginning of a meal; conversely, a savoury smell can be aversive if smelt at the end of a meal. Our experiment involved bringing subjects into the laboratory over a lunchtime period and testing their ERPs to odours before and after being given a standard meal consisting of fruit juice, soup, roll, butter and cheese. The ERPs were recorded in the Warwick low-odour room with subjects under conditions of perception isolation as described earlier. Headphones served both to eliminate external sound cues and also allowed the tones to be played to the subjects. The procedure involved the subjects coming to the laboratory and being fitted with the EEG headcap and prepared for an EEG recording. Subjects were then taken into the ventilated low-odour room and seated in a comfortable armchair where they listened to the tones.

The odours presented to the subjects were: (i) beef broth cube dissolved in water, to give a brothlike smell; (ii) a chocolate smell, created by using baby food chocolate pudding; (iii) a lemon smell; and (iv) a fishy smell, using a solution of trimethylamine. The material generating the smell was placed in capped plastic cups that allowed a large smell headspace to develop. The amounts and volumes used were adjusted to create isointensive smell levels. In addition to the five odours, we used an empty cup as a control. During the ERP trials the uncapped cups were held below the noses of the subjects to create an ambient food odour. Subjects were instructed to ignore any odours they might feel they could smell and to concentrate on detecting the tones. The odours were presented in a shuffled sequence. After the first set of trials, subjects were taken into an adjacent room and ate the provided lunch. Fifteen minutes after eating the lunch they were taken back into the experimental setting where the experimenters recorded a second set of ERP trials under similar conditions. The results are still being analysed but they reveal differences in ERP, both in terms of latencies and amplitudes, between what we have called preprandial and the postprandial stages.

4.11 Conclusions

Psychophysiological techniques involving EEG and BEAM techniques now being used at Warwick University and other research laboratories throughout the world are giving us valuable new perspectives and starting to shed light on how the cortex of the brain processes olfactory and taste sensations.

References

1. Booth, D.A., Objective measurement of determinants of food acceptance: sensory, physiological and psychosocial, in *Food Acceptance and Nutrition*, (eds J. Solms, D.A. Botth, R.M. Pangborn and O. Raunhardt), Academic Press, London, 1987.
2. Boakes, R.A., Popplewell, D.A. and Burton, J. (eds), *Eating Habits: Food Physiology and Learned Behaviour*. Wiley & Sons, Chichester, 1987.
3. Oatley, K., Brain mechanisms and motivation. *Nature*, 1970, **225**, 797–799.
4. Schachter, S., Some extraordinary facts about obese human and rats. *American Psychologist*, 1971, **26**, 129–144.
5. Logue, A.W., *The Psychology of Eating and Drinking*. W.H. Freeman, New York, 1986.
6. Zellner, D.A. and Kautz, M.A., Color affects perceived odor intensity. *Journal of Experimental Psychology*, 1990, **16**, 391–397.
7. Van Toller, S., Dodd, G.H. and Billing, A. *Ageing and the Sense of Smell*. C.C. Thomas, Springfield, Illinois, 1985 .
8. Cain, W.S., Taste versus smell in the organization of perceptual experience, in *Food Acceptance and Nutrition*, (eds J. Solms, D.A. Booth, R.M. Pangborn, and O. Raunhardt), Academic Press, London, 1987.
9. McBurney, D.H., Taste, smell and flavor terminology: taking the confusion out of fusion, in *Clinical Measurement of Taste and Smell*, (eds H.L. Meiselman and R.S. Rivlin) Macmillan, New York, 1986.
10. O'Mahoney, M., Descriptive analysis and concept alignment, in *Advances in Sensory Science*, (eds H.T. Lawless and B.P. Klein), Marcel Dekker, New York, 1900.
11. Moncrieff, R.W., A technique for comparing the threshold concentrations for olfactory trigeminal and ocular irritants. *Quarterly Journal of Experimental Psychology*, 1955, **7**, 128–132.
12. Moncrieff, R.W., Psycho-galvanic reflexes to odours. *Perfumery and Essential Oil Record*, 1963, **54**, 313–316.
13. Moncrieff, R.W., Effects of odours on EEG records. *Perfumery and Essential Oil Record*, 1962, **53**, 757–760, 825–828.
14. Moncrieff, R.W., Emotional responses to odours. *Soap, Perfumery and Cosmetics*, 1977, **50**, 24–25.
15. Moncrieff, R.W., *Odour Preferences*, Leonard Hill, London, 1966.
16. Alison T. and Goff, W.R., Human cerebral evoked responses to odorous stimuli. *Electroencephalography and Clinical Neurology*, 1967, **23**, 558–560.
17. Lorig, T.S. Human EEG and odor reponse. *Progress in Neurobiology*, 1989, **33**, 387–398.
18. Kobal, G. and Hummel, C. Cerebral chemosensory evoked potentials elicited by chemical stimulation of the human olfactory and respiratory nasal mucosa. *Electro-encephalography and Clinical Neurophysiology*, 1988, **71**, 241–250.
19. Funakoshi, M. and Kawamura, Y. Summated cerebral evoked responses to taste stimuli in man. *Electroencephalography and Clinical Neuropsychology*, 1971, **30**, 205–209.
20. Torii, S., Fukuda, H., Kanemoto, H., Miyanachi, R., Hamauzu, Y., and Kawasaki, M. Contingent negative variation (CNV) and the psychological effects of odour, in *Perfumery: The Psychology and Biology of Fragrance*, eds S. Van Toller and G.H. Dodd), Chapman & Hall, London, 1988.

21. Van Toller, S. *The Function of Human Emotion*. University of Warwick public lecture, 1976. Personal circulation.
22. Van Toller, S. Odours, emotion and psychophysiology. *Internat. Journal of Cosmetic Science*, 1988 **10**, 171–197.
23. Van Toller, S. Emotion and the brain, in *Perfumery: The Psychology and Biology of Fragrance*, (eds S. Van Toller and G.H. Dodd), Routledge, Chapman & Hall, London, 1988.
24. Van Toller, S., Kirk-Smith, M., Wood, N., Lombard, J. and Dodd, G.H. Skin conductance and subjective assessments associated with the odour of 5-alpha-androstan-3-one. *Biological Psychology*, 1983, **16**, 85–107.
25. Duffy, F. (ed.) *Topographic Mapping of Brain Electrical Activity*. Butterworths, Guildford, 1986.
26. Kolb, B. and Wishaw, I.Q. *Foundations of Human Neuropsychology*. Freeman, New York, 1990.
27. Dykes, R.W., and Ruesty, A. What makes a map in somatosensory cortex? in *Cerebral Cortex*. Vol. 5: *Sensory-motor Areas and Aspects of Cortical Connectivity*, (eds E.G. Jones and A. Peters) Plenum Press, New York, 1986.
28. Van Toller, S. and Kendal-Reed, M. BEAM activity to olfactory and chemosensory stimulation: qualitative and quantitative results. *Psychiatric Research*, 1989, **9**, 429–430.
29. Van Toller, S., Behan, J., Howells, P., Kendal-Reed, M. and Richardson, A. An analysis of spontaneous human cortical EEG activity to odours. *Chemical Senses*, 1993, **18**, 1–16.
30. Kendal-Reed, M. Human infant olfaction: responses to food odours measured by brain electrical activity mapping (BEAM). PhD Thesis, University of Warwick, 1990.
31. Kendal-Reed, M. and Van Toller, S. *Infant Responses to Food Odours: Does This Indicate Early Preference?* Proceedings of the Twelth Annual Conference of the British Nutrition Foundation, London, 27th June, 1990.
32. Hotson, S. and Van Toller, S. A longitudinal study of EEG responses to odours. *Chemical Senses*, 1990, **4**, 384–385.
33. Donkin, E., Ritter, W. and McCallum, W.C. Cognitive psychophysiology: the endogenous components of the ERP, in *Event-Related Potentials in Man*, (eds E. Calloway, P. Tueting and S. Koslow) Academic Press, New York, 1978.

5 Matching sensory and instrumental analyses

M. MARTENS, E. RISVIK and H. MARTENS

Abstract

Goals, methods and philosophy with respect to flavour assessment are discussed in this chapter. The intention is to raise 'old' fundamental questions in the context of recent developments in sensory analyses, instrumental analyses and data modelling, hopefully, to stimulate thoughts for the future.

Flavour quality may be looked upon as an interaction between product and man: an event at which awareness of both subject (here a multitude of sensory responses) and object (here a multitude of product stimuli) is made possible. Multivariate psychophysics in combination with multivariate data analysis ensure rapid, relevant and reliable flavour analyses that accommodate the complexity, latency and subjectivity in the total system. Instrumental analyses ought to be calibrated towards human responses before they are acknowledged as flavour assessments.

Qualimetrics is introduced as multivariate systems analysis for matching not only sensory and instrumental analyses but also matching the 'cultures' of humanists and technologists.

5.1 Introduction

The aim of this book is to present recent developments in the origin, assessment and application of biologically produced flavours. An understanding of flavours, aromas and fragrances, and their general use, is no less relevant today than 20 years ago when one of the 'original' flavourists, W.R. Littlejohn [1], strongly stressed that '. . . the nose is the best gas chromatograph . . .'. In the Littlejohn memorial lecture *Art and Science in the Understanding of Flavour* by another great flavourist, R. Harper [2], Harper lists 10 relevant topics for progress in the flavour field, among which developments in computers and statistics are expressed.

The theme 'use of computers in flavour research' has indeed rapidly progressed in the last decades, with respect to both the hardware and the software. When Martens and Harries wrote a bibliography of multivariate statistical methods in food science and technology in 1983 [3], they found that the average yearly number of publications showed:

1945–1969 : 1 paper/year
1970–1974 : 9 papers/year
1975–1979 : 26 papers/year
1980–1982 : 47 papers/year

Although the tendency for an increasing number of papers on matching sensory with instrumental analyses may be found in the literature since 1983, the main impression is that the academic 'sensory' and 'chemistry' worlds are still strongly separated. Most of the chemical flavour research seems to be dazzled by the increasingly more advanced instrumental techniques, while the sensory work turns increasingly towards market research and psychology; for example, see the book by Martens *et al.* [4].

Matching sensory with instrumental analyses, the topic of this chapter, seems to be an obvious topic when dealing with flavour. According to definitions of 'flavour' [5], and emphasis in several papers [6], flavour is '. . . a mingled but unitary experience that includes sensations of taste, smell, pressure, and other cutaneous sensations such as warmth, cold, mild pain' [5]; and '. . . an attribute of foods, beverages, and seasonings resulting from stimulation of the sense ends that are grouped together at the entrance of the alimentary and respiratory tracts – especially odour and taste . . .' [6].

Then, how can we defend the most advanced instrumental flavour work without matching it to the senses? How can we avoid expressing assessment of flavour as an interaction between a multitude of human variables and product variables – as a multivariate system? How can a multivariate system be simplified for mental interpretation and understanding?

The purpose of this chapter is to raise a discussion of goals, methods and philosophy behind flavour assessment. In Figure 5.1, an overview of the various topics to be presented is given. Starting at the top of the figure, flavour and flavour quality is considered as an event occurring in the interaction between product and man, that is, a multivariate system. Qualimetrics will here be used as a concept for matching sensory perception with instrumental analyses in a multivariate way.

5.2 Goals

One needs to match sensory with instrumental analyses for the following reasons:

1. In research: to reveal systematic structures in the real world complexity, using modern technology as well as human resources.
2. In R&D: to interpret underlying, latent phenomena in simplified terms without losing essential information.
3. In production and marketing: to acknowledge the human subjectivity

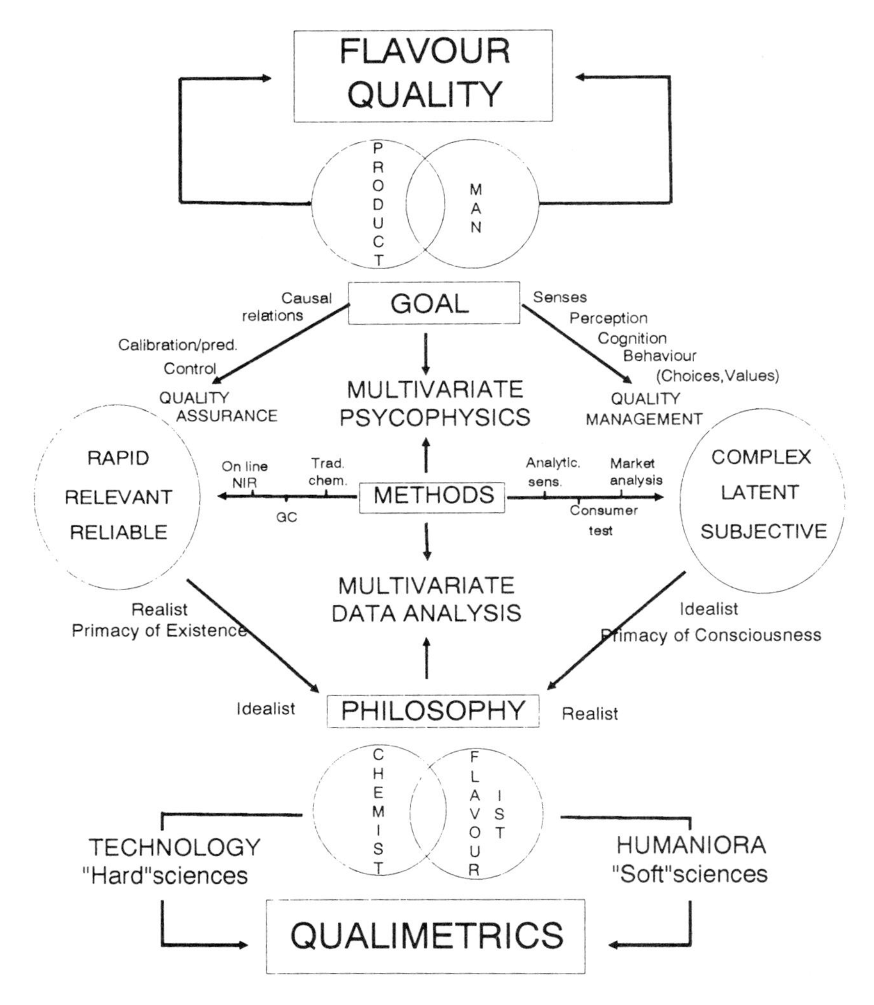

Figure 5.1 Matching sensory with instrumental analyses: an overview.

embedded in experiments, yet strive for objectivity through applying our knowledge in experimental planning and practical work.

5.2.1 *Understanding causal relationships*

To understand the formation and changes in different flavours and aromas, one needs to appreciate basic research in long-term goals. Since 1950, traditional psychophysics has encompassed mathematical formulae to relate stimulus to response (Weber's, Fechner's and Steven's laws [7]). According to Engen:

> In order to understand the behaviour in relation to physical energies that may elicit or control that behaviour, it is valuable to know the relationship between perceived magnitude and physical stimulus magnitude. This is the problem of psycophysical scaling and measurements. (*Engen, 1972*)

Recent reports in the psychophysical literature include time-intensity registrations in the theoretical models [8] and perceived patterns [9] as an interesting additional variable for deeper understanding of perceived taste and smell.

However, this field of flavour research is still, in general, heavily univariate in thinking. In the search for causal relations between molecules and man the more one focuses on the chemical/physical components, on judged performance, on consumer behaviour, choices and values, on culture, health and environment, the stronger is the requirement for multivariate thinking. Multivariate psychophysics is needed simply because the real world is multivariate.

5.2.2 Finding calibration and prediction models

The goals for most of the studies of sensory instrumental relationships related to R&D are:

1. To find relationships between sensory and instrumental (chemical, physical, optical) analyses for, for example, product optimization and research purposes.
2. To find predictive models for rapid on-line instruments for, for example, screening in breeding programmes and process control.

For research and product development purposes a pragmatic approach is often necessary since one may not understand in advance all the phenomena affecting the relationships. Foods and beverages are complex materials; however, they are not infinitely complex. For a given problem there usually exist latent phenomena constituting systematic variations in the complexity, which a multitude of sensory as well as chemical, physical and optical measurements reflect. This common, latent space is to be revealed by multivariate soft modelling.

5.2.3 Application in quality assurance and management

The final, common type of goals for matching sensory with instrumental analyses in production and marketing are:

1. To find rapid methods to control product quality to conform with requirements from producer to consumer.
2. To link sensory and instrumental analyses to other data (economical, production, market data, etc.).

The latter goal points at an extremely important current issue within total quality management, namely to value man as a resource in the production. It is somehow obvious that an operator either in the production or in the laboratory, possesses much useful information about the product in process. This information is gained mainly by sensory observation, thus a link between sensory science and total quality systems is a challenge: i.e. a total quality system that includes necessary sensory analyses and consumer tests and, in addition, methods for registration of operator's observations. The various methods described in the next section may be the means of obtaining the goals that harmonize with consumers, producers, authorities, economical and political considerations, in other words the idea of total quality management.

5.3 Methods

When choosing which sensory, instrumental and computer methods to use, one should endeavour to:

- match the right methods to achieve the goals in a rapid, relevant and reliable way;
- use data analytical methods that let 'nature' talk without forcing unneccesary assumptions from theoretical models upon the data, or without overloading our mental system; and
- overcome the tyranny of methods with our experience, intuition and the craftsman's pride.

5.3.1 Sensory methods

Depending on the goals for matching sensory and instrumental analyses, various sensory methods are available [10]:

1. *Analytical sensory methods* that directly use a group of trained people (a panel) to describe 'objectively' and evaluate properties like colour, flavour and texture, analogous to an instrument (e.g. conventional descriptive profiling and magnitude estimation).
2. *Consumer methods* that directly ask the consumers about their preference or degree of liking of different products (e.g. hedonic, preference or acceptance tests).
3. *Laboratory–consumer methods* that aim at bridging the gap between laboratory flavour research and the consumer [11]. Of recent interest are free-choice profiling [12], dissimilarity scaling [13] and projective mapping [14].
4. *Locus of control* is a measure of orientation toward life [15] and is thus a

value measurement. It is used to describe an individual's belief that a specific issue, such as health, is contingent upon specific behaviour [16].

From a sensory methodological point of view it was once necessary to add knowledge about the psychology of perception to the 'pure' physiological approach, resulting in a concept change from 'organoleptic' to 'sensory' analysis. Now, the question seems to be to what extent, and in which way recent knowledge from marketing research and cognitive science are to influence the sensory methods.

5.3.2 Instrumental methods

While analytical sensory methods are relevant in studies of causal relations, and laboratory consumer methods are relevant in quality analyses, correspondingly, various instrumental methods should be chosen to match the same goals:

1. *Direct methods* for studies of causal relations are used to understand the formation and changes in different chemical components in biological materials (e.g. traditional 'wet' chemical analyses, chromatography, NMR, MS).
2. *Indirect methods* may bypass much of traditional chemistry in product development and quality assurance: e.g. multichannel sensors based on spectrophotometry or chromatography may be calibrated towards sensory profiles, and the model patterns obtained used for future prediction of new samples.
3. *Interference problems* are important in conversion of nonselective measurements to selective information. These problems can be solved by multivariate calibration, which is discussed elsewhere [17].

5.3.3 Computers and data analytical methods

As mentioned in the introduction, use of computers in flavour research has rapidly progressed in the last decades both with respect to hardware and software.

Development of data automation in analytical chemistry first focused on data generation (pre-1960), then changed focus to data acquisition (1960s) to data reduction (1970s) to data management (1980s) for handling technical data [18]. The main breakthrough for data automation of sensory data came in the middle of the 1980s.

Flavour data-sets originated from biological materials may be characterized as multivariate (i.e. with many variables involved), often having more variables than objects, having strong multicollinearity, and possibly containing missing values. Traditional linear regression and discriminant methods cannot easily handle this type of real-world data. Bilinear soft modelling was developed to meet the necessary requirements, and was implied in sensory instrumental work in the early 1980s.

An overview of the various multivariate methods goes beyond the scope of this chapter. The many methods available for matching sensory and instrumental analyses are referred to in other papers in this volume and to literature on applied statistics [19]. However, there are three different classes of data analytical techniques that should be distinguished in the present context: (i) using one data block; (ii) using two data blocks; and (iii) using multiway data blocks.

5.3.3.1 Methods using one data block. 'One data block' refers to one data table: *p* variables (e.g. sensory or chemical) measured in *n* objects (X-block). Before matching sensory to instrumental data it is advisable to look for outliers and for main tendencies of variation in each data block separately by, for example, cluster analysis, principal component analysis (PCA), correspondence analysis or factor analysis (or multidimensional scaling for symmetrical data tables). If there are distinct classes in the dataset as seen by classification analysis (e.g. SIMCA), discriminant analysis or canonical (variate) analysis, then this may be helpful in interpreting nonlinear data sets. Also, during experimental planning, it is important to know the main 'structure' of the data set so that samples representing the different types of variation are obtained [20]. Experimental design based on so-called *principal properties*, as developed recently in pharmaceutical research within quantitative structure–activity relationships [21], should hold an interesting potential in flavour research.

5.3.3.2 Methods using two data blocks. 'Two data blocks' refers to *p* variables (e.g. instrumental data) measured in *n* objects (X-block) and *q* variables (e.g. sensory) measured in the same *n* objects (Y-block). Traditionally, relationships between only one *q*-variable and several *p*-variables have been studied by different versions of multiple linear regression. Response surface methods (power functions and polynomial models) have been developed to fit nonlinear relationships. However, one quickly runs into trouble when: (i) the X-block contains more variables than objects; and (ii) the Y-block contains more than one variable. Partial least squares (PLS) regression is designed to handle these problems by extracting the main, systematic variation in the X-block and Y-block into estimates of a few latent phenomena. By introducing indicator variables (1/0, or design variables) in either the Y-block or the X-block, the PLS methods are, in principle, a discriminant analysis or an analysis of variance, respectively. Operator's registrations or prior knowledge about the materials may constitute an X-block or be embedded in the X- or Y-block.

Thus, valuable experience is included in the analyses. This will be illustrated in the jam example below.

Theory and applications of PLS regression are described in several papers [22,23]. PLS regression can be seen as a linear back-propagation neural net [24]. The use of nonlinear neural nets seems to have potential if used carefully.

5.3.3.3 Methods using multiway data blocks. 'Multiway data blocks' refers to p_1 variables (e.g. sensory) evaluated by p_2 variables (e.g. judges) in n objects. Several interesting studies add to the development in this field, for example, advances in Procrustes analysis [12], consensus analysis [25] and multiway data analysis [26,27].

Good data analysis. The multivariate data analytical models for matching two blocks (X,Y) must be based on:

1. Good problem formulation: What is expected to affect the data? What is the noise level of the sensory and instrumental measurements, and what will be 'good enough' precision of results?
2. Good experimental design: Relevant choice of variables and objects to ensure sufficient variability and all potential phenomena spanned in both blocks. One-block methods may be used for X and Y respectively.
3. Preprocessing of the data: Multiway data methods and unfolding multiblock analyses are relevant to check judge performance. Outliers in X and Y revealed by one-block methods, both with respect to variables and objects, are to be interpreted and taken out, if necessary.
4. Matching X and Y require two-block methods with conservative validation routines and useful graphics.
5. Interpretation of the results both statistically and with the use of experience and common sense.

A user-friendly software program. This must allow an interactive data process with the following elements:

- enter real-world data (see 'flavour data sets');
- look for outliers and handle them;
- ensure predictive validity;
- check residuals for unexpected patterns;
- look for interpretable structures in variables and in objects;
- be conscious about *a priori* weights and linearization;
- check resulting interpretation back to raw data;
- good presentation of results; and
- test hypotheses generated with new data (i.e. enter new real world data).

5.4 Philosophy

5.4.1 To what extent can flavour be measured?

Matching sensory with instrumental analyses has an epistemological value; it is one way to study the limits for what can be measured as an interaction between *a priori* knowledge and the physical world. A hermeneutic sensory method is suggested to indicate flavour assessment as a dynamic process.

5.4.2 Philosophical concepts of relevance in flavour assessment

What is the relation between a chemical profile of 'peach' and 'peach' perceived by the human senses? Is the resonance of my brain to the flow of nerve impulses and energies only a world for me; is 'peach' for me the same as 'peach' for you? These are classical questions in philosophy since the year 500 BC. Philosophy is not the topic of this paper. Nevertheless, it is interesting to consider some classical philosophical principles relevant to the methodology of flavour assessment.

5.4.2.1 Primary and secondary qualities. These have traditionally referred to the distinction between primary properties possessed by the materials (the objects), such as external size and shape, and secondary properties assigned only to the consciousness of man, such as colour, taste and smell [28]. From this theory, flavour should not at all be measured by instruments. Is the time ripe to discuss the primacy of existence versus the primacy of consciousness – theories about the relation between sense perception and objects in cognition, how to understand phenomenology and flavour? Cain [29] touches this discussion when questioning the differences in classification of taste and odour. He indicates that taste categorization is 'built in' and in contrast to this, odour categorization is not 'built in', and thus seems to be constructed from experience.

5.4.2.2 'Hard' versus 'soft' sciences. These have for a long time been at issue in academic circles. In Figure 5.1 the authors have tried to indicate that the technologist or chemist is originally realist with focus on the primacy of existence, and that a technologist interested in relating chemistry to the senses turns to become an idealist (relatively). Conversely, a humanist or psychologist who originally is phenomenalist or idealist with focus on the primacy of consciousness, turns somehow realist when he or she starts working in the field of matching sensory with instrumental analyses. The flavour industry needs both categories at the end-points.

5.4.2.3 Objectivity versus subjectivity. This has always been confusing words when it comes to measuring sensory qualities. People outside the field consider measurements performed by an instrument as objective, while evaluations by man are subjective. People inside the field distinguish between analytical sensory measurements as objective, analogous to instrumental analyses, and consumer responses as subjective. Objectivity is then understood as intersubjectivity or panel consensus.

5.4.2.4 Measuring the unmeasurable. This is the title of a book by researchers engaged in measurement and mathematical and statistical modelling of human phenomena in time and space [30]. Interesting in this connection is what is meant by 'unmeasurable'. Matching sensory with instrumental analyses by data analytical techniques always gives a residual variance not explained by the model. Technologists quickly reject this as noise and focus on the signal. Sociologists and psychologists are more interested in the noise that may reflect some underlying individual differences between persons. Sensory scientists have to concentrate on both the signal and the noise in the measurements and, at the same time, accept a limit for what can be measured either instrumentally or by sensory analysis.

5.4.2.5 A hermeneutic sensory methodology. This will capture much of the more or less theoretical considerations described above, and suggest a preliminary dynamic flavour assessment. Starting with *a priori* knowledge, researchers design and perform experiments (choose objects and variables), do data modelling, interpret the results, which again form new knowledge to be embedded in the new design, and so forth. Thus, we get a knowledge spiral in time, as a reflective equilibrium; acknowledging interpretative information both from 'the masters of wine' and from the on-line instrumental sensors.

5.5 Qualimetrics

5.5.1 Why qualimetrics?

Qualimetrics may be considered as a field in which the physical world (stimulus) and the sensory world (response) are linked in a multivariate psychophysical model. This field has developed as a practical, problem-oriented and goal-oriented multivariate systems analysis of relevance in flavour assessment. Qualimetrics links quality and the art of quantitative measurements.

5.5.2 *Definitions*

Definitions according to the European Organization for Quality Control (EOQC, 1989):

> Quality: the totality of features and characteristics of a product or service that bear on its ability to satisfy stated or implied needs.
>
> Qualimetrics: the discipline dealing with quantitative methods of quality evaluation.

Other definitions of quality and reflections upon consumer quality perception are discussed in an article by Schutz and Judge [31].

5.5.3 *Multivariate psychophysics*

The discussions about goals, methods and philosophy above are to fall out in an empirical driven modelling. In practical flavour assessment:

- stimuli are several components from biological materials,
- human responses are a multitude of sensations, perceptions, cognitions (and behaviour); and
- instrumental responses are a multitude of measured signals.

These two-block (or multi-block) multivariate response spaces require multivariate data analysis.

5.5.4 *Multivariate data analysis*

Quality analysis is characterized with a multitude of variables to cover the complexity of biological materials. Multivariate data analysis is developed to treat the many data. Soft multivariate data techniques (e.g. PLS regression) reveal the systematic covariation between sensory and instrumental analyses by concentrating the complexity onto a few, latent phenomena. Thus, according to Tschudi ‘one can indeed regard much of the history of multivariate methods as attempts to replace “subjective” with “objective” criteria of decisions’ [32].

5.5.5 *Multivariate calibration and prediction*

Multivariate calibration means to make a mathematical formula that optimally converts, for example, a multitude of chemical data X to sensory response data Y in an interpretable model.

Multivariate prediction means to use the obtained model to predict certain variables from other, more easily available variables, for example to predict sensory response data Y in new, unknown samples from chemical data X.

This topic is discussed in depth elsewhere [17].

5.6 Examples

5.6.1 Goals

These were as follows:

1. To find methods for rapid, relevant and reliable quality determination of raspberry jam.
2. To make use of prior knowledge in the multivariate sensory–instrumental system.

5.6.2 Materials and methods

Twelve raspberry jam samples were selected to span certain relevant quality variation. The samples were from four cultivars and three harvesting times.

Twelve sensory colour, flavour, and texture attributes were evaluated along a 1–9 point intensity scale (four replicates) by a trained sensory panel. Consumers evaluated jam preference.

Six chemical/instrumental analyses (i.e. spectrophotometric colour measurements – *L,a,b* and absorbance, percentage soluble solids and percentage total titratable acidity) were performed on the same material as used for the sensory analyses.

The PCA and PLS regression were validated by cross-validation; the UNSCRAMBLER program was used (CAMO A/S, Trondheim, Norway).

The experimental method is described in detail elsewhere [33].

5.6.3 Results: rapid, relevant and reliable methods (goal 1)

Results of the raspberry jam experiment are reported in detail elsewhere [33]. The results confirm the argument that analytical sensory profiling is a rapid, relevant and reliable method for quality determination. It is *rapid* because often no time-consuming sample preparation is needed. Further, many product properties, such as colour, flavour and texture, are measured by one instrument (i.e. the human senses) and on the same samples. By using PCA, the number of variables may be reduced to a few but with a sufficient number of variables to cover the main tendencies of variation in a rational way. In the raspberry jam experiment, the 12 sensory attributes were reduced to seven variables based on (i) the ability to discriminate between the jam samples; (ii) minimum redundancy (i.e. a few variables from each of the three PCA-factors were representative for the sensory space); and (iii) cognitive clarity.

Sensory profiling is *relevant* since food quality usually refers to the perception of the product by consumers, and analytical sensory methods use people as instruments, thus it is more likely to be relevant for the final aim. In the raspberry jam experiment, a PLS regression revealed which instrumental analyses covaried with the sensory variables and to what extent. A two-factor analysis of variance showed that 85% of the colour/viscosity variation related to spectrophotometric colour measurements, while 70% of the variation in perceived sweetness could be explained by the chemical sugar and acidity analyses. A review showing to what degree sensory and instrumental analyses are related is given in a thesis by Martens [34]. The importance of calibrating instruments towards human responses before it is valid as relevant sensory quality measure cannot be over-emphasized.

Sensory profiling is also *reliable*, both in statistical terms (i.e. reproducible) and interpretatively when performed in a correct way [34,35]. In the raspberry jam experiment, sensory analytical results showed no greater experimental 'noise' than comparable chemical and instrumental analyses. Procrustes analysis and PCA-unfolding ensured panel precision and consensus. Furthermore, a systematic error in the instrumental analyses was discovered through interpretation against the sensory results: the spectrophotometric values *L,a,b* 'blindly' measured the sensory redness as green (i.e. a negative *a* value) because the samples were out-of-range for the instrument and no warnings were given. In fact, there is an intelligent self-correction in sensory panels that is often lacking in other instruments. More research into this issue is needed.

5.6.4 *Results: use of prior knowledge (goal 2)*

Based on the results already mentioned, the number of variables could be reduced to seven sensory variables (i.e. redness, raspberry flavour, sweet, sour, bitter, off-flavour and juicy), and three instrumental variables (i.e. absorbance, percentage soluble solids and percentage total titratable acidity). PLS regression on the three instrumental analyses (X1-block) and seven sensory variables (Y-block) showed two significant results, explaining sweetness (factor 1) and redness (factor 2); Table 5.1.

However, when prior information about the materials, expressed as 1/0 indicator variables for the four cultivars and the three harvesting times (i.e. seven design variables), were included in the X2-block, the percentage explained variance in Y after three significant factors increased (Table 5.1).

Table 5.1 Results from PLS regression: percentage explained variance of the seven sensory variables (Y-block) versus three instrumental variables (X1-block) and three instrumental + seven design variables (X2-block) with an optimal number of factors

Y-block	X1-block	X2-block
Redness	80	88
Raspberry flavour	0	45
Sweet	90	91
Sour	19	60
Bitter	45	68
Off-flavour	0	49
Juiciness	10	41

Figure 5.2 shows results from the latter PLS regression. In factor 1, sweet (negatively correlated), bitter and sour (positive correlated) were related to percentage titratable acidity (positive correlated) and percentage soluble solids (negatively correlated); a factor spanned by variation in cultivar 1 versus cultivar 4, and time 1 versus 3. The second colour factor shows positive relation between redness and absorbancy; a factor spanned by cultivar 3 versus the rest, and time 3 versus 1. By including design variables in the X-block, information about the 12 jam samples is seen in the loading plot, and makes the score plot (Figure 5.2 b) superfluous.

The flavour variation constituted a third factor, which was not related to the instrumental/chemical analyses, but strongly spanned by differences between the cultivars (not shown here).

5.6.5 Discussion

The results with respect to goals 1 and 2 indicate that a combination of instrumental analyses and prior knowledge about the materials (expressed as 1/0 variables) is best when predicting sensory quality.

However, it is a long step forward to be able to predict whether the consumer will buy jam sample 1 or 12. Recently researchers have focused on factors that could be more important in predicting consumer behaviour than product attributes [15,16]. Let us assume that the raspberry jams had been irradiated or that they were produced from gene-manipulated raw materials, for example to enhance the bioflavour; perhaps the consumers would avoid the jam samples even if they were optimally produced with respect to redness, sweetness and raspberry flavour. Would all these sensory and instrumental analyses then be in vain?

5.7 Conclusions

On one hand, we may repeat Littlejohn's conclusion from 1968: '. . . the nose is the best gas chromatograph . . .'. At least it is the most relevant

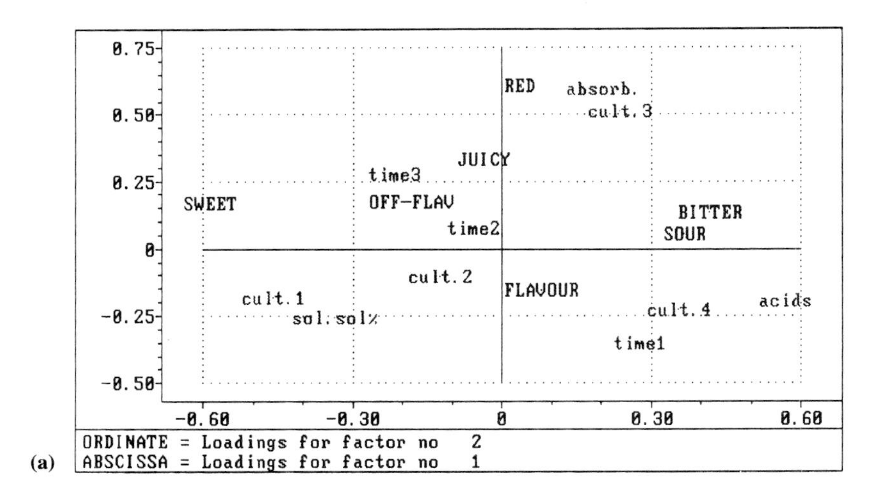

(a)

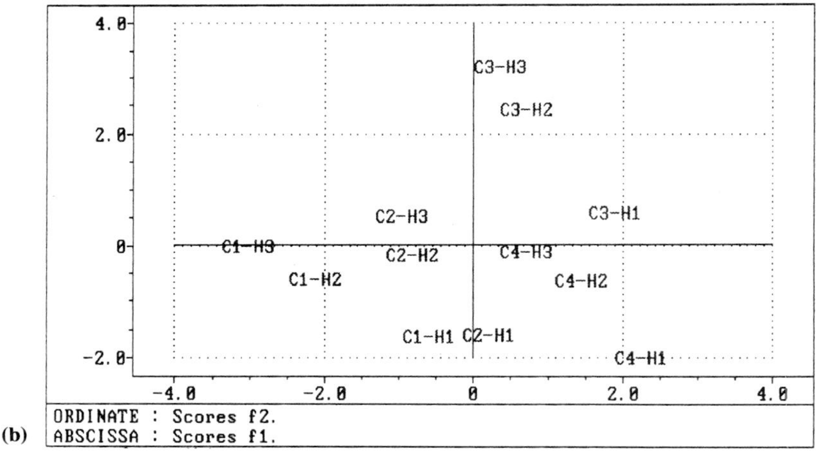

(b)

Figure 5.2 Results from PLS regression on three instrumental plus seven design variables (X) and seven sensory variables (Y) on 12 jams: (a) PLS loadings for factors 1 and 2; (b) PLS scores for factors 1 and 2 (C = cultivars 1–4; H = harvesting times 1–3).

tool in flavour assessment. On the other hand, making this obvious conclusion more scientific, which has been the prime consideration over the last 20 years, has been a valuable process in itself. Realizing that flavour is more than chemistry is one important step; realizing that flavour is a multivariate system needing multivariate approaches, is the next important step. Finally, realizing that sensory science is a good arena for understanding future developments both in modern intelligent instrumentation *and* in increased knowledge about the human brain and behaviour, justifies the strengthening of this field both academically and in the process industry.

Acknowledgement

Thanks are due to the Sensory Group at MATFORSK for preparing the data for analyses, and to Bjørg Narum Nilsen for technical drawings.

References

1. Littlejohn, W.R., Editorial comments. *Perfumery and Essential Oil Record*, 1968, **59**, 1.
2. Harper, R., Art and Science in the understanding of flavour. *Food Flavours, Ingredients, Packaging and Processing*, January 1982.
3. Martens, M. and Harries, J.M., A bibliography of multivariate statistical methods in food science and technology, in *Food Research and Data Analysis*, (eds. H. Martens and H. Russwurm Jr), Applied Science Publishers, London, 1983, pp. 493–518.
4. Martens, M., Dalen, G.A. and Russwurm, H. Jr, *Flavour Science and Technology*, Wiley & Sons, Chichester, 1987.
5. Amerine, M.A., Pangborn, R.M. and Roessler, E.B., *Principles of Sensory Evaluation of Food*, Academic Press, New York, 1965.
6. Pangborn, R.M., Sensory science in flavour research: Achievements, needs and perspectives, in *Flavour Science and Technology*, (eds. M. Martens, G.A. Dalen, H. Russwurm Jr.), Wiley & Sons, Chichester, 1987, pp. 275–289.
7. Engen, T., Psychophysics, (parts 1 and 2), in *Experimental Psychology*, (eds. J.W. Kling and L.A. Riggs), Methuen & Co., London, 1972, pp. 11–86.
8. Overbosch, P., A theoretical model for perceived intensity in human taste and smell as a function of time. *Chemical Senses*, 1986, pp. 315–329.
9. Kuo, Y-L., Pangborn, R.M. and Noble, A.C., Temporal patterns of nasal, oral, and retronasal perception of citral and vanillin and interaction of these odorants with selected tastants. *Int. J. Food Sci. Technol.*, 1993, **28**, 127–137.
10. Risvik, E., *Sensory Analysis,* Tecator AB, Høganeas, Sweden, 1986.
11. Piggott, J.R., *Bridging the Gap Between Laboratory Flavour Research and the Consumer*. Proceedings of the Sixth International Flavour Conference, (ed. G. Charalambous), Amsterdam, July 1989.
12. Arnold, G.M. and Williams, A.A., The use of generalized procrustes techniques in sensory analysis, in *Statistical Procedures in Food Research*, (ed. Piggott, J.R.), Elsevier, London, 1986, pp. 233–253.
13. Risvik, E., Colwill, J.S., McEwan, J.A. and Lyon, D.H., Multivariate analysis of conventional profiling: a comparison of British and Norwegian trained panel. *Journal of Sensory Studies*, 1992, **7**, 97–118.
14. Risvik, E., McEwan, J.A., Colwill, J.S., Lyon, D.H. and Rogers, R., Projective mapping: A new tool for linking sensory profile and consumer research data. *Food Quality and Preference*, submitted.
15. Bruhn, C.H., Schutz, H.G. and Sommer, R., Food irradiation and consumer values. *Ecology of Food and Nutrition*, **21**, 219–235.
16. Pangborn, R.M., Relationship of personal traits and attitudes to acceptance of food attributes, in *Food Acceptance and Nutrition*, (eds. J. Solms, D.A. Booth, R.M. Pangborn and O. Raunhardt), Academic Press, London, 1987, pp. 353–370.
17. Martens, H. and Næs T., *Multivariate Calibration*, Wiley & Sons, Chichester, 1989.
18. Betteridge, D.J., Towards intelligent automation, *Analytical Proceedings*, 1987, **24**, 106–108.

19. Piggott, J.R., *Statistical Procedures in Food Research*, Elsevier, London, 1986.
20. Næs, T., The design of calibration in near infra-red reflectance analysis by cluster analysis, *Journal of Chemometrics*, 1987, **1**, 121–134.
21. Skagerberg, B., Principal properties in design and structural description in QSAR, Doctorate Thesis, University of Umeå, Sweden, 1989.
22. Martens, M. and Martens, H., Partial least squares regression, in *Statistical Procedures in Food Research*, (ed. J.R. Piggott), Elsevier, London, 1986, pp. 293–359.
23. Williams, A.A., Rogers, C.A. and Collins, A.J., Relating chemical/physical and sensory data in food acceptance studies. *Food Quality and Preference*, 1988, **1** (1), 25–31.
24. Alsberg, B., Doctoral Thesis (in preparation).
25. Geladi, P., Martens, H., Martens, M., Kalvenes, S. and Esbensen, K., *Multivariate Comparison of Laboratory Measurements*. Proceedings of the Symposium in Applied Statistics, (ed. P. Thorbøll), UNI-C, Copenhagen, January 1988, pp. 15–30.
26. van der Burg, E. and Dijksterhuis, G., Nonlinear canonical correlation analysis of multiway data, in *Multiway Data Analysis*, (eds R. Coppi and S. Bolasco), Elsevier, The Netherlands, 1989, pp. 245–255.
27. Næs, T. and Kowalski, B., Predicting sensory profiles from external instrumental measurements. *Food Quality and Preference*, 1989, **1** (4/5), 135–147.
28. Kelley, D., *The Evidence of the Senses: A Realist Theory of Perception*, Louisiana State University Press, 1986.
29. Cain, W.S., Taste versus smell in the organization of perceptual experience, in *Food Acceptance and Nutrition*, (eds J. Solms, D.A. Booth, R.M. Pangborn and O. Raunhardt), Academic Press, London, 1987, pp. 63–77.
30. Nijkamp, P., Leitner, H. and Wrigley, N., *Measuring the Unmeasurable*, Martinus Nijhoff, Dordrecht, 1985.
31. Schutz, H.G. and Judge, D.S., Consumer perception and food quality, in *Research in Food Science and Nutrition*, (eds J.V. McLoughlin and B.M. McKenna), Boole Press, Dublin, 1984, pp. 229–242.
32. Tschudi, F., The latent, the manifest and the reconstructed in multivariate data reduction models. Doctor of Philosophy Thesis, University of Oslo, 1972.
33. Martens, M., Risvik, E., Rødbotten, M., Thomassen, M. and Redalen, G., Sensory analysis – A rapid, relevant and precise method for food quality determination, in *Rapid analysis in Food Processing and Food Control*, (Proceedings of EUROFOOD CHEM IV), (eds W. Baltes, P. Baardseth, R. Norang, K. Søyland), 1987, vol. 2, pp. 290–301.
34. Martens, M., Determining sensory quality of vegetables: A multivariate study. Doctor of Agriculture Thesis, Agricultural University of Norway, 1986.
35. Pangborn, R.M., Individuality in responses to sensory stimuli, in *Criteria of Food Acceptance*, (eds J. Solms and R.L. Hall), Forster Verlag, Zurich, 1981, pp. 177–219.

6 Product optimization

M. BERTUCCIOLI and I. ROSI

Abstract

This chapter reviews the strategies of interest for the food industry in order to establish a statistical model for food-quality optimization. Two available statistical procedures for response surface optimization based on multiple regression analysis and partial least squares modellings are discussed. Three examples of recent applications of response surface optimization are illustrated: (i) the concentration of sulphur dioxide (SO_2) in white wine production; (ii) wine quality obtained by immobilized yeasts, and (iii) cheese preference as a function of sensory attributes.

6.1 Introduction

The success of food products depends on the results of studies attempting to identify factors contributing to their outstanding performance. Technological development has improved ways for prolonging the life of perishable food products, thus improving the quality of raw materials, packing and finished products, and by developing standardized products, employing a wide variety of new technologies. In order to put a product on the market, food technologists need to know what makes a product acceptable. A food product is expected to be healthy, nutritive, palatable and easy to use, it should be able to be kept for a certain period of time and be readily available, and all of these for a certain price [1]. Attention is currently being focused on the routine use of statistical experimental design in the improvement of production processes and product characteristics.

The food industry is interested mainly in the description of the products by sets of attributes that may be quantitative and qualitative. As a result of this, any kind of data analysis, aimed at discovering relationships between analytical measurements and organoleptic responses, amounts generally to comparing two sets of multivariate data [2]. Analytical food data, sensory acceptability, consumer testing and complex statistical procedures for relating product attributes to consumer preferences represent important changes to the traditional approaches for defining and identifying desirable product attributes.

Over the past few years multivariate statistical analysis techniques have been used increasingly by food chemists and technologists to reach a better understanding of food characteristics [3]. There are two main reasons for this: (i) awareness that very few problems in this area are univariate; (ii) the increasing availability of suitable software. The task of multivariate analysis is to establish patterns within data sets or relationships between different sets of data: for example, flavour judgements (descriptive or hedonic) versus chemical measurements and/or product properties versus technological parameters. Often such studies are carried out with a precise objective, such as modelling a desired response (consumer acceptability or process productivity) in terms of a few parameters already recognized to be relevant to the process. This optimization approach consists of determining the mathematical model of a response surface as a function of a few casual variables. The response surface model permits us to determine optimal conditions for process and/or the level of individual parameters required to reach a particular response level [4–6].

The purpose of this paper is to outline the basic principles of food product optimization and describe three recent applications in evaluating: (i) the concentration of SO_2 and thiamine in white wine production; (ii) wine quality obtained by immobilized yeasts; and (iii) cheese preference as a function of sensory attributes.

6.1.1. *Optimization procedure*

The optimum conditions of a process or optimal properties of a product are usually determined by measuring the desired response, while varying the causal variables one at a time. This strategy is inadequate to explore the true situation, since it gives no information on the interactive effects of the examined variables. Fractional and composite design [4] can be used successfully to collect required information, with the minimum number of experiments and coupling such information with surface-response modelling enables all effects to be investigated at once.

Mathematically speaking, a response surface has the form of a polynomial equation in which the factors are present in their linear and quadratic terms as well as their bifactorial cross products. The polynomial equation of three causal variables is the following:

$$Y = b_0 + b_1x_1 + b_2x_2 + b_3x_3 + b_{11}x_1^2 + b_{22}x_2^2 + b_{33}x_3^2 + b_{12}x_1x_2 + b_{13}x_1x_3 + b_{23}x_2x_3$$

The polynomial coefficients are usually computed as the regression coefficients obtained by multiple regression analysis (MRA) on an appropriate data set. These data sets are orthogonal, that is, collected according to a factorial design or a central composite design, in the sense that the

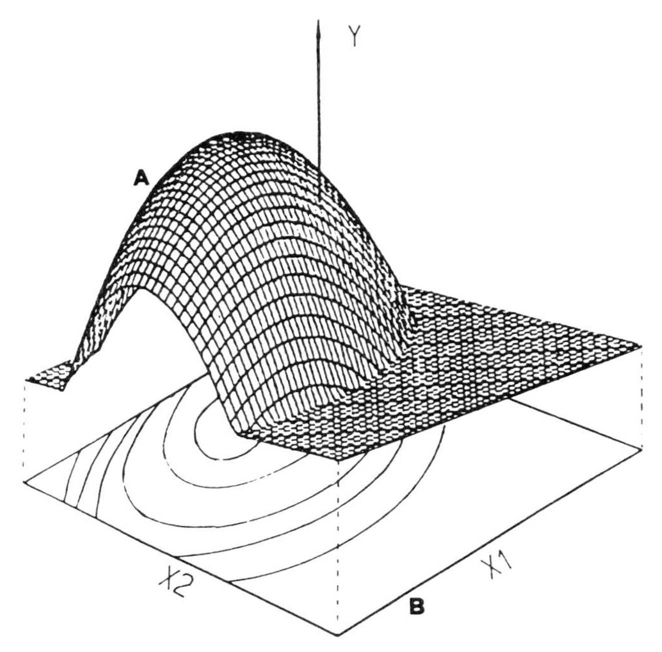

Figure 6.1 A response surface (A) with the relative response diagram (B).

experimental points span the domain in a symmetrical way with respect to its centre [4]. When this is the case, the response surface can be studied by means of canonical analysis [7].

A response surface with the relative isoresponse plot is illustrated in Figure 6.1. For example, central composite design in three causal variables, by adding experiments along the co-ordinate axes at a distance equal to half of the diagonal of the cube, provides the best sample sets for such analysis (Figure 6.2).

In the preliminary part of every study the variables most relevant to the response are unknown, and the number of experiments required become impractically large. If n is the number of causal variables we would have $2^n + 2_n + 1$ experiments in a central composite design (Figure 6.3). Consequently, a preliminary study is often carried out with a fractional factorial design, which permits control of a larger number of variables with fewer experiments [4].

The results of a fractional factorial design, which gives insufficient information to distinguish between the effect of an individual variable and the interaction of other variables (confoundings), can properly be used to select the most important effects, being a conservative procedure. The data can be analysed again using multiple regression (in this case the polynomial equation does not contain the quadratic terms but contains the higher

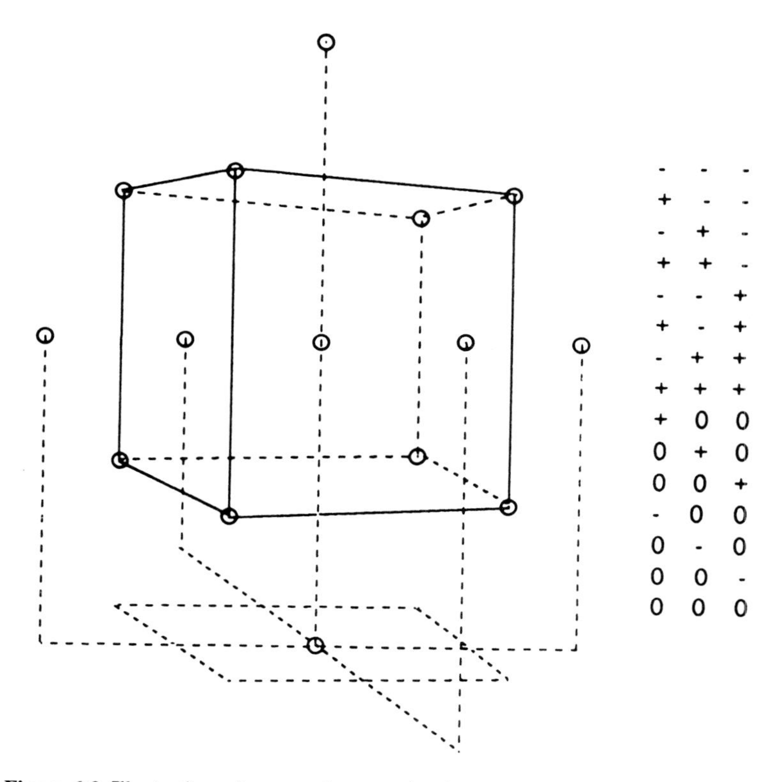

Figure 6.2 Illustration of a central composite design using three causal variables.

order interaction terms) or by simpler procedures such as the Yates algorithm [8].

Quite often, in several food product problems, such as the description of quality in terms of analytical parameters, the collection of experimental data according to an orthogonal design is not possible, hence it is impossible for such an investigation to follow a factorial design. In order to obtain a response surface in the absence of an appropriate design a new procedure, developed to obtain response surface by partial least squares (PLS) modelling, called CARSO (computer-aided response surface optimization) [9], can be applied. Using CARSO, the coefficients of the polynomial describing the surface are obtained by PLS instead of MRA, followed by the subsequent collapse of the PLS loadings. Accordingly, the strategy of CARSO procedure can be briefly outlined as follows:

1. The minimum number of explanatory variables is selected, in order to describe all relevant features by variables as orthogonal as possible, by a preliminary 'normal' PLS modelling.

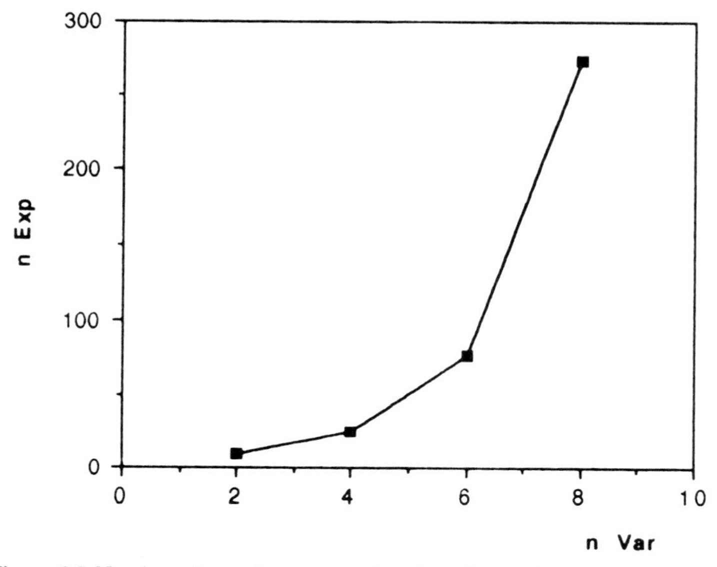

Figure 6.3 Number of experiments as a function of a number of variables.

2. The X-matrix is coded and expanded to include squared and cross-product terms.
3. PLS loadings are transformed into polynomial coefficients for each of the terms; the polynomial equation now describes the response hyper-surface mathematically and can be used for its characterization.
4. The surface is studied to find out the maximum response value within the experimental domain.
5. For this 'best point', the ranges, within which the response value is higher than a certain value, are determined by means of analytical or graphical tools.

6.1.2 Definition of the desired response

Optimizing a process means determining experimental conditions that give rise to certain properties within the product. Often more properties acting in opposition to each other have to be optimized, so that a final product is a compromise. A typical example is the improvement of the yield of a process accompanied by a suppression of undesiderable side-products [10].

As already mentioned, the PLS method can handle different responses; however, the knowledge that two responses vary in an opposite way does not yet tell us the best compromise between them. One must introduce an external evaluation of the 'goodness' (desirability) of each response. To

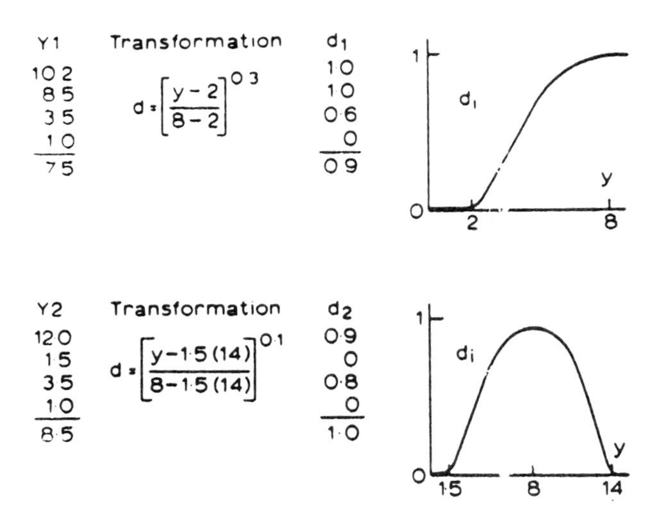

Figure 6.4 Types of transformation associated with desirability function [11].

achieve this, each response is allowed to assume values between zero and one by means of suitable transformation of the actual y values [11]. Each such transformation is called a desirability function (d_i) and it is defined as a dimensionless scale between zero (value of d_i given for an actual y value corresponding to a very poor result) and one (value of d_i given for an actual y value corresponding to an extremely good result). The y values corresponding to zero and one are a subjective choice taking into account the knowledge of process and the expected characteristic of product. As an example, if a wine fermentation process may be accepted when the alcohol formed reaches at least 7% and may be considered fully satisfactory when it reaches 10%, the values assumed by the desirability function will be zero for $y = 7\%$ and one for $y = 10\%$.

Desirability function can exhibit linear, nonlinear or bell-shaped profiles, as reviewed by Derringer and Suich [11], and illustrated in Figure 6.4. Again, the form of the transformation is selected subjectively according to the level of knowledge of process and the desired optimal response.

An overall desirability (D) can thereby be calculated as the geometric mean of the individual d_i values:

$$D = (d_1 d_2 d_3 \ . \ . \ . \ d_n)^{1/n}$$

The geometric mean is appropriate to express the overall desirability (D), which also varies between zero and one, since it increases as the balance of all required properties becomes more favourable and it decreases to zero as soon as one of the properties is out of range.

Although desirability functions have been used successfully in organic synthesis [10] and in our previous studies [12], it is clear that this approach is equal to taking the mean of the logarithms of the individual d_i values. Using such averages of overall desirability may be seen as destroying a significant amount of information contained in the individual responses. Despite this, the desirability approach has the following advantages:

1. External evaluation of the 'goodness' of the actual response values could not have been taken into account if the y values were not transformed into desirabilities.
2. Handling several responses at the same time would have necessitated the superimposition of many contour plots (in our example, with four responses and three explanatory variables, this would have been as many as 12, thus rendering the procedure very impractical).
3. The overall desirability is a good compromise between conflicting responses and it becomes zero as soon as any one response is abnormal and consequently dominates the acceptance of the product. Therefore it can immediately detect the parameters required for a good product or an efficient process.

6.2 Applications of response surface optimization

6.2.1 Thiamine and SO_2 content in quality white wine production

SO_2 exhibits antioxidant and antimicrobial properties and is therefore used as a preservative for wine. Health and organoleptic reasons suggest that the use of this additive during the wine-making process should be limited. Among the practical possibilities to reduce the level of SO_2 in wine is the addition of thiamine. Thiamine is a vitamin that acts as growth factor for the yeast population and as a co-enzyme in the decarboxylation of ketoacids (pyruvic and α-ketoglutaric acids). These compounds are able to bind to SO_2 and therefore to reduce the effect of this additive in wine. When the thiamine is added to grape must, it allows a more regular fermentation and reduces the production of ketoacids by yeast cells.

This study was undertaken to verify the opportunity to reduce the amount of SO_2 during the wine-making process of white wine by adding a defined amount of thiamine. The desired measured response was the amount of iso-amyl acetate as a 'marker' of white wine quality. Studies of the correlation between sensory quality and chemical and physical composition of white wines have demontrated clearly that only some components of the aroma are sufficient to characterize the quality of white wines [13] and, among these, iso-amyl acetate represents one of the most important compounds associated with wine quality. Thus its concentration can

Table 6.1 Matrix of central composite design and relative response

	Variables			Response
Fermentations (*n*)	pH	Thiamine (mg/1)	SO_2 (mg/1)	iso-amyl acetate (mg/1)
1	2.8	0.7	20	1.12
2	3.2	0.3	20	2.25
3	3.2	0.3	40	2.67
4	2.8	0.7	40	2.32
5	3.2	0.3	20	1.86
6	2.8	0.7	20	1.52
7	2.8	0.3	40	2.38
8	3.2	0.7	40	2.35
9	3.4	0.5	30	2.74
10	2.6	0.5	30	1.96
11	3.0	0.9	30	3.05
12	3.0	0.1	30	2.13
13	3.0	0.5	50	2.82
14	3.0	0.5	10	1.24
15	3.0	0.5	30	2.51
16	3.0	0.5	30	2.52
17	3.0	0.5	30	2.65
18	3.0	0.5	30	2.48
19	3.0	0.5	30	2.41
20	3.0	0.5	30	2.55

represent a useful means of evaluating the technological incidence of different operations carried out during the vinification procedures [14].

A series of batch fermentations, under simulated vinification conditions and based on a central composite design with three variables (the amount of SO_2, thiamine and pH) was set up. The data were to be used for determining response surface by multiple regression analysis. The total number of fermentations was 20 and six replicates were collected at the central points. Table 6.1 summarizes the fermentations according to the central composite design. The coefficients of the polynomial equation describing the response surface, calculated by multiple regression analysis, are reported in Table 6.2 with their statistical significance. The response surface showed a maximum point at the following variable values: SO_2 = 48 mg/1, thiamine 0.52 mg/1 and pH = 3.2.

Figure 6.5 shows the appropriate graphical output from a surface (isoresponse plot) where the co-ordinates (x_1 = pH and x_2 = thiamine) represent the coded value of the original variables in such a way as to give − 1 and + 1 to the lowest and the highest raw data. The third variable (SO_2) was maintained at constant coded value 0 (30 mg/1). The iso-response plot permitted modelling of the ester production and to predict the area where the ester should be formed in a larger amount when the content of SO_2 was maintained constant while pH and thiamine content increased or decreased.

Table 6.2 List of regression coefficients (β_1)[a,b]

β_1	$P(0.05)$[c]	$F(1,10)$[d]
$\beta_0 = 25.10$	±1.46	4462.32
$\beta_1 = 2.27$	±0.99	26.06
$\beta_2 = 1.08$	±1.03	5.37
$\beta_3 = 4.31$	±0.97	96.31
$\beta_{11} = -1.00$	±0.97	5.28
$\beta_{22} = -0.66$	±1.01	2.14
$\beta_{33} = -1.97$	±0.99	19.80
$\beta_{12} = -0.34$	±1.29	0.33
$\beta_{13} = -1.43$	±1.29	6.12
$\beta_{23} = -0.92$	±1.21	2.86

[a] Standard error of regression S(E) = 1.6440
[b] Multiple correlation coefficient R^2 = 0.94168
[c] Confidence interval for the 95% level.
[d] Significance of the coefficient by an F test (95% level).

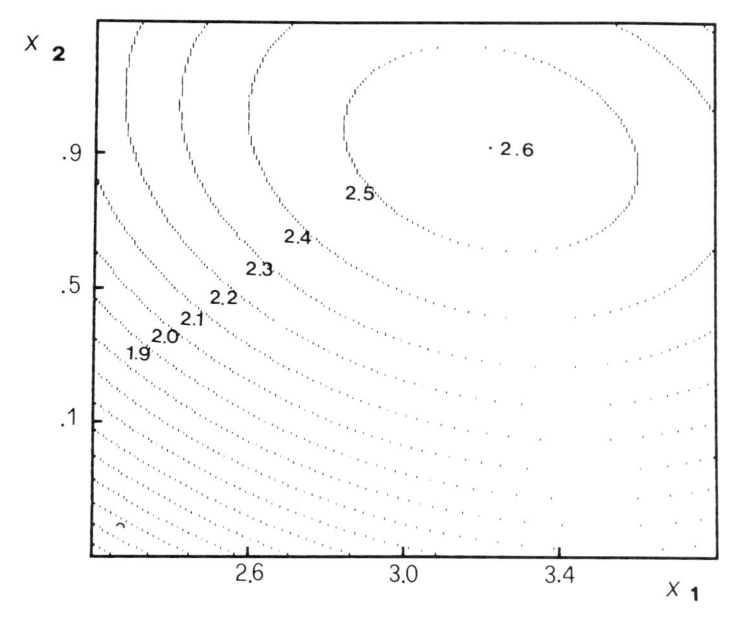

Figure 6.5 Isoresponse plot obtained by MRA for describing ester production (mg/l) in terms of thiamine (x_2) and pH (x_1).

6.2.2 *Optimization of wine fermentation by using immobilized yeast cells*

Cell immobilization is considered as the physical confinement or localization of a microorganism, either alive or dead, that permits economical reuse of the microorganism in either continuous flow processes or in repeated batch contacts.

The immobilization of whole cells has become a promising area of biotechnological research chiefly because this technique offers several advantages:

- obviates enzyme extraction/purification;
- lower effective enzyme cost;
- generally higher operational stability;
- co-factor regeneration;
- greater potential space for catalysing a series of sequential reactions;
- greater resistance to environmental perturbations;
- capability of recycling or reusing the microorganisms;
- accelerated reaction rate due to increased cell density;
- higher specific product yield; and
- better provision and easier control of the fermentation process.

The principal disadvantages can be associated with undesirable side-reactions, catalysed by many enzymes, and prevention of substrate permeation into and out of cells by the cell wall and the cell membrane of intact cells.

The immobilization of yeast cells for application in wine production have been described [12, 15–19].

This study was undertaken to explore the applicability of an optimization procedure to a simulated wine fermentation by immobilized yeast cells. The study began with a fractional factorial design in which nine variables were controlled in 16 experiments: 2^{9-5}. The variables taken into account were: temperature (15–25°C), pH (2.7–3.3), the amount of living cells in the gel (15–20% wet wt/v), the amount of gel beads (10–40% in g/ml), time (2–4 days), glucose (16–22%), nitrogen content (as Casaminoacids Difco, 0.6–1.0 mg/1), ergosterol (15–25 mg/1), and SO_2 (50–100 mg/1). All these variables were used at two levels according to the design matrix reported in Table 6.3.

The amount of ethyl alcohol and residual glucose was determined as a measure of process productivity, acetic acid as an indicator of volatile acidity, and iso-amyl acetate as an indicator of sensory quality. Each response was linearly converted into a desirability function so that individual responses and total desirability function could be evaluated. The results of the fractional factorial design were converted, by the Yates algorithm, into the relative importance of individual variables in determining each response (Table 6.4). Increasing temperature was always effective, increasing the desirability of all responses, while increasing pH improved the desirability of alcohol and ester, and increasing time increased all desirabilities except the volatile acidity. On the other hand, no significant effect was produced by increasing the content of nitrogen, ergosterol and

Table 6.3 Matrix of the fractional factorial design[b]

Experiment	Variables[a] 1	2	3	4	5	6	7	8	9
1	−	−	−	−	−	−	−	−	+
2	+	−	−	−	+	−	+	+	−
3	−	+	−	−	+	+	−	+	−
4	+	+	−	−	−	+	+	−	+
5	−	−	+	−	+	+	+	−	−
6	+	−	+	−	−	+	−	+	+
7	−	+	+	−	−	−	+	+	+
8	+	+	+	−	+	−	−	−	−
9	−	−	−	+	−	+	+	+	−
10	+	−	−	+	+	+	−	−	+
11	−	+	−	+	+	−	+	−	+
12	+	+	−	+	−	−	−	+	−
13	−	−	+	+	+	−	−	+	+
14	+	−	+	+	−	−	+	−	−
15	−	+	+	+	−	+	−	−	−
16	+	+	+	+	+	+	+	+	+

[a] 1 = temperature (15–25°C); 2 = pH (2.7–3.3); 3 = the amount of living cells in the gel (15–20% wet wt/v); 4 = the amount of gel beads (10–40% in g/ml); 5 = time (2–4 days); 6 = glucose (16–22%); 7 = nitrogen content (as Casaminoacids Difco, 0.6–1.0 mg/l); 8 = ergosterol (15–25 mg/l); 9 = SO_2 (50–100 mg/l).
[b] Fermentation condition: yeast strain was *S. cerevisiae* – commercial active dry yeast.

Table 6.4 Result of fractional factorial design

Response	Variables[a] 1	2	3	4	5	6	7	8	9
Ethanol (v/v%)	+	+	+	+	+	0	−	0	0
Volatile acidity (g/l)	+	0	−	−	0	−	0	0	0
Residual sugar (g/l)	+	0	0	+	+	−	0	0	0
iso-Amyl acetate (mg/l)	+	+	−	0	+	0	0	0	0
D_{total} (by Yates)	+	+	0	+	+	−	0	0	0
D_{total} (by PLS)[b]	0.48	0.24	0.08	0.57	0.45	−0.31	−0.17	−0.18	0.11

[a] + = the response increases when variable increases; − = the response decreases when variable increases; 0 = no significant effect on the response.
[b] Loadings of first latent variable (explained variance = 86%).

SO_2. Increasing glucose would decrease the desirability in terms of volatile acidity and residual glucose, while the effect of the other two variables was opposite on diverse responses.

The effects on the overall desirability showed that five variables had a

Table 6.5 Central composite design: Matrix and results

	Variables			
Experiment	Glucose	T(°C)	pH	D_{total}
1	20.0	16	2.7	0.37
2	30.0	16	2.7	0.41
3	20.0	22	2.7	0.72
4	30.0	22	2.7	0.65
5	20.0	16	3.3	0.00
6	30.0	22	3.3	0.39
7	20.0	22	3.3	0.72
8	30.0	16	3.3	0.85
9	33.4	19	3.0	0.00
10	16.5	19	3.0	0.00
11	25.0	24	3.0	0.81
12	25.0	14	3.0	0.00
13	25.0	19	3.5	0.69
14	25.0	19	2.5	0.61
15	25.0	19	3.0	0.76
16	25.0	19	3.0	0.76
17	25.0	19	3.0	0.61
18	25.0	19	3.0	0.69
19	25.0	19	3.0	0.74
20	25.0	19	3.0	0.69

significant effect: a positive effect was given by an increase in temperature, pH, time and the amount of gel beads, and a negative effect was given by glucose. However, since it is obvious that the amount of gel beads will improve the process and that the percentage of cells in the gel should be kept as low as possible to avoid saturation, the second part of the study was carried out using only temperature, pH and glucose as controlled variables. Besides the traditional interpretation of a set of experiments for a fractional factorial design, it is possible to carry out a PLS analysis where the total desirability is related to the design matrix of the nine variables. The PLS model required one latent variable only, which explained 86% of the D variance, and which pointed out the same type of effect seen by the Yates algorithm (Table 6.4).

A new series of experiments was then set up based on a central composite design in three variables. The total numbers of experiments was 20 as six replicates were collected at the central point (temperature = 19°C, pH = 3.0, glucose = 25%). Again the four quality responses were measured and the actual values were transformed into individual desirability functions from which a total desirability was calculated. The values are listed in Table 6.5. Since the experimental design was appropriate, the coefficients of the polynomial equation describing the response surface have been computed by multiple regression. The response surface showed a saddle point, canonical analysis giving two positive eigenvalues and one

Table 6.6 List of regression coefficients (β_1)

β_1	MRA[a]	CARSO[a]
β_0	0.70	0.70
β_1	0.03	0.06
β_2	0.16	0.27
β_3	−0.04	−0.01
β_{11}	−0.21	−0.59
β_{22}	−0.06	−0.19
β_{33}	−0.02	0.05
β_{12}	−0.16	−0.45
β_{13}	0.07	0.19
β_{23}	−0.04	−0.11

[a] Standard deviation: MRA = 0.13 and PLS = 0.22

negative eigenvalue. The procedure based on PLS algorithm (CARSO) gave polynomial coefficients almost identical to those obtained by MRA (Table 6.6), and the limit values for the experimental range expected to generate a desirability 10% higher than the best experimental result that could be found. These were: temperature higher than 24.7°C, glucose between 24.4% and 26.6% and pH lower than 2.5 or higher than 3.9; the lower pH increasing the amount of alcohol and the higher favouring ester production and a low volatile acidity. These results are illustrated in Figure 6.6. The validation of these predictions has been confirmed by three experiments carried out in that area [20].

6.2.3 Optimization of Provolone cheese quality (preference) by sensory descriptive analysis

This study was undertaken to evaluate the sensory descriptive profile of Provolone cheese and to optimize the consumer preference as function of sensory characteristics. The study was carried out on 22 samples of commercial Provolone cheeses. A consumer laboratory panel of about 50 people was used for preference testing. Panellists were asked to sample up to 22 products and assign to each one a score for overall preference using a 1–9 point scale. The flavour sensory profile of Provolone cheese was determined by a trained panel of 10 assessors. Each product attribute was scored on a point scale. The 13 selected flavour and texture terms (within four categories) were:

1. *Odour*:
   (1) Pungent
   (2) Butter
   (3) Curd

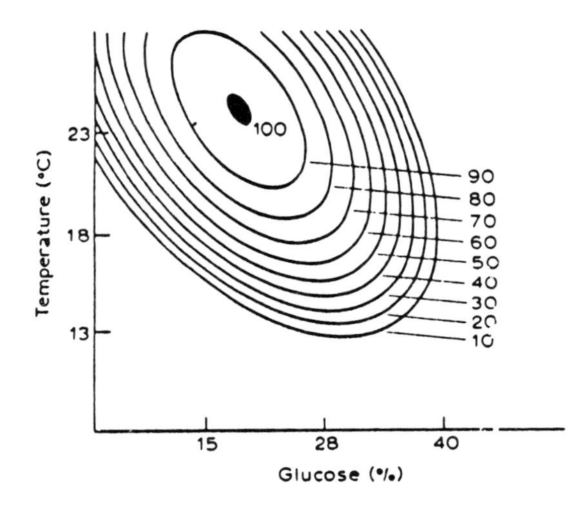

Figure 6.6 Isoresponse plot obtained by the CARSO procedure for describing the desirability of the process in terms of glucose and temperature at pH = 2.5 [20].

2. *Taste*:
   (4) Salty
   (5) Sweet
   (6) Bitter
3. *Flavour by mouth*:
   (7) Pungent
   (8) Butter
   (9) Curd
4. *Texture*:
   (10) Hardness
   (11) Friability/cohesiveness
   (12) Elasticity
   (13) Chewiness

The score for each attribute for each cheese sample was then taken as the average of the scores given by all judges. Analysis of variance (three-factor design) was performed to verify the panel consistence. The sensory data collected in this way were then analysed by the principal component analysis using the SIMCA-3B package developed by Wold *et al.* [21]. The results of the PC analysis of the sensory data are illustrated in Figure 6.7. The two-component model explained 77% of the variance and the diagram showed the association between groups of attributes.

In order to check whether the data of consumer preference was homogeneous, PCA of the preference data of the cheeses was carried out.

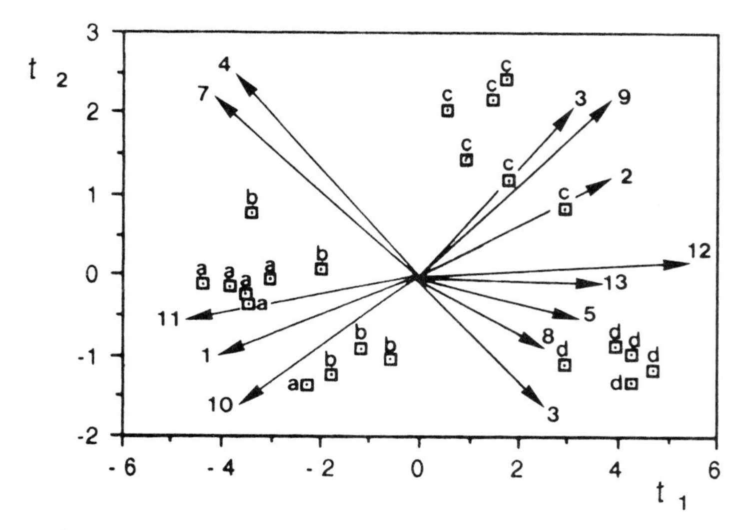

Figure 6.7 Superimposition of score and loading plot obtained by PC analysis on sensory data from Provolone cheese (letters represent samples and vectors represent attributes – numbers indicate which flavour or texture term is being described – see list in text).

Each consumer panellist was described in terms of the scores assigned to each of 22 cheeses. The plot of the first (t_1) versus the second (t_2) component shows a clear separation of people into two groups (Figure 6.8). To obtain information about which descriptive sensory variables were important in determining the preference of cheeses, each group was analysed separately by PLS. The results are shown in the Table 6.7.

The preference scores of the cheeses are reported in Table 6.8 together with the score for attributes, which were recognized to be the most relevant by a one component PLS model (Table 6.7) for the two groups of consumers.

The CARSO procedure was considered more appropriate to determine the relationship between preference and the sensory attributes selected by PLS analysis. In fact this package permitted modelling of the consumer preference by an appropriate surface in terms of the selected attributes. The polynomial coefficients obtained by CARSO procedure (Table 6.9) allowed the modelling of the preference of two groups of cheese consumers in terms of two sensory descriptive variables.

Figure 6.9 shows the isoresponse plots made to detect the variable ranges and the variable thresholds required for a preference greater than an established value. The isoresponse plot that described group A preference (Figure 6.9a) showed a saddle point in the experimental domain and Lagrange analysis [9] indicated that the best constrained maximum had co-ordinates of 1.0 and 0.6. The analytical solution corresponding to this

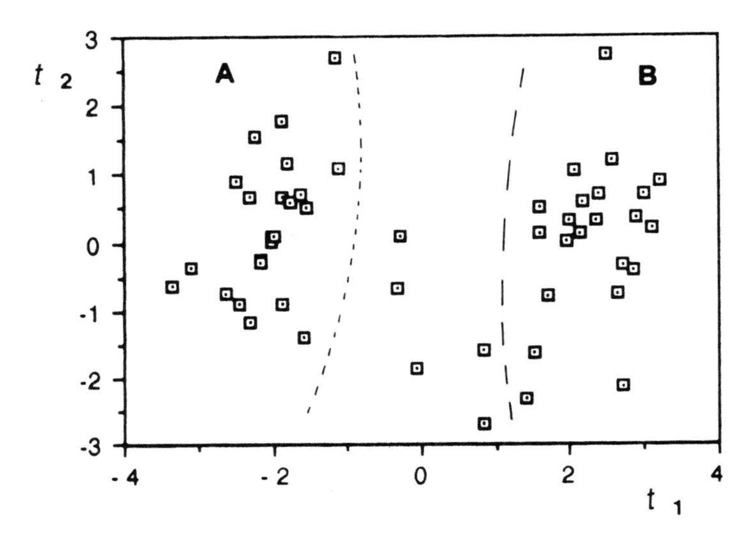

Figure 6.8 PCA of consumer preference data of several samples of Provolone cheese.

Table 6.7 Statistical results of PLS analysis of the preference versus sensory attributes

Variables	Loadings	
	Group A	Group B
1 Pungent	−0.31	0.32
2 Butter	0.25	−0.25
3 Curd	0.29	−0.29
4 Salty	−0.24	0.24
5 Sweet	0.24	−0.24
6 Bitter	0.17	−0.17
7 Pungent (by mouth)	−0.25	0.24
8 Butter (by mouth)	0.27	−0.27
9 Curd (by mouth)	0.28	−0.27
10 Hardness	−0.30	0.30
11 Friability/choesiveness	−0.32	0.32
12 Elasticity	0.31	−0.31
13 Chewiness	0.31	−0.31
Percentage explained variance (*Y*)	70%	83%

point revealed that the maximum preference for elasticity was over 4.4 and friability less than 3.8.

The isoresponse plot that described the group B preference (Figure

Table 6.8 Data matrix for preference optimization

Cheese sample	Sensory score					
	Group A			Group B		
	Number of variables			Number of variables		
	11[a]	12[a]	Preference	1[a]	12[a]	Preference
1	4.1	2.6	3.2	4.5	2.6	6.6
2	1.8	3.9	6.6	2.6	3.9	3.7
3	1.1	4.4	6.4	2.1	4.4	3.8
4	4.3	2.1	3.4	5.1	2.1	6.7
5	2.5	3.0	3.6	3.8	3.0	6.0
6	1.8	3.4	6.8	2.5	3.4	3.7
7	1.7	3.6	7.2	2.5	3.6	3.6
8	4.4	2.0	2.4	5.2	2.0	7.7
9	3.2	2.9	2.5	4.9	2.9	7.5
10	2.2	3.3	6.1	3.0	3.3	4.1
11	1.3	3.9	5.8	2.1	3.9	4.3
12	4.5	2.1	3.1	5.6	2.1	7.5
13	4.1	2.6	3.1	5.1	2.6	7.3
14	2.0	3.7	6.2	2.9	3.7	3.9
15	1.1	3.8	6.4	1.9	3.8	4.1
16	4.4	2.5	2.1	4.6	2.5	7.9
17	3.2	2.7	2.2	4.3	2.7	7.7
18	2.0	3.0	4.6	2.3	3.0	4.3
19	1.4	3.8	4.4	2.3	3.8	4.1
20	4.2	2.2	1.9	5.6	2.2	7.8
21	3.4	2.7	1.8	4.5	2.7	7.6
22	2.2	3.0	4.4	2.4	3.0	4.2

[a] 1 = pungency; 11 = cohesiveness; 12 = elasticity.

Table 6.9 List of the regression coefficients (β_1) of two preference models

β_1	Group A	Group B
β_0	4.06	5.67
β_1	−1.21	1.73
β_2	1.18	−0.93
β_{11}	1.10	0.21
β_{22}	−0.54	−1.22
β_{12}	0.07	−0.78

6.9b) allowed the definition of values of pungency and elasticity attributes to have the highest score preference. For instance, the value of 7 for the preference score could be attained by a pungency value above 5.6 and by an elasticity value of less than 2.8.

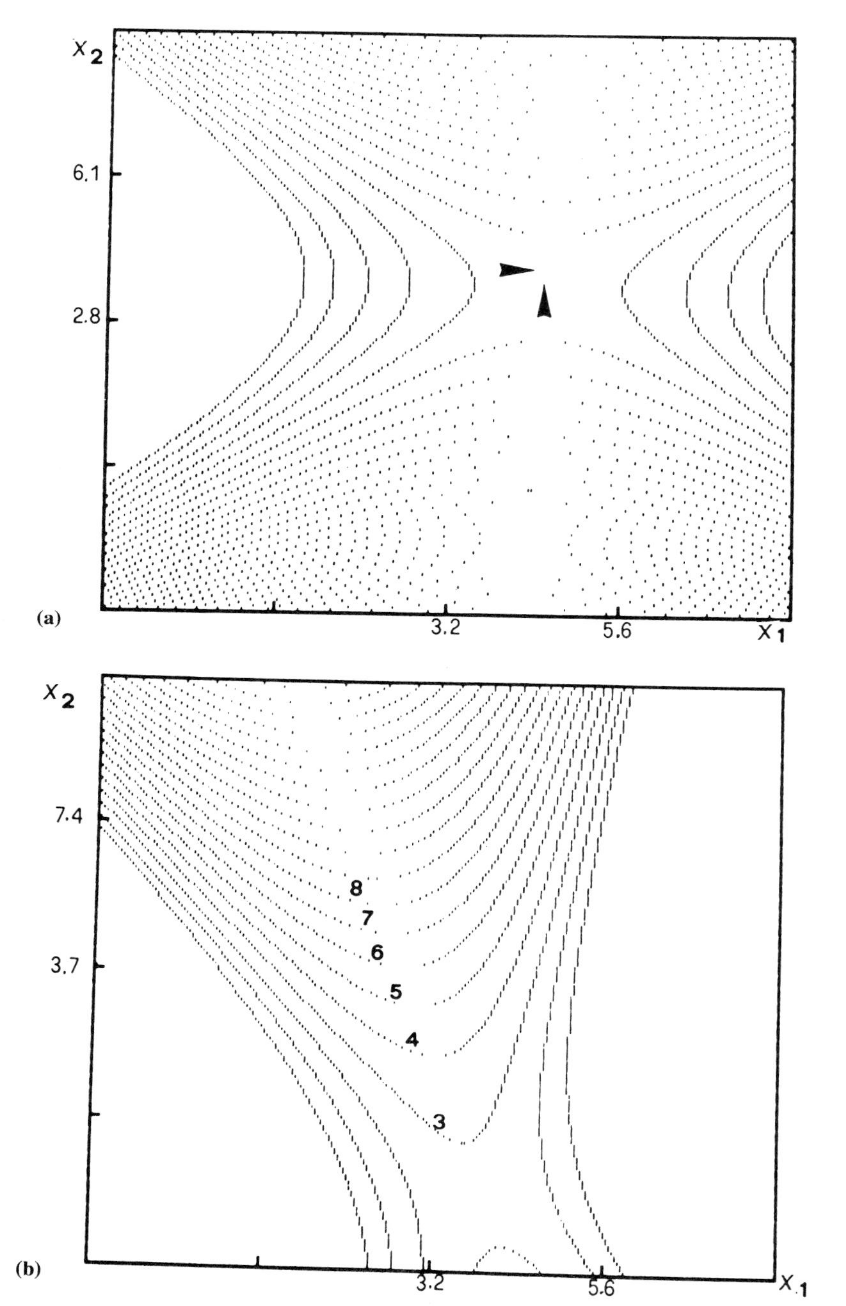

Figure 6.9 Isoresponse plot obtained by the CARSO procedure for describing the preference of the consumers in terms of: (a) friability (x_2) and elasticity (x_1); and (b) pungency (x_2) and elasticity (x_1).

6.3 Conclusion

This chapter was directed at suggesting strategies for handling the optimization of some characteristics of food products. The procedure to collect data following experimental designs has been illustrated and the results obtained by different statistical approaches in defining the causal variables relevant to optimize the response have been discussed. Moreover the response surface methodology can be applied successfully to optimize the consumer preference in terms of sensory characteristics of food products.

References

1. McEwan, J.A. and Thomson, D.M.H., A behavioural interpretation of food acceptability. *Food Quality and Preference*, 1988, **1**, 3–9.
2. Martens, M. and Martens, H., Partial least squares regression, in *Statistical procedures in food research*, (ed. J.R. Piggott), Elsevier, London, 1986, pp. 293–359.
3. Williams, A.A., Rogers, C.A. and Collins, A.J., Relating chemical/physical and sensory data in food acceptance studies. *Food Quality and Preference*, 1988, **1**, 25–31.
4. Box, G.E.P., Hunter, W.G. and Unter J.S., *Statistics for Experimenters: An Introduction to Design, Data Analysis, and Model Building*, John Wiley & Sons, New York, 1978, p. 639.
5. Box, G.E.P. and Draper, N.R., *Empirical Model-Building and Response Surface*, John Wiley & Sons, New York, 1987, p. 734.
6. Gacula, M.C., Jr. and Singh, J., *Statistical Methods in Food and Consumer Research*, Academic Press, Orlando, Florida, 1984, p. 489.
7. Mardia, K.V., Kent, J.T. and Bibby, J.M., *Multivariate analysis*, Academic Press, London, 1979, p. 452.
8. Yates, F., *The Design and Analysis of Factorial experiments*, Bulletin 35, Imperial Bureau of Soil Science, Harpenden, Herts, England, Hafner (Macmillan), 1937.
9. Clementi, S., Cruciani, G., Curti, G. and Skagerberg, B., PLS response surface optimization: the CARSO procedure, *Journal of Chemometrics*, 1989, **3**, 499–509.
10. Carlson, R., Hansson, L. and Lundstedt, T., Optimization in organic syntesis. Strategies when the desired reaction is accompanied by parasitic side reactions. An example with enamine synthesis, *Acta Chemica Scandinavica*, 1986, **b40**, 444–452.
11. Derringer, G. and Suich R., Simultaneous optimization of several response variables, *Journal of Quality Technology*, 1980, **12**, 214–219.
12. Rosi, I., Costamagna, L., Bertuccioli, M., Clementi, S. and Cruciani, G., Wine fermentation by immobilized yeasts: an Optimization study, in *Flavour Science and Technology*, (eds M. Martens, G.A. Dalen and H. Russwurm Jr.), John Wiley & Sons, New York, 1987, pp. 239–252.
13. Bertuccioli, M., Clementi, S., and Giulietti G., Relazione fra dati sensoriali ed analitici dei vini. *Vini d'Italia*, 1984, **XXVI** (3), 27–36.
14. Van Rooyen, P.C. and Tromp A., The effect of fermentation time (as induced by fermentation and must conditions) on the chemical profile and quality of a Chenin blanc wine. *South African Journal of Enology and Viticulture*, 1982, **3**(2) 75–80.
15. Bidan, P., Divies, C. and Dupuy, P., Procedé perfectionné de préparation des vins mousseux. ANVAR French Patent No. 822131, 1978.
16. Divies, C., Les possibilitées d'emploi des germes fixés en oenologie. *Bulletin O.I.V.*, 1981, **54**(608), 843–857.
17. Cantarelli, C., Factors affecting the behaviour of yeast in wine fermentation, in *Biotechnology in Beverage Production*, (eds C. Cantarelli and G. Lanzarini), Elsevier, London, 1989, pp. 127–150.

18. Pardonova, B., Polednikova, M. and Sedova M., Application of immobilized yeasts in the production of sparkling wines. (transl.), *Kvasny Premysl.*, 1986, **32**, 232–235.
19. Fumi, M.D., Trioli, G. and Colagrande, O., Alcuni aspetti dell'impiego di lieviti immobilizzati in spumatizzazione. Proceedings of the Third International Symposium on *Progressi scientifici e tecnologici nella produzione dello spumante*. Chiriotti Editori, 1987, pp. 94–102.
20. Clementi, S., Cruciani, G., Giulietti, G., Bertuccioli, M. and Rosi, I. Food quality optimisation. *Food Quality and Preference*, 1990, *2*, 1–12.
21. Wold, S., *SIMCA 3B %3F: Soft Independent Modelling of Class Analogy*. Brief program description, Umeå University, Umeå, Sweden, 1985.

7 Software for data collection and processing

P.H. PUNTER

Abstract

Future generations will without a doubt refer to the last part of the 20th century as the computer revolution. The first PCs were very simple and slow, and in those days no one would have predicted that PCs would take over most tasks from the mainframe computers. Software is the necessary tool to make computers do anything. Without the proper software the most expensive and sophisticated computer will be useless. But with the growing number of people using PCs the need for simple and 'user friendly' software arose. The first important change toward the user was the introduction of the Apple Macintosh (1984), which introduced the desktop metaphor. This is a so called graphical user interface (GUI). The availability of very cheap computing power, the miniaturization of computers and the introduction of GUIs has had a tremendous influence upon the way we work and upon the way we collect and handle our data. In this chapter the discussion will be limited to the application of personal computers in sensory analysis with special reference to data collection and processing.

7.1 Introduction

Future generations will no doubt refer to the last part of the 20th century as the computer revolution. It is only some 15 years ago that the personal computer was introduced. Before that time, calculations were carried out by hand or on complex mainframe computers. The introduction of the pocket calculators (the well known Texas Instruments and Hewlett Packard) changed our dependency upon the mainframe drastically. Although it took some time, many things could now be performed manually. The simple pocket calculators computed means and standard deviations, the more expensive ones could even do linear regressions. Shortly after that, the first personal computers (PCs) appeared. Owing to their limited capabilities and memory they were not really considered as something serious but more as sophisticated pocket calculators.

This changed when IBM decided to enter the PC market in 1981 with the introduction of the IBM-PC. The first PCs (like the Apple computers who preceded IBM) were very simple and slow, no one would have predicted in those days that PCs would take over most tasks from the mainframe computers. The situation really changed. Now, almost 10 years later, PCs have the power and capacities of minicomputers at a fraction of the price. Computing has become completely decentralized, everyone has the hardware to do most computing alone, without the need of a mainframe. The popularization of computing had also a very important consequence for the software. In the past, there was a small group of specialists (the programmers) who could work with the mainframe (locked away in air-conditioned rooms). Like ancient priests, they were the only ones who could communicate with the computer. When increasingly more people were able to do some things with there own PC, it became necessary to develop a simple language to interact with it. A whole new range of software had to be developed with special emphasis on the inexperience of the user.

Software is the necessary tool to make computers do anything. Without the proper software the most expensive and sophisticated computer will be useless. In principle, software is the language one must use to tell the hardware (the computer) what to do. Not yet in words (although the moment that one can just shout 'PCA' to the computer is near) but in some kind of language the computer understands. In 'computerland' there are almost as many languages as in the human world: for example, Algol, FORTRAN, BASIC, Cobol, Pascal, APL, Modula-2, C, and so on. As in the real world, if you speak different languages you cannot talk to each other. In the early computer days these languages were very difficult to learn and to use. For many years, software for PCs was very similar to software for mainframe computers, in other words, difficult to use. The word 'friendly' was completely unknown: the most clear example of this is the EDLIN editor for MS-DOS computers. However, with the growing number of people using PCs the need for simple and 'user friendly' software increased.

The first important change toward the user was the introduction of the Apple Macintosh in 1984. The Macintosh (based on a different central processing unit or CPU than the MS-DOS computers) changed the way the user interacts with the computer (the 'user interface') fundamentally. Instead of the classical 'type an instruction approach' where the user typed commands, they introduced the desktop metaphor (the idea originated from Rank Xerox and was developed in 1973 [1]). The computer screen is like a desktop and the user can manipulate things by moving them on the desktop. This is a so-called graphical user interface (GUI). Although you still could not talk to the computer to let it do something, you could simply point to a symbol or picture to activate it. Since Apple used a different CPU (the Motorala 680x0 series instead of Intel's 80x86 chip) new software had to be developed specifically for the Macintosh. The emphasis on graphic symbols and the fact that Apple provided programmers with both the programming tools and very strict guidelines resulted in a very uniform 'user interface' for all Macintosh software. The strict guidelines for a very uniform protocol for software developers (and the fact that it could be enforced) was and is a very important innovation in the computer world. The effects are that every Macintosh user already knows how to open files, edit them, print out results for every program that appears. In the early years of the Macintosh, the emphasis was on text and desktop publishing, in recent years increasingly more statistical applications have been developed for the Macintosh. In combination with GUIs this has resulted in a fundamentally different approach to statistical analysis and visualization of data and/or results.

The availability of very cheap computing power, the miniaturization of computers and the introduction of GUIs has had a tremendous influence upon the way we work and upon the way we collect and handle our data [2,3]. These developments are by no means stagnant, the last decade of this century will certainly have a lot more in stock. In this paper, the discussion will be limited to the application of personal computers in sensory analysis with special reference to data collection and processing. The subject matter will be divided into the following categories: (i) software for data collection; (ii) software for data analysis; and (iii) software for data exploration. It is not the intention of this chapter to give a full account of every existing software package that can be used in sensory analysis. The perfect statistical package does not exist, each one has its own (dis)advantages for different users. Reviews of these packages can be found in the different software magazines [4–6]. The emphasis of this paper will be on important developments and future trends. The selection made by the author is personal and not exhaustive. Inclusion or omission should not be regarded as approval or disapproval.

Since software is dependent upon hardware, any discussion about software should first specify for which operating system it is meant: MS–DOS or Macintosh. The basic difference between these two is the way the user communicates with the computer (the user interface). IBM-compatibles still use the classical user interface. You have to enter commands as text strings and communicate in that way. Although the use of a uniform screen layout, colours and pull-down menus has strongly improved the user interface of MS-DOS computers (a nice example is PC-Tools version 6.0) it is not a GUI. The Apple-Macintosh type of computers and IBM-compatibles running Windows use a fundamentally different approach, the desktop metaphor. The screen resembles a desktop on which objects can be moved, activated or put away. Most statistical software originates from mainframe computers and is adapted for MS-DOS. However, in recent years increasingly more statistical software is written for the Macintosh.

7.2 Software for data collection

Computers can be of great assistance in sensory analysis where large quantities of data are collected. Owing to the complexity of sensory analysis and the peculiar measuring instrument (humans), automation is still in an early stage.

The first stages in sensory analysis are test preparation, data collection and data entry. The classical approach is using the paper and pen. Questions are printed on a sheet, the subjects write down the response and the sensory analyst enters these data into the computer. This is a very error-prone and time-consuming procedure. In the late 1970s and early 1980s several systems have been developed to facilitate these activities. The first important step forward was the use of card readers. The subject used a specific score sheet to write down the answers and some kind of optical reader could transform these answers into the proper format for a computer. Optical mark readers are still used today [7,8], although we strongly support the belief that card readers and digitizers are becoming outdated [9].

Another simple input device is the digitizer or graphics tablet. The scoring sheet is placed on the surface of the tablet and the responses are entered by touching the correct answers. The necessary hardware and software is cheap and, with the use of laser printers, it is relatively easy to design the necessary score sheets. However, for several tests (e.g. paired comparison or ranking) it is not so easy to make standard forms. Furthermore, when many different tests are carried out it is rather time-consuming to adapt the score sheets. With these systems, the sensory analyst still has to prepare the layout and text of the scoring sheets and has to collect and check the responses before feeding them to the optical or electromechanical reader. It is much more efficient when the subject interacts directly with the computer to enter its responses.

In marketing research, several software programs exist for the administration of questionnaires. This software is not integrated, every subject uses a simple stand-alone PC or laptop computer and is guided through the questionnaire. Most programs make conditional answers possible (i.e. the routing through the questionnaire is dependent upon the answer). After finishing the test, the results have to be removed from the different computers to a central computer to be processed. Several programs make it possible to design your own forms (electronic form processing), which can be used on laptop computers or in a networking environment (for instance PerFORMER from Delrina Technology Inc. and Horizon from FormMaker Software).

In sensory analysis, a more integrated approach is often needed [9–13]. Most tests are carried out using eight or more subjects, this means that some kind of network must be used. Several such systems are available

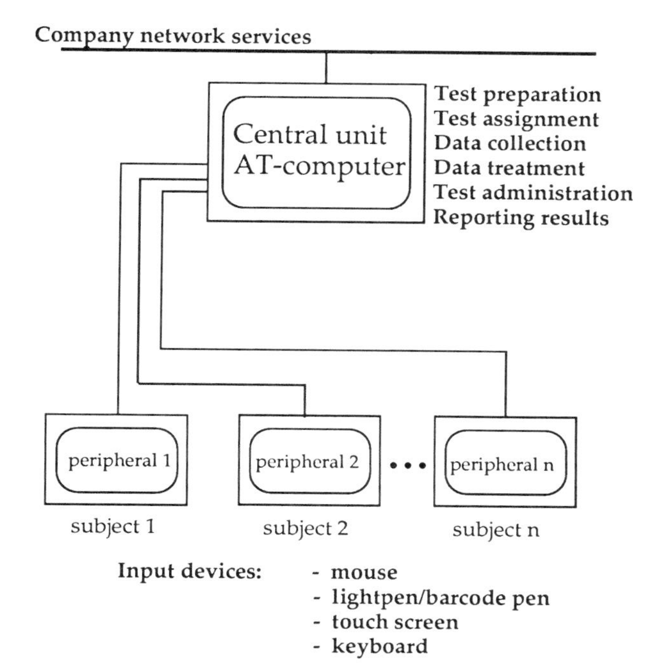

Figure 7.1 Layout of computer-aided sensory analysis (CASA). The influence of the user interface depends strongly on the type of interface. The subjects can have different tasks: marking a scale (e.g. line or category), moving objects on the screen, entering numerical or textual information. Marking scales can be performed in different ways, such as with touch-screens [14,15], light-pens [10], mouse- or trackball-controlled cursor movement or keyboard-controlled cursor movement. Touch-screens work excellently with children and inexperienced subjects, the disadvantage is the relatively low resolution of the screen. Mouse-controlled movement is probably the most accurate and efficient method, even completely naive subjects master it in minutes.

today, ranging from single-purpose systems to complete systems for computer-aided sensory analysis (CASA) [14–16]. Some examples of commercially available systems for CASA are Actis [17], Biosystèmes [18], Compusense [19], the PSA System [20], Tastel [21] and Taste [22]. All CASA systems use some kind of network where peripheral units in the tasting booths are connected to a central computer. A diagram of this is shown in Figure 7.1. Several peripheral units (in general up to 32) are connected to one central computer. The central computer (generally an AT or 386) is used to prepare the experiments, assign them to peripheral units and collect the results. Most systems allow different tests to be assigned to the peripherals, subjects can be assigned from the central computer or subjects can identify themselves at the peripheral. The resulting output files should be in such a format that statistical analysis is

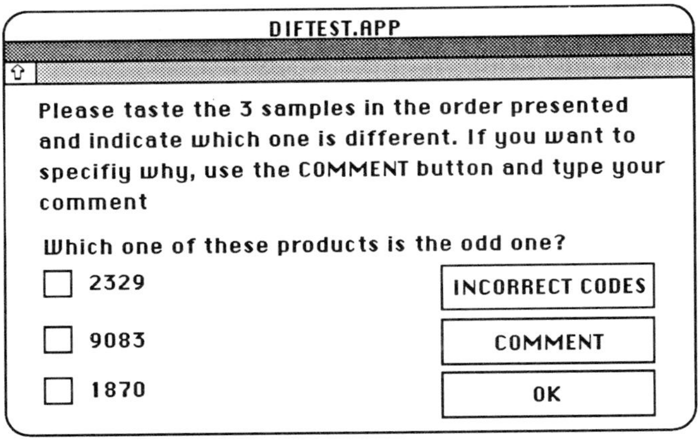

DIFTEST.APP

Please taste the 3 samples in the order presented and indicate which one is different. If you want to specifiy why, use the COMMENT button and type your comment

Which one of these products is the odd one?

2329

9083

1870

INCORRECT CODES

COMMENT

OK

Figure 7.2 Example of the user interface for a difference-test item from the PSA system (redrawn in *MacDraw*). Responses are entered with the mouse-controlled cursor.

possible, indeed simple summary statistics should be available immediately after the test.

The following peripheral computers are used: Atari, Minitel (France), BBC (England) or standard AT computers. Recently, there has been a demand for flat screens because of space limitations in most tasting booths. One disadvantage of flat screens is the low-resolution and the critical viewing angle. The disadvantage is the cost, at this moment VGA flat screens are as expensive as a complete AT computer. Although flat screens look good in tasting booths, the major criterion should be the resolution and quality of the text on the screen for the subjects.

In most cases an AT or 386 computer is used as the central unit. The advantages of CASA are the increased efficiency and speed: less personnel can run more tests in a shorter time. Results (at least summary data) are available immediately. Both experienced and inexperienced subjects like to work with the computer, they use the mouse or keypad without any problem (Figures 7.2 and 7.3). Once they are used to CASA, they often refuse to return to paper-and-pencil tests. There is some discussion about this. The main questions are whether the way the subject interacts with the computer influences the response and whether the glare of the screen influences the test environment [9]. The answer to the last question is simple, when special lighting conditions are necessary a coloured screen can be used. The first question is more difficult to answer.

Keyboard-controlled cursor movement should not be allowed since it is too difficult and unnatural. Furthermore, most people dislike entering free-format input through the keyboard. It is slow, error-prone and very

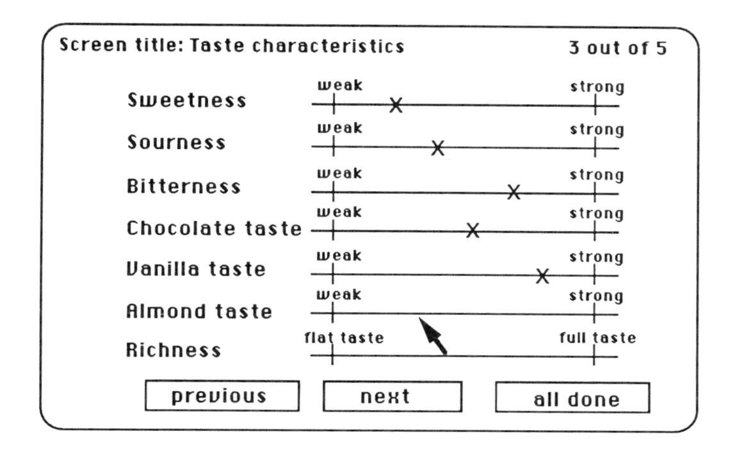

Figure 7.3 Example of the user-interface for a profiling test from the PSA system (redrawn in *MacDraw*). Responses are entered with the mouse-controlled cursor, the cross only appears after selection.

difficult to analyse. At this moment, voice recognition is not possible but it certainly will be in the near future. Considering the layout of most sensory facilities it probably will not be of much use, unless the subjects can whisper.

A special kind of computer-aided sensory test is the time-intensity (TI) experiment. In TI experiments the subjects have to express their sensation as it evolves in time. TI experiments have been carried out since the early 1950s but the availability of the microcomputer made the method much more popular [23,24]. In general, TI experiments are carried out as shown in Figure 7.4. The data collection is relatively easy. For each subject, the perceived intensity as expressed by the slider is measured n times/second. Data treatment is more complex, in most cases the following parameters are computed from the TI curves: total area, maximum value, time to maximum value and slope of the rising and falling part of the curve (Figure 7.5). Data analysis of these results is still in its infancy, although several ways have been suggested [25,26].

Although the software and hardware costs for CASA are relatively high, it is generally assumed that the advantages outweigh the investments. Systems for CASA should enable the sensory analyst to prepare sensory tests, assign these tests to the peripheral units, collect and summarize the results. In future, it should also be possible to get on-line help for experimental design and selection of the proper test form (as an addition to test preparation) and to prepare the output files for further statistical analysis. Test preparation should work according to the WYSIWYG

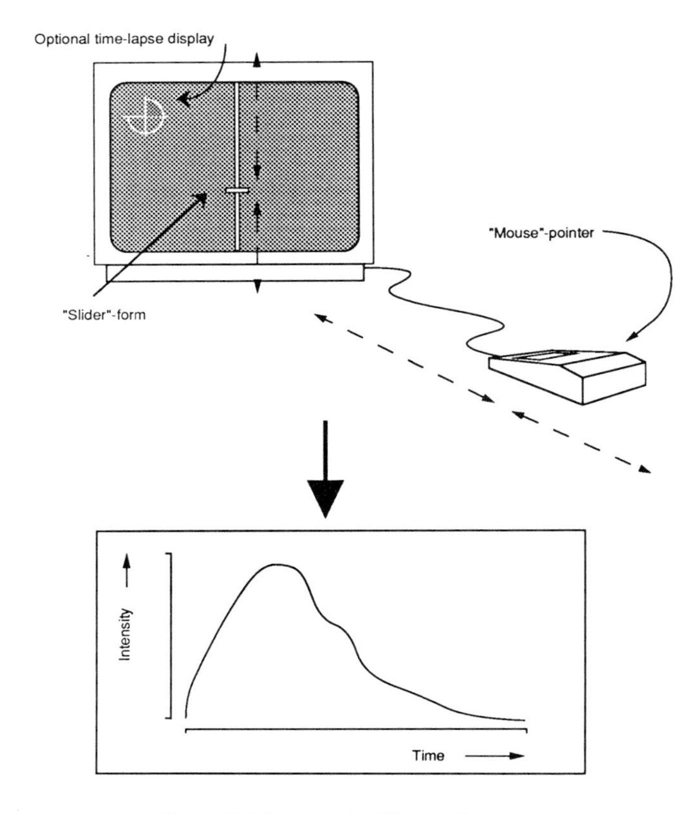

Figure 7.4 Layout of a TI experiment.

principle (what you see is what you get), enabling the sensory analyst to design the questions just as they appear in front of the subjects. Finally, it should be possible to work interactively with the subjects, showing on-line results during test sessions. Development of CASA systems is a very time-consuming and expensive activity, companies who have tried to develop an adequate system on their own have often experienced great difficulties because of the difficult nature of sensory analysis [9]. An alternative is the purchase of one of the commercially available systems.

7.3 Software for data analysis (traditional statistical analysis)

With the advancement of statistical methods and computing power, more sophisticated and complex tools became available for the sensory analyst [27–32]. In the early days of sensory analysis, simple tests were carried out

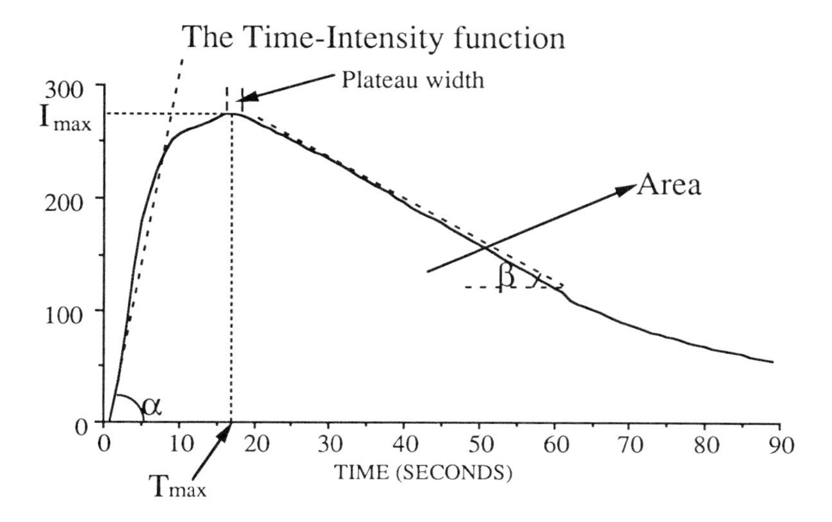

Figure 7.5 Typical measures from the TI curve.

and statistical treatment was relatively easy. For example, to analyse the difference test the only things needed were a table and the number of correct responses. More complex designs required more complex methods. In the 1950s and the 1960s the most widely used forms of analyses were analysis of variance (ANOVA), *t*-tests, factor analysis, regression analysis. Some of these analyses could be performed by hand (although two- or three-factor ANOVA could take several days), while others needed a computer (factor analysis on 30 attributes requires 30 × 29 correlation coefficients). The most important (mainframe) packages for statistical analyses then were SPSS [33], SAS [34], Genstat [35] and BMDP [36]. With the introduction of PCs it became possible to do the same analysis at your own desk. At first, there were clear limitations on size and speed but now most of the sensory analysis can be performed on PCs.

Today, a large number of PC packages is available, ranging from simple, single-purpose packages to packages that cover most statistical methods, with or without graphics. Although the statistical procedures incorporated in these packages do not differ (mostly in the size of the data matrices they can process), there are large differences in the way the user interacts with these programs. The most important distinction is the 'command language' user interface (from the mainframe tradition) in which the user has to know or learn the specific command language and the graphical, menu- and icon-driven command structure. Although several MS-DOS packages use a menu-driven structure, they are by no means comparable with specific features of the GUI as used by the Macintosh computers. These differences are most clearly demonstrated by software that is developed specifically for the Macintosh. Most databases and spreadsheets also have some statistical features, ranging from simple descriptive statistics to complex analyses. Since their primary function is not statistical analysis, they will not be discussed here.

Although it is difficult to indicate the quality of individual software, some classification is possible. An important point to remember is that ideal software does not exist – each package has its strong and weak points. Switching between different statistical packages for MS-DOS is much more difficult then switching between packages for the Macintosh. The strictly defined user interface and the extensive use of the GUI simplify the use of different packages considerably. The most extensive 'general purpose' statistical packages for PCs are SPSS-PC and SAS. SPSS-PC is probably the most widely used PC package, SAS the most extensive and expensive. Besides general purpose packages, there are specific programs for sensory analysis. Some examples are PC-MDS (Brigham Young University, Utah, USA), Senstat (OP&P, Utrecht, The Netherlands), Senpak (Reading University, Reading, England), Procrustes-PC (OP&P) and the Unscrambler (Camo, Trondheim, Norway). Statistical packages should not only be able to compute but should also be able to present the data in graphical form. Not every package available has the same capabilities for graphical output.

Table 7.1 gives an overview of several statistical packages for both MS-DOS and the Macintosh. They range from 'general purpose' packages that can do almost anything to dedicated programs for one specific application. Indicated for each package is the type of computer (i.e. MS-DOS or Mac), whether it is a general-purpose package (e.g. descriptive statistic, regression, ANOVA) or a specific one, whether it can be used for multivariate analysis (MVA), nonmetric MVA (ALS), MDS and whether is has extensive graphical possibilities. For more detailed information about these programs see references 4–6.

In the last few years, several important new developments have been implemented in the statistical software packages. The department of Datatheory (University of Leiden) has developed several techniques of the alternating least squares (ALS) type. These programs perform nonlinear multivariate analyses and have recently been incorporated into SPSS-PC (Categories). The most important programs are Homals, Princals and Overals. Other important new developments are models for multivariate calibration (PLS models as in the Unscrambler package) and generalized Procrustes techniques (Procrustes-PC, the Genstat macros and the SAS macros) [37–43]. Since these products will be updated or extended at regular intervals, this table should be interpreted with caution. Check the recent literature or your local dealer.

Table 7.1 Statistical software for MS-DOS and Macintosh computers[a]

Package	MS-DOS	Mac	General	MVA	ALS	MDS	Graphics[b]
SPSS-PC	x	x	x	x	x	x	x
SAS	x	–	x	x	–	–	x
BMDP	x	–	x	x	–	–	x
Genstat	x	–	x	x	–	–	x
Systat	x	x	x	x	–	x	x
Stata/stage	x	–	x	–	–	–	x
Statgraphics	x	–	x	x	–	–	x
CSS	x	–	x	x	–	x	x
Microstat	x	–	x	x	–	–	x
Stat80	x	x	x	x	–	–	x
Statpak	x	x	x	–	–	–	x
Statfast	x	x	x	–	–	–	x
Statistix	x	–	x	–	–	–	x
Statview	–	x	x	–	–	–	
SCA	x	–	x	–	–	–	x
Statworks	–	x	x	–	–	–	x
Unscrambler	x	–	–	x	–	–	x
PC-MDS	x	–	–	x	–	x	–
Procrustes-PC	x	–	–	x	–	–	–
Senstat2	x	–	–	–	–	–	–
Senpak	x	–	–	–	–	–	–
Data desk	–	x	x	x	–	–	x
JMP	–	x	x	x	–	–	x

[a] x = available; – = not available.
[b] Every package for the Macintosh has excellent and extremely user-friendly graphical possibilities.

7.4 Software for data exploration

One of the basic problems in statistical analysis is the extraction of useful information from raw data. The availability of very sophisticated and complex software prompted sensory analysts to collect larger and more complex datasets. The emphasis moved away from hypothesis testing to exploration of the relationships between the data, not only within one domain but also between different domains (e.g. instrumental and sensory data). An interesting introduction to the importance of data exploration can be found in a discussion between Molenaar, de Leeuw and de Gruijter [44–46]. Without the guidance of a hypothesis the amount of information is overwhelming. To impose structure upon the data, it became necessary to be able to manipulate and explore the data dynamically. This is impossible in the classical, instruction-oriented approach to statistics and has led to a completely new kind of (statistical) software: explorative statistical software. Explorative data analysis is about looking at data to see what it seems to say. It concentrates on simple arithmetic and easy-to-draw pictures. It regards whatever appearances we have recognized as partial descriptions, and tries to look beneath them for new insights. Its concern is with appearance, not with confirmation. (*Tukey, 1977*)

Dynamic graphical methods [48] are important aspects of data exploration. The essential characteristic of a dynamic graphical method is the direct manipulation of elements of a graph on a computer screen and the instantaneous change of the elements. The data analyst takes an action through manual manipulation of an input device (mouse) and something happens in real time on the screen. Linked with these techniques is a collection of new methods: identification, deletion, linking, brushing, scaling and rotation. Identification refers to the labelling of points on the screen (i.e. what does this point represent?) and to the location of a point or variable (i.e. where are these values or variables located?). Deletion refers to the removal of points or variables and to the immediate redrawing of the screen (i.e. what happens if an outlier is removed?). Linking and brushing refer to the possibility of showing interrelated points in different plots (e.g. with more than two variables). Scaling can be used to change the ratio aspect of graphs (the relationship between the physical length of x and y axis). Changing the ratio can dramatically change the information. The last method is rotation – the movement of three-dimensional plots on the screen.

Visual analysis of the data is essential in explorative statistics. There should be a figure for each and every result. Since this is only possible with a very flexible and graphical user interface, the available software is exclusively for the Macintosh at present. Two different programs are available: Data Desk and JMP. Data Desk has been developed in 1984, recently version 3.0 has been released (Odessa Corporation, Northbrook Ill, USA). JMP is developed by the SAS Institute (Cary, NC, USA), version 1.01 was released recently. Both programs use the same desktop metaphor as the Macintosh. The desktop is an integrated statistical working environment for data analysis, data transformation, visualization and reporting. At every stage, context-sensitive expertise is available to suggest possible further actions. Since all information is linked dynamically, results can be changed interactively by changing or transforming data. Figure 7.6 is an example of the desktop from Data Desk. To quote Forest Young [48]:

writing about dynamic graphics is a very frustrating business. After all, writing is linear, static, and nongraphical. What could be further from the subject we are addressing here? (*Forest Young, 1988*)

Both Data Desk and JMP have the possibility of displaying rotating three-dimensional plots and even using colour as a fourth dimension. The plot-rotation technique is one of the most valuable and interesting

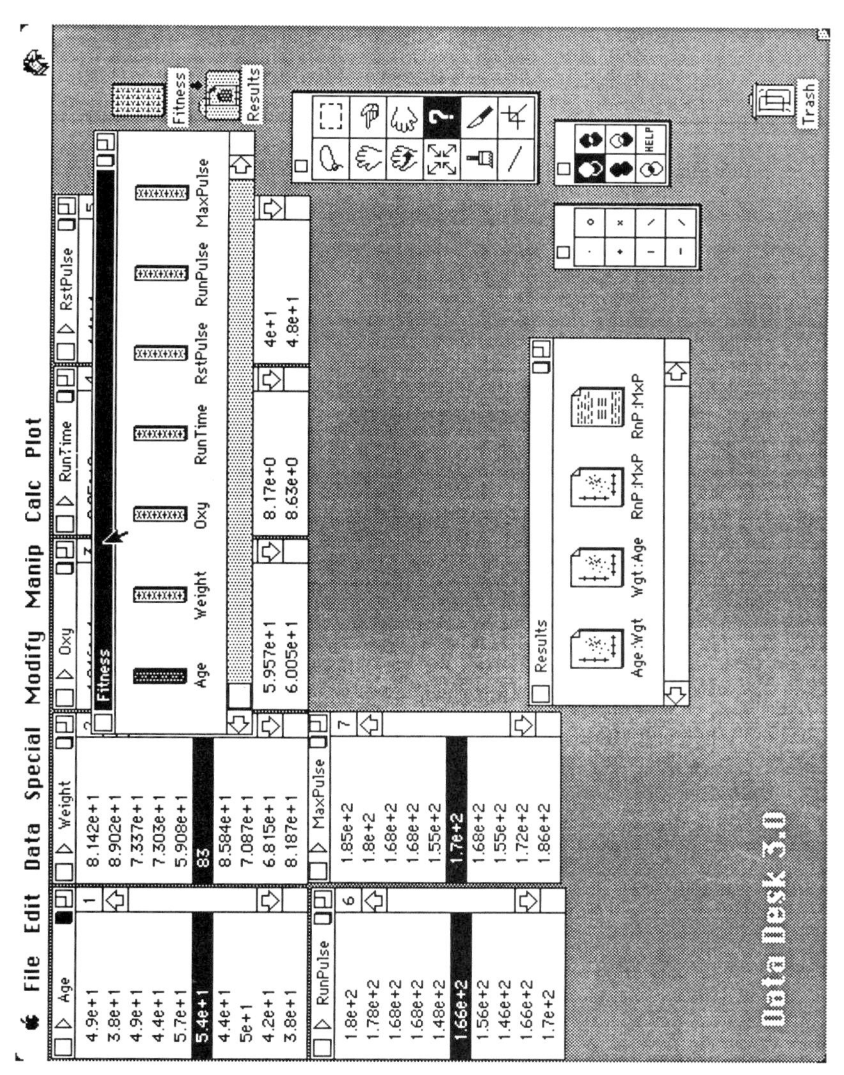

Figure 7.6 An example of the screen layout from *Data Desk* (printout of the actual screen).

elements of these programs. Rotating, three-dimensional plots have been introduced in the early 1970s (PRIM-9, running on an IBM mainframe [48]). In 1986, at the Annual Meeting of the American Statistical Society, the program MacSpin was demonstrated, running on a simple Apple Macintosh. Another program for manipulating multidimensional spaces is VISUALS, developed by Forest Young and colleagues [49].

These programs do not offer statistical procedures but are mainly used to explore data in a dynamic way. For further information, the reader is referred to ref. 48.

7.5 The future

The development of software for data collection, data analysis and data exploration is by no means finished or slowing down. Computer-aided sensory analysis (CASA) will become more sophisticated and flexible. The sensory analyst will be guided by the selection of the proper experimental design and screen layout of the tests and by the data analysis. For many applications, standard report formats will be generated automatically from the raw data files. Data from experiments will be portable to any other program, without difficult transformation steps (e.g. DBMS/COPY, from BMDP Statistical software, which converts data from approximately 100 different programs). There is also a trend to standard data formats in the MS-DOS environment and compatibility with Macintosh formats.

As the amount of data increases and the information it contains becomes increasingly complex, the need for explorative analysis will increase. Presentation of results will also change dramatically, rotating three-dimensional data representation and on-line interactive computation ('what if . . .' questions) will replace the classical report. This type of analysis and data treatment is made possible by the powerful PCs and sophisticated GUIs. The problem in the future will not be whether it is possible but which one to use. There is an overwhelming amount of software and one must try to decide what is most suitable. From the other side, automated sensory analysis will generate increasingly more data, so one needs sophisticated ways of data storage. One of the most important steps towards easy control of data is some kind of standardized data-format and standard procedures in sensory analysis. To quote Risvik and Rogers [9]:

> We will have to deal with problems created by computer technology, faced decades ago by more classical sciences. We may also be able to learn from the destructive discussions they [the more classical sciences] have already experienced. To achieve progress we will have to agree to standards for the basic sensory methodology . . . (*Risvik and Rogers, 1989*)

If this standardization coincides with a standard and uniform user interface, most of our problems will be solved. In the near future, statistical analysis might finally become fun for all of us.

Acknowledgments

The contribution of the following colleagues from OP&P is gratefully acknowledged: Garmt B. Dijksterhuis, Michael Janse, Paul Krabbe and Peter E. Roos.

References

1. Smith, D.K. and Alexander, R.C., *Fumbling the future*, William Morrow, New York, 1988.
2. Roos, P.E., Impact of computers on sensory evaluation: past, present and future. *Food Quality and Preference*, 1989, **1**, 165–170.
3. Schmid, J.P., Computerization in flavour research., *Flavour Science and Technology*, (eds Y. Bessière and A.F. Thomas), Wiley, New York, 1990.
4. Francis, I., *Statistical Software; A Comparative Review*, Elsevier, Amsterdam, 1981.
5. Lehman, R.S., Statistics on the Macintosh. *Byte*, 1987, 207–214.
6. Raskin, R., Statistical software for the PC. *PC Magazine*, 1989, **8**, 5.
7. Piggott, J.R., Automated data collection in sensory analysis, in *Flavour of Distilled Beverages*, (ed J.R. Piggott), Ellis Horwood, Chichester, 1983.
8. McLellan, M.R., Hoo, A.F. and Peck, V., A low-cost computerized card system for the collection of sensory data. *Food Technology*, 1987, **41**, 68–72.
9. Risvik, E. and Rogers, R., Sensory analysis: a view on the use of computers. *Food Quality and Preference*, 1989, **1**, 81–85.
10. Brady, P.L., Computers in sensory research. *Food Technology*, 1984, **38**, 81–83.
11. Lyon, D.H., Sensory analysis by computer, Part 1., *Food Manufacture*, 1986, November, 40–42.
12. Arnott, M.L., Computerised collection and statistical analysis of descriptive sensory profiling data, in *Distilled Beverage Flavour*, (eds J.R. Piggott and A. Paterson), VCH Publishers, New York, 1989, pp. 65–75.
13. Williams, A.A. and Brain, P., The scope of the microcomputer in sensory analysis. *Chemistry and Industry*, 1986, **4**, 118–122.
14. Daget, N., Voirol, E., Resenterra, P. and Cabi-Akman, R., A system of data acquisition for sensory testing: the tactile plasma screen. *Nestlé Research News*, 1986/1987, 211–213.
15. Findlay, C.J., Gulett, E.A. and Genner, D., Integrated computerized sensory analysis. *Journal of Sensory Studies*, 1986, **1**, 307–315.
16. Winn, R.L., Touch screen systems for sensory evaluation. *Food Technology*, 1988, **42**, 98–100.
17. Actis, ISHA, BoxPostale 138, 91163 Longjumeaux Cedex, France.
18. Biosystèmes, 3 Rue de la Breuchillière, Dijon, France.
19. Compusense, 173 Woolwich Street, Suite 201, Guelph, Ontario, Canada.
20. PSA-System, OP&P Inc., PO Box 14167, 3508 SG Utrecht, The Netherlands.
21. Taste, Software package for sensory analysis, Reading Scientific Services, Lord Zuckerman Research Centre, Reading, UK.
22. Tastel, Pernod-Ricard, 120 Avenue du Marechal-Foch, 94015 Creteil Cedex, France.
23. Lee. W.E. III, Evaluation of time-intensity sensory responses using a personal computer. *Journal of Food Science*, 1985, **50**, 1750–1751.
24. Yoshida, M., A microcomputer (PC 9801/MS mouse) system to record and analyse time-intensity curves of sweetness. *Chemical Senses*, 1986, **11**, 105–118.

25. Overbosch, P., A theoretical model for perceived intensity in human taste and smell as a function of time, *Chemical Senses*, 1986, **11**, 315–329.
26. Dijksterhuis, G.B. and Roos, P.E., *Collection and Processing of Time-Intensity Data, Lecture for the Nordic Workshop, 'Advanced Sensory Analysis III, the Consumer in Focus'*, February 1990, Aarhus, Denmark (reprint available upon request from OP&P).
27. Gacula, M.C. and Singh, J., *Statistical Methods in Food and Consumer Research*, Academic Press, New York, 1984.
28. Powers, J.E., Using general statistical programs to evaluate sensory data. *Food Technology*, 1984, **38**, 74–84.
29. O'Mahony, M., *Sensory Evaluation of Food*, Marcel Dekker, New York, 1986.
30. Stevens, J., *Applied Multivariate Statistics for the Social Sciences*, Lawrence Erlbaum, New Jersey, 1986.
31. Piggott, J.R. (ed.), *Statistical Procedures in Food Research*, Elsevier, London, 1986.
32. MacFie, H.J.M., Data analysis in flavour research: Achievements, needs and perspectives, *Flavour Science and Technology*, (eds M. Martens, G.A., Dalen and H. Russwurm Jr.), Wiley, New York, 1987.
33. SPSS, SPSS International BV, PO Box 115, 4200 AC Gorinchem, The Netherlands.
34. SAS, SAS Institute GmbH, PO Box 105307 Heidelberg, West Germany.
35. BMDP Statistical Software, Cork Technology Park, Cork, Ireland.
36. Genstat, Numerical Algorithms Group Ltd., England.
37. Gower, J.C., Generalized Procrustes Analysis, *Psychometrica*, 1975, **40**, 33–51.
38. Tyssø, V., Esbensen, K. and Martens, H., UNSCRAMBLER, an interactive program for multivariate calibration and prediction, *Chemometrics and Intelligent Laboratory Systems*, 1987, **2**, 239–243.
39. Burg, E. van der, *Nonlinear Canonical Correlation and some related techniques*, Thesis, DSWO Press, Leiden, 1988.
40. Martens, H. and Naes, T., *Multivariate calibration*, Wiley, New York, 1989.
41. Burg, E. van der and Dijksterhuis, G.B., Nonlinear canonical correlation analysis of multiway data, in *Multiway Data Analysis*, eds R. Coppi, and S. Bolasco, North Holland, 1989.
42. Gifi, *Nonlinear Multivariate Analysis*, Wiley & Sons, New York, 1990.
43. Dijksterhuis, G.B. and Punter, P.H., Interpreting Generalized Procrustes 'Analysis of Variance' Tables, *Food Quality and Preference*, 1990, **2**, 255–265.
44. Gruijter, D.N.M. de, Data analysis and statistics, report of a discussion. *Statistica Neerlandica*, 1988, **42**, 99–102.
45. Leeuw, J. de, Models and techniques. *Statistica Neerlandica*, 1988, **42**, 91–98.
46. Molenaar, I.W., Formal statistics and informal data analysis, or why laziness should be discouraged. *Statistica Neerlandica*, 1988, **42**, 83–90.
47. Tukey, J.W., *Exploratory data analysis*, Addison-Wesley, Reading, Massachusetts, 1977.
48. Cleveland, W.S. and McGill, M.E. (eds), *Dynamic Graphics for Statistics*, Wadsworth & Brooks/Cole, Belmont, 1988.
49. Visuals, F.W Young, UNC Psychom., CB-3270 Davie Hall, Chapel Hill, NC, USA.

8 Citrus breeding and flavour

R. ROUSEFF, F. GMITTER and J. GROSSER

Abstract

An original review of the information pertaining to citrus breeding and flavour is presented. Nonvolatile flavour component, such as sugars, acids and bitter components, in *Citrus* can be modified through breeding. Low-acid and nonbitter grapefruit hybrids have been developed. Whereas nonvolatile flavours of many progenies can often be predicted, volatile flavours are impossible to foretell. Aroma quality from progeny of commercial quality parents has, in many cases, been definitely inferior to either parent. The quantitative and qualitative differences in the volatile components that produce the characteristic aroma of traditional hybrids such as orange, grapefruit, lemon and lime are presented. New interspecific and intergeneric somatic hybrids involving *Citrus* have been developed using protoplast fusion. Owing to the long juvenile period in *Citrus*, only one hybrid has produced fruit to date. The flavour characteristics of this new hybrid remain to be evaluated.

8.1 Introduction

The importance of citrus to the world economy is demonstrated by its wide distribution and extensive production. Citrus is grown throughout the world in tropical and subtropical areas where suitable soils and climates are found. It is a major component in the economies of many countries such as Brazil, Israel, South Africa, Spain and part of the USA.

In 1987, citrus accounted for 58% of the USA commercial fruit production [1]. Citrus growing is Florida's largest agricultural industry and it generated 70% of the total USA production of citrus in 1988 [2]. During the 1987–1988 season, 12.4 millon metric tons of citrus were harvested in Florida. In that season, approximately 92% of the crop was processed. Of the portion that was processed 85% was converted into concentrates, 12% went into chilled juices and 3% went into canned juice. The on-tree value of this crop was over a billion dollars and the value added from processing was an additional two billion dollars.

Citrus fruits have distinctive flavours that are esteemed by people throughout the world. The diversity of citrus flavours ranges from the acidic, zesty and distinctive light aroma of limes (*C. aurantifolia*) to the rich sweet taste and full-bodied aroma of sweet oranges (*C. sinensis*) to the pungent aroma and astringent taste of the citron (*C. medica*). Orange flavour is probably the most widely recognized and accepted flavour in the worldwide food and beverage industry; it is widely used to flavour or aromatize foods and beverages because of its distinctive flavour and aroma. Lemon flavour is probably second to orange flavour in overall popularity. Lemon oils are used to flavour both carbonated and non-carbonated beverages, which have become increasingly popular world-wide, and to aromatize household products to produce a clean, light fragrance. To a lesser extent, lime and tangerine oils are also used as flavourings.

South-east Asia, particularly China, Cochin China, India, and the Malay Archipelago are commonly considered as places where citrus originated. Citrus has been used as a food source for thousands of years. The cultivation of citrus predates recorded human history. In the ancient Confucian writings called the *Book of History* it is recorded that citrus fruit were considered prized tributes during the reign of Ta Yü (approximately 2205–2197 BC). The two species specifically mentioned are the chü and the wu. Today they are known as ancestors of the mandarin and pummelo, respectively. Gradually over the centuries, variations in size and sweetness of the original chü were developed and the name kan, meaning 'tree of sweet' was used in the names of several popular cultivars. The sweet orange (*C. sinensis*) is thought to have originated from natural hybridization between a kan mandarin and a pummelo in southeast China [3,4]. The Chinese never attached much significance to this new cultivar, considering it merely as a tight-skinned kan selection. These early horticulturists apparently selected wild types that produced the most desirable fruit and took them back to their villages for cultivation. Later writings would demonstrate a rather high degree of horticultural sophistication in that they were familiar with asexual reproduction techniques and the use of dissimilar rootstocks to increase yield and reduce juvenility.

8.2 Botanical considerations

Botanically, a citrus fruit is classified as a hesperidium, a particular sort of berry with a leathery rind that is internally divided into segments. There are six closely related genera (*Citrus, Fortunella, Poncirus, Microcitrus, Eremocitrus* and *Clymenia*) and seven less-related genera (see Table 8.1).

8.2.1 Taxonomic systems

Citrus fruit and plants are botanically diverse and have been difficult to classify. Two of the most widely used taxonomic systems are those

Table 8.1 Swingle classification of *Citrus* spp.: family – Rutaceae; orange subfamily – Aurantiodeae[a,b]

Tribe I: Clauseneae: very remote and remote citroid fruit trees	
Subtribe 1. Micromelinae: very remote citroid fruit trees	
Micromelum (9)	
Subtribe 2. Clauseninae: remote citroid fruit trees	
Glysosmis (35)	*Murraya* (11)
Clausena (23)	
Subtribe 3. Merrillinae: large-fruited remote citroid fruit trees	
Merillia	
Tribe II: Citreae: citrus and citroid fruit trees	
Subtribe 1. Triphasiinae: minor citroid fruit trees	
Wenzelia (9)	*Triphasia* (3)
Monanthocitrus	*Pamburus*
Oxanthera (4)	*Luvunga* (12)
Merope	*Paramignya* (15)
Subtribe 2. Citriniae: citrus fruit trees	
Severinia (6)	***Fortunella*** (4)
Pleiospermium (5)	*Eremocitrus*
Burkillanthus	***Poncirus*** (2)
Limnocitrus	*Clymenia*
Hesperathusa	*Microcitrus* (6)
Citropsis (11)	***Citrus*** (16)
Atalantia (11)	
Subtribe 3. Balsamocitrinae: hard-shelled citroid fruit trees	
Swinglea	*Balsamocitrus*
Aegle	***Feronia***
Afraegle (4)	*Feroniella* (3)
Aeglopsis (2)	

[a] Number in parentheses indicates number of species per genus.
[b] Genera in bold type indicate success in somatic hybridization with *Citrus*.

developed by Swingle [5] and Tanaka [6,7]. Tanaka initially proposed the existence of 145 and later 159 citrus species. Swingle [5] suggested there were only 16 species. Without judging the merits of either classification system we will use the Swingle system (Table 8.1) in this discussion.

8.2.1.1 Citrus *and closely related genera.* The Citrinae, subtribe 2 of tribe II Citreae, subfamily Aurantiodiae, family Rutaceae, of which *Citrus* is a member, can be divided into additional subtribal groups, consisting of 28 genera and 125 species. *Citrus* spp. belong to the true *Citrus* fruit tree subtribal group. Even though many of the genera and species related to *Citrus* have been characterized and evaluated for potential value to *Citrus* improvement, intraspecies and intragenus variability has not been adequately assessed. *Citrus* has remarkable sexual compatibility within its own subtribal group, but species that contain many desirable traits exist outside this subtribal group. Unfortunately, these species are sexually incompatible with *Citrus*.

Contemporary *Citrus* taxonomists propose that there are only three true citrus species, namely mandarin (*C. reticulata*), citron (*C. medica*) and pummelo (*C. grandis*). The extensive diversity of citrus flavours is thought to arise from various combinations and successive back-crosses of these three progenitors and some gene exchange with close, sexually compatible relatives [3,8,9].

8.2.2 Breeding problems

Breeding for specific characteristics is difficult with citrus because of its nucellar embryony and long periods of juvenility. *Citrus* is generally cross-pollinated and highly heterozygous, and hybrid offspring can be exceedingly variable. Facultative apomixis via nucellar embryony is a unique feature of citrus reproductive biology, common to most commercial cultivars. Both sexual and asexual seedlings can be produced from the same seed, but usually the nucellar embryos outcompete the zygotic embryos for nutrients and space within the developing seed. The percentage of nucellar seedlings from polyembryonic clones can range from 20% to 100%, depending on environment and pollen sources. It can be difficult, and sometimes impossible, to determine if the resulting progeny are true sexual hybrids. Further, the two species of greatest economic significance in Florida, *C. sinensis* and *C. paradisi* (grapefruit), are considered more correctly as hybrid forms, and not true biological species. They are highly heterozygous, and intraspecific crosses yield few (if any) sexual hybrids that resemble the parental types or exhibit the typical characteristics of the parents. Self- and cross-incompatibility, together with partial to complete pollen and/or ovule sterility, frequently prevent hybridization among most commercial *Citrus* selections such as navel orange, satsuma mandarin, etc. The reproductive cycle is lengthy, requiring 3 to 10 years to obtain fruit from seeds. Consequently, improvement of these important types will likely be achieved by methods other than sexual hybridization. This has been one of the major problems in citrus breeding.

Breeders have established evaluation criteria for new hybrids based on horticultural vigour and disease resistance, as well as marketing characteristics such as fruit size, shape, seediness, peel thickness, colour and season/ maturity. Unfortunately flavour is usually regarded as a secondary objective in evaluating new hybrids. Nevertheless, some interesting flavours have been observed from the products of established breeding programmes.

8.3 Breeding of citrus flavours

Citrus flavours arise from a variety of sources, both volatile and non-volatile. However, in evaluating flavours of citrus hybrids several factors must be considered. First the flavour of the hybrid is usually evaluated as fresh fruit, rather than juice. Second, the amount of fruit from hybrid trees will be limited and subject to the usual horticultural variables that influence flavour such as maturity, climate, tree position, rootstock and so on. Unfortunately the sensory procedures such as quantitative descriptive analysis and other flavour profiling or category scaling tests that have been developed to evaluate processed products in general, have not been used for fresh fruit. Most sensory evaluations have involved a single evaluator who usually has not had sensory training. This evaluator is usually the breeder, and assessments tend to be subjective. Since it takes an average of 5–6 years before fruit from sexual crosses can be evaluated, breeders must make some early subjective evaluations as to the value of each hybrid. By the time the trees bear fruit, they are fairly large (2–4 m tall). Owing to space limitations and sheer number of hybrid trees it is usually neither possible nor desirable to evaluate rigorously the fruit of all hybrids for flavour.

8.3.1 Sensory analysis

In one of the uncommon uses of a taste panel for the sensory evaluation of fresh fruit, Genzi and Cohen [10] evaluated sensory characteristics and chemical composition of 50 different ‘Minneola’ tangelos. They employed a three- to five-member panel to evaluate fruit segments on a scale of 1–3 for sweetness and sourness and estimate aroma and bitterness on a scale of 0–2. In addition, panellists were asked to grade each fruit on overall flavour. The forced-choice categories were: (i) tasty; (ii) edible; or (iii) inedible. One half of each fruit was used for chemical analysis and individual segments from the other half were given to taste panellists for evaluation. Even with limited forced choices the panel was in complete agreement only 8% of the time when each fruit was considered individually. In 46% of the fruit tasted, the panel was in serious disagreement, indeed, some, panellists ranked the fruit as inedible, whereas others thought it was tasty. However, there was considerably less divergence in the evaluations for sourness, aroma and bitterness.

Correlation coefficients between chemical measurements and taste-panel assessments were low. The highest correlations were between: (i) inedibility and the sugar:acid ratio ($r = -0.56$); (ii) sourness and acid content ($r = 0.76$); and (iii) sourness with total soluble solids, (TSS):acid ratio ($r = -0.79$). The results were also scrutinized using canonical correlations. Linear combinations were sought within both sensory and analytical measurement groups. Subgroups within these two groups were also evaluated to discard variables that did not contribute to the correlation. Individual canonical correlations are shown in Table 8.2.

Table 8.2 Canonical correlations from 'Minneola' tangelos[a]

Variable	I: 'Maturity'	II: 'Taste acceptability'
Aroma	0.76	0.86
Sweetness	0.74	0.82
Sourness	−0.77	−0.87
Edibility	0.66	0.73
TSS : acid ratio	0.89	0.80
Acid	−0.89	−0.79
Ethanol	0.40	0.36
Weight	0.49	0.44

[a] Source: Genzi and Cohen [10].

The highest correlations were obtained from the following canonical variables:

I: 0.41 (TSS:acid) − 0.60 (acid) + 0.0021 (ethanol) + 7.5 (weight in kg)

II: 1.08 (aroma) + 0.52 (sweetness) − 0.48 (sourness) + 0.0086 (edibility)

Canonical variable I was termed 'maturation' because the terms included correlated in general with accepted measurements for maturity. In a similar fashion, canonical variable II was termed 'taste acceptability'. The correlation between the two canonical variables was very good ($R = 0.90$) and is shown in Figure 8.1.

The lines dividing the figure separate three different stages of the development of 'taste acceptability,' which can be separated with very few misclassifications using the 'maturation' variable. For example, if the highest taste acceptability were desired then the analytical values used to calculate the 'maturation' variable would have to be greater than or equal to 0.62.

8.3.2 *Nonvolatile flavours*

8.3.2.1 Sweetness. The primary nonvolatile flavour perception is the blend of sweetness and sourness due to organic acids and sugars. Lemons and limes are high-acid fruits, whereas oranges and tangerines typically contain more sugars than acids. The relative sugar-acid distribution between high-sugar low-acid species such as sweet orange, and low-sugar high-acid types such as limes is shown in Figure 8.2. Since the relative blend of sugars to acids is maturity related, breeders have sought to develop cultivars that produce favourable sugar acid blends early.

8.3.2.2 Sourness. High acidity causing sourness is usually a negative flavour attribute in cultivars that normally contain primarily sugars, such as sweet oranges and grapefruit (Figure 8.2). However, acidity is esteemed in

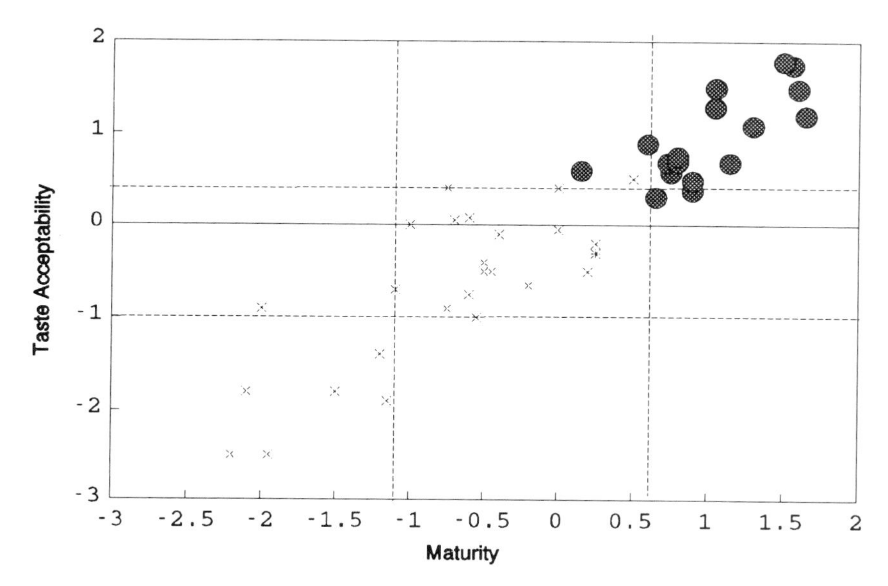

Figure 8.1 Canonical correlation between I, 'maturity', and II, 'taste acceptability', for 'Minneola' tangelos. (× = total soluble solids, TSS; acid <6.1:1 and/or acid >2.2%; ● = TSS:acid >6.1:1; acid ≤2.0%). Source: Genzi and Cohen [10].

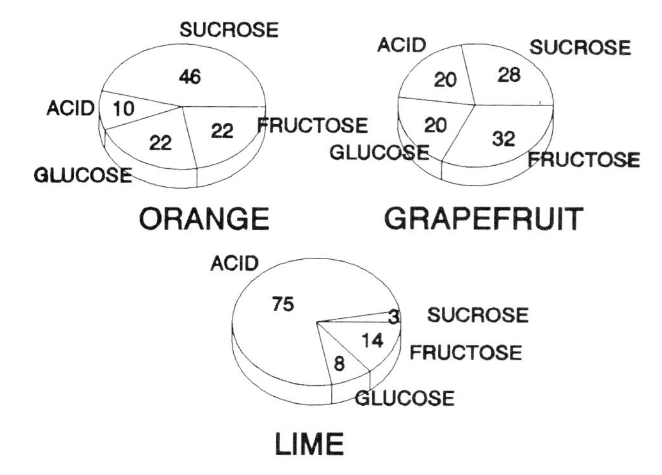

Figure 8.2 Relative sugar-acid composition of sweet orange *C. sinensis*, grapefruit *C. paradisi* and lime *C. aurantifolia*.

high-acid fruit such as limes; indeed growers are paid more for high-acid fruit. Breeders have found that in sexual crosses, high acidity can be a persistent trait. Barrett [11] made several crosses with the high-acid, cold-tolerant, inedible citrus relative, *P. trifoliata*. The hybrids of *P. trifoliata* × *C. sinensis* and *P. trifoliata* × *C. paradisi* were back-crossed with *C. sinensis*. The acid levels in the first hybrids (F_1 progenies) were very high, and none was fit to eat as a dessert fruit. All fruit from F_1 hybrids involving *P. trifoliata* were highly acidic, acrid and bitter. With the exception of one, acid levels in the back cross progenies (BC_1) were still much higher than is commercially acceptable. However one of the back cross progenies, (*C. paradisi* × *P. trifoliata*) × *C. sinensis*, produced selections whose fruit acid levels are similar to *C. sinensis*. This work suggests that high acidity is a dominant trait with *P. trifoliata* hybrids and requires successive back crosses with low acid cultivars before it can be reduced to commercially acceptable levels.

8.3.2.3 Bitterness. Bitterness is also a major nonvolatile flavour attribute that arises from one or both of two classes of chemical compounds – liminoids and flavanone glycosides. Citrus bitterness arises from flavanone neohesperidosides and/or liminoids. Interestingly, the bitterness caused by flavanone neohesperidosides is perceived immediately (if present above taste thresholds), whereas liminoid bitterness is not perceived in fresh fruit or freshly squeezed juice. Liminoid bitterness is detected only after the juice is allowed to stand or is heated.

The key to flavanone glycoside bitterness lies in the manner in which the two sugars, rhamnose and glucose, are linked. Linked 1–2 and the resulting flavanone neohesperidoside is bitter, but linked 1–6 and the resulting structural isomer (flavanone rutinoside) is tasteless. Horowitz [12] was one of the first to recognize that citrus cultivars usually contained entirely one or the other form. Most of the major commercial citrus cultivars, that is, sweet oranges (*C. sinensis*), mandarins (*C. reticulata*), lemons (*C. lemon*) and citrons (*C. medica*) contained only flavanone rutinosides. Those containing the bitter form were pummelo (*C. grandis*), sour orange (*C. aurantium*), and the trifoliate orange (*Poncirus trifoliata*). Albach and Redman [13] found that in crosses between cultivars containing only flavanone rutinosides, only flavanone rutinosides and no flavanone neohesperidosides were observed in the progeny. Conversely, in crosses between cultivars containing only flavanone neohesperidosides, no flavanone rutinosides were observed. However, in crosses between cultivars containing only flavanone rutinosides with cultivars containing only flavanone neohesperidosides, the progeny usually contained both forms. Thus the alleles responsible for the production of the two rhamnosyl glucoside isomers often exhibit additive inheritance in the F_1 generation. Grapefruit is perhaps the best known example of this process. Grapefruit (*C. paradisi*) contains both flavanone rutinosides and flavanone neohesperidosides, adding further evidence to the supposition that this cultivar is actually a hybrid of pummelo (*C. grandis*) and, possibly, sweet orange (*C. sinensis*). It is also interesting to note that when hybrids containing both bitter and nonbitter forms were back-crossed with a nonbitter cultivar, the resulting progeny usually contained only the nonbitter forms. Other workers found similar results when working with leaves [14, 15]. More recently high-pressure liquid chromatography (HPLC) has been used to obtain quantitative information as to the absolute amounts of each flavanone glycoside form present in fruit [16].

8.3.3 Volatile flavours

8.3.3.1 Unpredictable aromas. Whereas the nonvolatile flavour qualities such as acidity or bitterness can be predicted using parents that have high or low acidity or bitterness, the volatile aroma quality cannot be predicted. For example Reece and Hearn [17] reported that the aromas from progeny of 'Clementine' mandarin × 'Mott' grapefruit were fishy, garlicky or had a 'sickeningly sweet perfumed quality'.

8.3.3.2 Oil-soluble components. Citrus volatile flavour components can be separated into one of two broad categories. The first is the oil-soluble constituents present in peel oil and in juice oil. The other flavour component consists of the water-soluble constituents present in the juice.

Citrus peel oil is located in small, ductless glands present in the flavedo, or outer section of the peel. Peel oil from each citrus cultivar provides much of the characteristic aroma and flavour of that cultivar. Ethyl acetate, ethyl butyrate, ethyl hexanonate, ethyl octanoate, ethanol and limonene have been identified in headspace gases from oranges stored in closed containers [18]. Juice sacs contain a small amount of oil that differs slightly in composition from that found in the peel. The fruit must be peeled carefully before the juice is extracted to eliminate peel oil in the juice. Since this is not commercially possible, the usual industrial practice is to remove most of the excess oil through evaporation either in a separate de-oiling step or in the evaporator used to concentrate the juice.

A capillary gas chromatogram of a 'Valencia' orange oil is shown in Figure 8.3. Over 90% of the oil consists of D-limonene, an essentially odourless terpene. However, it is the oxygenated terpenes (e.g. esters, aldehydes and alcohols) that are the primary aroma-generating compounds. The general odour characteristic of these components is indicated above each identified peak or group of peaks. For example, linalool produces a floral fragrance, neral produces a lemon-like aroma and ethyl butyrate produces a fruity fragrance.

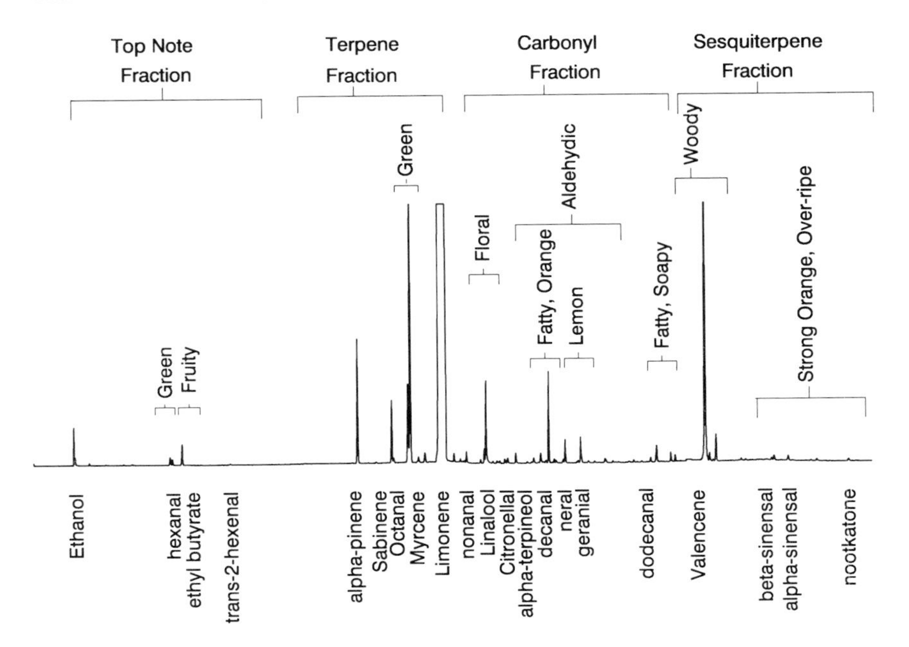

Figure 8.3 Capillary gas chromatogram of a Valencia orange oil.

Citrus oils have distinctly different aromas arising from both qualitative and quantitative compositional differences. Time and space limitations will not allow for an adequate discussion of this complex topic. As a brief example, the chromatograms from 'Valencia' orange and 'Duncan' grapefruit are shown in Figure 8.4.

The aroma of orange oil is characterized by the mixture of acetaldehyde, ethyl butyrate, decanal, neral, geranial, valencene and linalool whereas grapefruit is characterized by having significant amounts of nootkatone, caryophyllene, small but highly significant amounts of *p*-methenethiol and comparatively low linalool and citral. Other cultivars have their own distinctive aroma active components. For example lemon is characterized by citral, neryl acetate and geranyl acetate. Tangerine has γ-terpinene, dimethylanthramilate, α-sinensal but not β-sinensal.

8.3.3.3 Water-soluble components. The majority of the characteristic flavour of fresh citrus fruit comes from the water-soluble components found in the juice sacs. Identification of these components has been accomplished from juice extracts or commercial aqueous essence prepared from juice. One of the more extensive studies of the volatile components of fresh orange fruit was carried out by Schreier and co-workers [19, 20]. The list of compounds identified and quantified is shown in Table 8.3

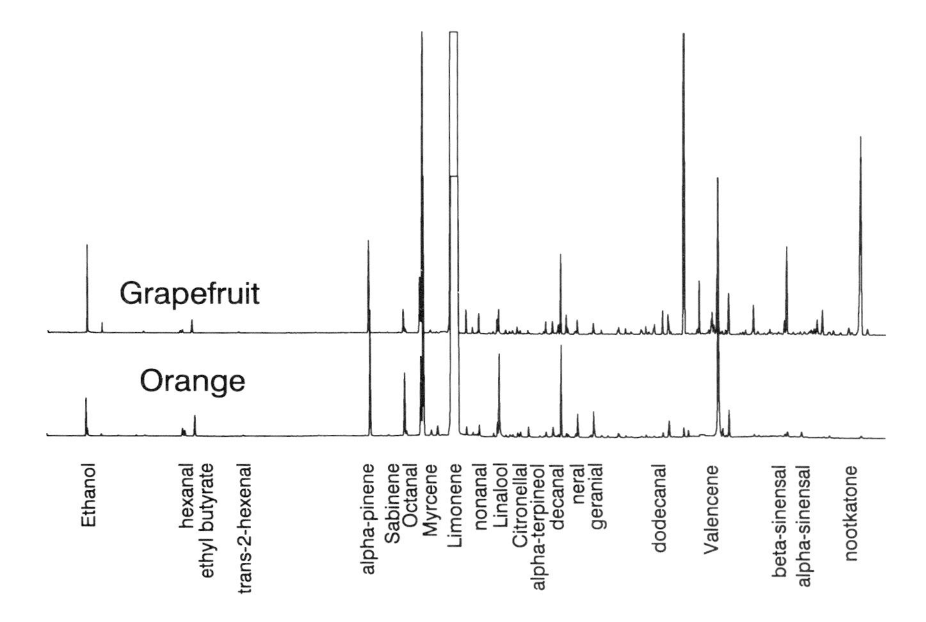

Figure 8.4 Comparison of a capillary gas chromatogram of orange and grapefruit oils.

8.3.4 Conventional breeding programmes

The majority of citrus scion cultivars in current production originated as chance seedling selections or bud sport mutations of existing cultivars and not from established breeding programmes. The limited impact of planned breeding has not been for lack of effort. Citrus breeding has been carried out by workers worldwide. However the earliest, extensive studies appear to have been carried out in the USA, primarily Florida and California [21]. W.T. Swingle and H.J. Webber carried out some of the first systematically planned hybridizations in Florida beginning in 1893. One of their major breeding objectives was to transmit cold-resistance into their hybrids. Interestingly, this continues to be a major breeding goal almost 100 years later. In the early 1900s this same group created citranges, citrangors, citrangedins, limequats and tangelos. Commercially successful mandarin hybrids such as 'Robinson', 'Lee', 'Page', and 'Nova' cultivars were developed in the 1940s and released several years later [22, 23] and 'Sunburst' in the 1960s by Hearn [24].

Soost and Cameron [25] reported that they had found an essentially acidless pummelo (*C. grandis* Osbeck) cultivar, CRC 2240. They crossed this cultivar with a seedy, white, tetraploid grapefruit and found that the triploid progeny contained low acidity. Two cultivars had particularly favourable characteristics and have been released as 'Oroblanco' and 'Melogold' [26].

Table 8.3 Components and concentrations from 'Sanguinello' orange flesh

Component	μg/l	Component	μg/l
α-Pinene	178	Butyl octanoate	36
β-Pinene	8	Linalyl acetate	12
Myrecene	690	Menthyl acetate	6
Limonene	70 000	Citronellyl acetate	12
γ-Terpinene	30	Neryl acetate	20
ρ-Cymene	17	Geranyl acetate	15
Terpinolene	10	Ethyl 3-hydroxybutanoate	650
allo-Ocimene	85	Methyl 3-hydroxyhexanoate	100
β-Caryophyllene	64	Ethyl 3-hydroxyhexanoate	700
Farnesene	48	Ethyl 3-hydroxyoctanoate	30
α-Humulene	30	Linalool	215
Valencene	5000	4-Terpineol	47
δ-Cadinene	31	α-Terpineol	55
Selinadiene	150	Nerolidol	10
Hexanal	20	Geraniol	11
Octanal	25	Citronellol	21
Nonanal	15	Nerol	20
Decanal	77	*trans*-ρ-Menthen-9-ol	22
Citral	58	*cis*-Linalool oxide	10
β-Ionone	25	*trans*-Linalool oxide	50
Perillaldehyde	30	3-Methylbutan-1-ol	569
Ethyl butanoate	900	Hexan-1-ol	30
Ethyl 2-methylbutanoate	22	*cis*-Hex-3-en-1-ol	54
Butyl butanoate	13	Octan-1-ol	17
Ethyl hexanoate	52	2-Phenylethanol	20
Ethyl octanoate	27		

Source: Schreier *et al.* [19].

8.3.5 New breeding approaches

Advantageous recombination within the genetic diversity of citrus is limited in large measure by sexual incompatibility. With the advent of certain biotechnological techniques such as protoplast fusion, additional genetic material can now be introduced [27, 28]. Protoplast fusion has been used to develop somatic hybrids in citrus. These hybrids contain the complete genetic information from both parents and can be used as breeding parents to introduce new genetic diversity into citrus hybrids. A schematic representation of a general approach for the production of citrus somatic hybrids is shown in Figure 8.5. Generally excised portions of leaf tissue are treated with enzymes to digest cell walls and connective tissue to produce spherical cells called protoplasts containing visible chloroplasts. These cells are mixed with identically treated cells from embryogenic suspension cultures that were developed from nucellar tissue. Mixed protoplasts are then treated with polyethylene glycol to induce fusion and incubated. If successful, each hybrid fusion product (heterokaryon) should contain the complete set of genes from each parent. These cells are said to be tetraploid (4X) as each parent is diploid (2X). (In normal sexual crosses

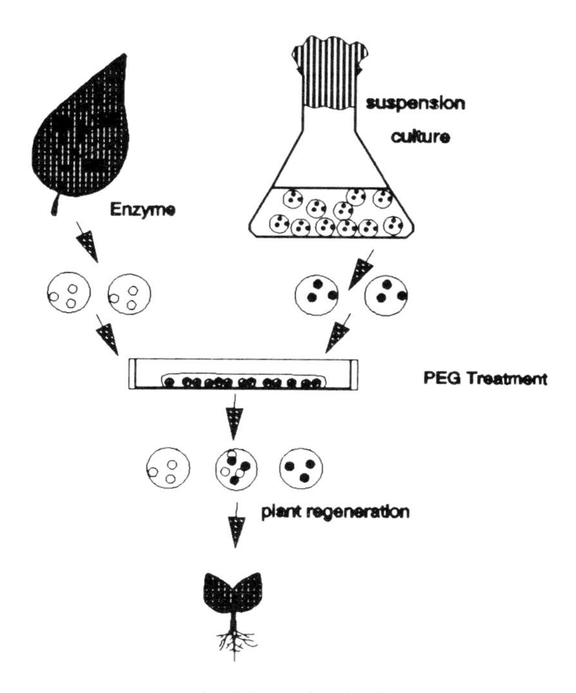

Figure 8.5 Somatic hybrid formation in *Citrus*.

the ovule and pollen are haploid (1X) which are produced as a result of meiosis.) The success rate of somatic hybrid production may range from 1–40%. However, the multiple steps required to regenerate a viable plant from somatic hybrid cells is the most difficult and critical step in the overall process. The processes involved in plant regeneration are not well understood and represent a combination of art and science.

Several somatic hybrids have been developed at the Citrus Research and Education Center of the University of Florida at Lake Alfred. Several hybrids of sexually incompatible *Citrus* cultivars have also been produced. One interspecific somatic hybrid, a 'Valencia' sweet orange + 'Key' lime hybrid, has produced fruit to date. Unfortunately there was not enough fruit to make a representative assessment of the flavour characteristics of this hybrid. It will take at least a further year before the tree will be large enough to bear a representative crop for evaluation.

8.3.5.1 Limitations. All of the current somatic *Citrus* hybrids were produced from protoplasts of one parent that had been isolated from nucellar derived embryogenic cultures. This is a necessary requirement for hybrid plant regeneration following protoplast fusion. One of the principal limiting factors in somatic hybrid production is the availability of embryogenic cultures amenable to protoplast culture and regeneration. Initiation and maintenance of such cell lines is exacting and time-consuming. Media requirements for many important genotypes are not completely understood and production of embryogenic cultures has been confined to polyembryonic genotypes. Additional research is needed to improve culture efficiency of responding genotypes, other than sweet orange, and to determine media requirements for more difficult genotypes [26]. Other limitations include lack of fruitfulness, negative effects of polyploidy on fruit size, texture, peel thickness and so on.

8.4 Conclusion

A summary of citrus flavour sources and components and how some of these components can be modified through conventional breeding programmes has been presented. The problems and limitations of conventional programmes are delineated. Newer biotechnological approaches such as protoplast fusion can be used to introduce new genetic material into *Citrus*. The progeny from this technique are usually tetraploid, having the full set of genes from each parent. This should eliminate the expression of undesirable recessive genes that are frequently unmasked in sexual crosses. Currently, only a few fruit from this process have been produced. Unfortunately it will be several years before most of the newly created hybrids will be sufficiently mature to produce enough representative fruit for flavour evaluations.

Acknowledgements

The authors would like to express their appreciation to Dr Tim Anglea for many helpful discussions.

University of Florida Agricultural Experiment Station Journal Series No. R- 01162.

References

1. Market Research Report, Florida Research Report, Florida Department of Citrus Lakeland, Florida, 1988.
2. Citrus Summary, Florida Crop and Livestock Reporting Service, Orlando, Florida, 1989.
3. Barrett, H.C. and Rhodes, A.M. A numerical taxonomic study of affinity relationships in cultivated *Citrus* and it close relatives. *Sys. Bot.*, 1976, **1**, 105–136.
4. Cooper, W.C. and Chapot, H. Fruit production – with special emphasis on fruit for

processing, in Citrus Science and Technology, Vol. 2 eds S. Nagy, P.E. Shaw and M.K. Veldhuis AVI Publishing, Westport, Connecticut pp. 1–127.
5. Swingle, W.T. The botany of *Citrus* and its wild relatives, In *Citrus Industry I*, (eds H.J. Webber and L.D. Batchelor), University of California Press, Berkeley, 1967, pp. 129–474.
6. Tanaka, T., Species problems in *Citrus* (Revisio Aurantiacerum, IX), *Jap. Soc. Prom. Sci.*, 1954, Ueno, Tokyo, p. 152.
7. Tanaka, T. Taxonomic problems of *Citrus* fruits in the Orient: Misunderstandings with respect (to) *Citrus* classifications and nomenclature. *Bull. Univ. Osaka Perf. Ser.*, 1969 **B21**, 133–169.
8. Scora, R.W. and Kumamoto, J. Chemotaxonomy of the genus *Citrus*, in *Chemistry and Chemical Taxonomy of the Rutales*, 1983 (eds P.G. Waterman and M.F. Grundon) Academic Press, London, p. 343–351.
9. Scora, R.W. On the history and origin of *Citrus Bull. Torrey Bot. Club*, 1975, **102**, 369–375.
10. Genzi, A. and Cohen, E. The chemical composition and sensory flavour quality of 'Mineola' tangerines. II Relationship between composition and sensory properties. *J. Hort. Sci.*, 1988, **63**, 179–182.
11. Barrett, H.C. *Breeding Cold-Hardy Citrus Scion Cultivars, Proceedings of the International Society of Citruculture*, 1981, pp. 61–66.
12. Horowitz, R.M. *in Biochemistry of Phenolic Compounds*, (ed. J.B. Harborne), Academic Press, New York, 1964, p. 545.
13. Albach, R.F. and Redman, G.H., Composition and inheritance of flavanones in citrus fruit. *Phytochem.* 1969, **8**, 127–143.
14. Kamiya, S., Esaki, S. and Konishi, F. Flavonoids in citrus hybrids. *Agric. Biol. Chem.*, 1979, **43**, 1529–1536.
15. Mohi, M. and Aminuddin, M. Flavonoid patterns of leaves of some citrus species and their hybrids. *J. Plant Biochem.* 1981, **8**, 56–60.
16. Rouseff, R.L., Martin, S.F. and Youtsey, C.O. Quantitative survey of naringin, narirutin, hesperidin and neohesperidin in *Citrus*. *J. Agric. Food Chem.*, 1987, **35**, 1027–1030.
17. Reece, P.C. and Hearn C.J. The breeding behavior of the 'Clementine' tangerine. *Proc. Fla. State Hortic. Soc.*, 1964, **77**, 76–83.
18. Attaway, J.A. and Oberbacher, M.F. Studies on the aroma of intact Hamlin oranges. *J. Food Sci.*, 1968, **33**, 287–289.
19. Schreier, P., Drawert, F. Junker, A. and Mick, W., The quantitative composition of natural and technologically changed aromas of plants. II Aroma compounds in oranges and their changes during juice processing. *Z. Lebensm.-Unters.-Forsch.*, 1977, **164**, 188–193.
20. Schreier, P. Changes of flavor compounds during the processing of fruit juices. *Proc. Long Ashton Symp.*, 1981, **7**, 355–371.
21. Soost, R.K. and Cameron, J.W. *Citrus, in Advances in Fruit Breeding'*, (eds J. Janick and J.N. Moore) Purdue University Press, West Lafayette, Indianapolis, 1975, pp. 507–540.
22. Reece, P.C., Hearn, C.J. and Gardner, F.E. 1964 *Nova Tangelo – An Early Ripening Hybrid. Proc. Fla. State Hortic. Soc.*, **77**, 109–110.
23. Reece, P.C. and Gardner, F.E. Robinson, Osceola and Lee – new early maturing tangerine hybrids. *Proc. Fla. State Hortic. Soc.*, 1959, **72**, 49–51.
24. Hearn, C.J. Performance of 'Sunburst', a new Citrus hybrid. *Proc. Fla. State Hortic. Soc.*, 1979, **92**, 1–3.
25. Soost, R.K. and Cameron, J.W. Contrasting effects of acid and non-acid pummelos on the acidity of hybrid citrus progenies. *Hilgardia*, 1961, **30**, 351–358.
26. Soost, R.K. and Cameron, J.W. 'Melogold', A triploid pummelo-grapefruit hybrid. *Hort. Sci.* 1985, **20**(6), 1134–1135.
27. Grosser, J.W. and Gmitter, F.G. Wide hybridization of citrus via protoplast fusion: Progress, strategies and limitations, in *Horticultural Biotechnology*, Wiley-Liss, New York, 1990a, pp. 31–41.
28. Grosser, J.W. and Gmitter, F.G. Protoplast fusion and *Citrus* improvement. *Plant Breed. Rev.*, 1990b, **8**, 339–374.

9 Cereal flavours

C. ERIKSSON

Abstract

Flavours in raw cereals originate from the general metabolism of the grain. Cereal products obtain their flavour enzymatically, mainly in treatments like dough-kneading and fermentation, and nonenzymically, in processing for instance by different types of heat treatment. The principal flavour-formation routes seem to be autoxidation of lipids and amino-carbonyl reactions. Special attention has been paid to the formation of alkyl-pyrazines in extrusion and roasting processes. The possible causes of less flavour formation in microwave heating of cereal products is discussed. Off-flavour in cereal and cereal products is caused by advanced lipid oxidation and growth of microorganisms on the grain.

9.1 Introduction

Cereals are consumed in enormous amounts throughout the world. They come in many different forms after many different types of processing. When reviewing the literature one finds that there is little attention paid among flavour scientists to the flavour chemistry of raw cereal grain. The reason for this is to be found in the fact that much of the flavour in cereal products is a function of the processing. Yet the aroma and taste representing the raw material might be of some importance for the flavour of the final product. In this chapter some of the flavour characteristics of both the raw material and the cereal products will be discussed, bearing in mind that the quantitative and qualitative results from different studies, to a large extent, depend on the sensitivity of the laboratory techniques available at the time of the investigation.

As usual in flavour chemistry the standard gas chromatographic/mass spectrometric methods (GC-MS) result in endless lists of volatile compounds, mostly chemically identified. Sometimes individual compounds are described in a sensory way by GC effluent sniffing; flavour threshold and suprathreshold intensity are also often determined. In order to avoid these long lists of compounds most of which are common to many different materials, in this chapter an attempt will be made to highlight

which flavour compounds are characteristic for a variety or a pro
product.

9.2 Principal flavour-formation routes in cereals and cereal products

There are three general major routes in the formation of volatile compounds, many of which can reach the status of contributing to the flavour of cereals and cereal products depending on their characteristics (e.g. threshold value, intensity).

The first route is the formation of secondary products in the normal metabolism of the plant and seed. Probably the main objective of the plant to produce such compounds is to control its environment. This can be either to repel other competing organisms, for instance by producing phenolic compounds, or to attract organisms for the purpose of pollination, as for example with volatile pheromones.

Second, flavour compounds are produced by enzymatic and microbiological processes. Enzymes can be either indigenous or added ones that affect primarily protein, starch and lipids. Indigenous enzymatic reactions can be induced by, for example, flour-milling, dough-kneading, or unintentional mechanical damage. Microbiological processes are started by adding yeast or sour dough, as in dough fermentation, or by inducing mould and bacterial growth, as in for instance tempeh fermentation.

Third, nonenzymatic reactions are introduced primarily by heat, producing flavour compounds from precursors occurring naturally in the raw material or formed in the enzymatic and microbiological processes mentioned. The flavour of flour and cereal products is formed, primarily, by the second and third routes even though the metabolic route can have an influence on the development and differences in flavour through variety, climate, harvesting and storage conditions.

Flavour compounds are formed directly in lipid degradation and in fermentation. Wheat flour contains about 2% lipids, among them unsaturated fatty acids as oleic and linoleic acid. These acids are degraded in a series of reactions into volatile flavour compounds (Figure 9.1). The extent to which these reactions occur depends on factors such as flour or flour mixture composition, moisture content, oxygen availability, catalysts, temperature during storage of the flour or kneading of a dough. The enzymatic lipid oxidation is essential for the dough formation in French bread baking since the peroxides interact with the proteins and ascorbic acid in a complex series of reactions to form the gluten network. Typical flavour compounds developed in lipid oxidation, both in enzymatic and autocatalytic reactions, are hexanal and decadienals. Lipid oxidation is a complex reaction. In processed cereals the autoxidation of unsaturated fatty acids is an important flavour-developing reaction. The first step is

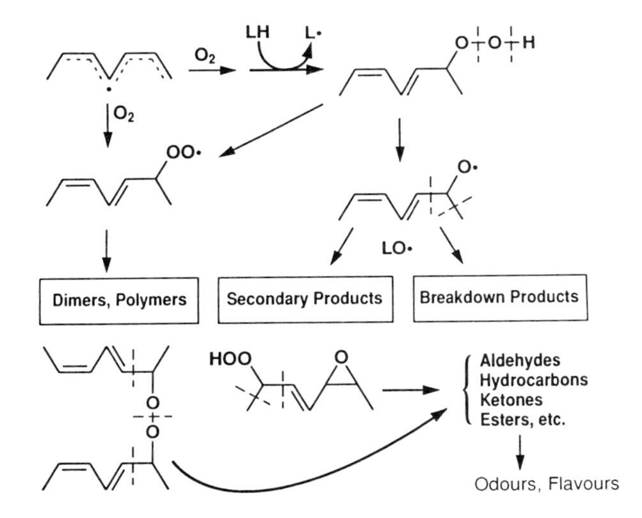

Figure 9.1 Decomposition of linoleate hydroperoxides [1].

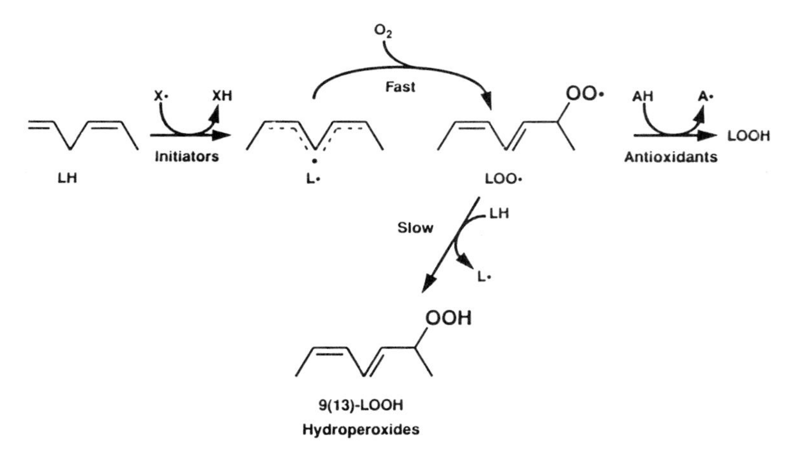

Figure 9.2 Free-radical autoxidation of linoleate [1].

either a radical (Figure 9.2) or a photosensitized reaction (Figure 9.3). The primary oxidation products, the hydroperoxides, can degrade as such (Figure 9.4), or after that they have either dimerized (Figure 9.5) or cyclized (Figure 9.6). The homolytic or heterolytic chain scission gives rise to different mixtures of volatile products as indicated in the figures. Detailed descriptions of these reactions can be found in a recent overview [1].

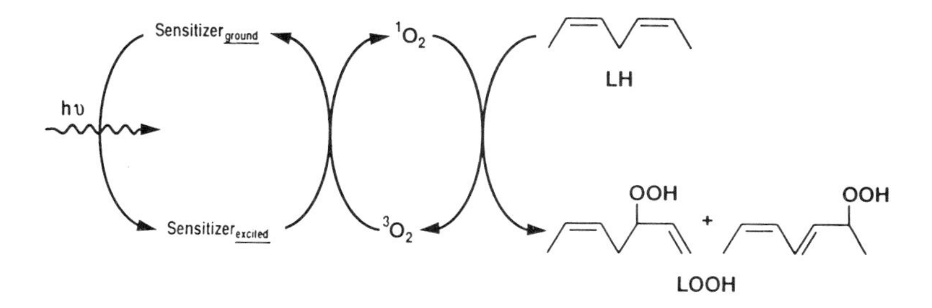

Figure 9.3 Mechanism of photosensitized oxidation of linoleate [1].

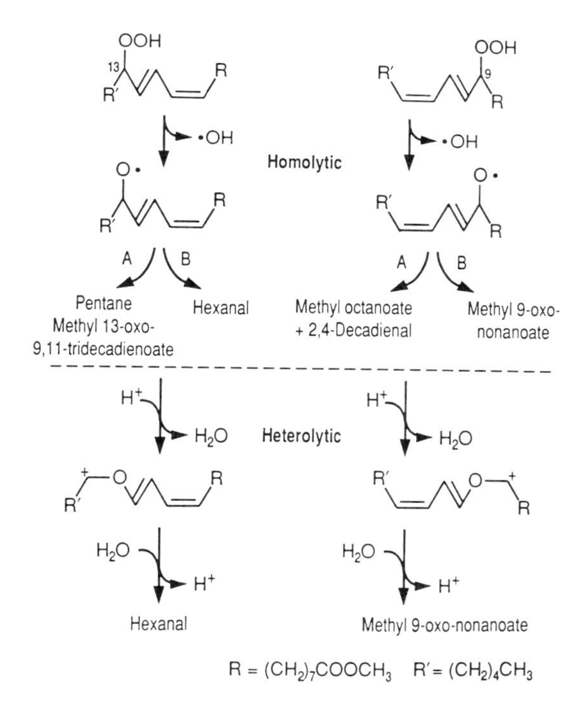

Figure 9.4 Homolytic and heterolytic scission mechanisms for the decomposition of hydroperoxides [1].

Volatile lipid degradation products are very common in the raw material as well as in different cereal products. Evidence whether or not they contribute to a positive flavour of some cereals or cereal products is lacking. More common is to look at these volatile products as the cause of a rancid type of off-flavour. It seems, however, that in some products as

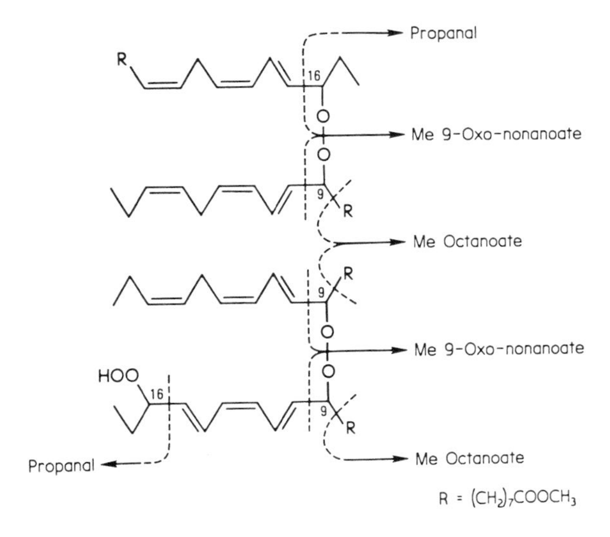

Figure 9.5 Some volatile decomposition products of linolenate dimers [1].

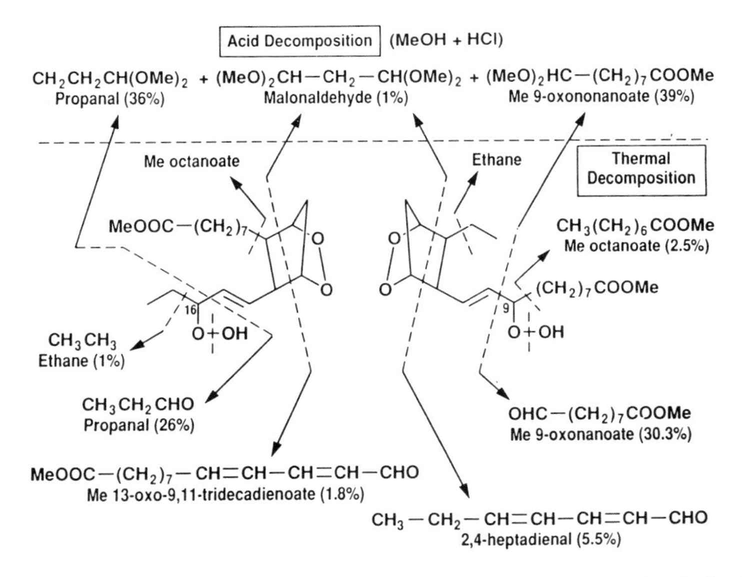

Figure 9.6 Acid and thermal decomposition of hydroperoxy endoperoxides derived from methyl linoleate [1].

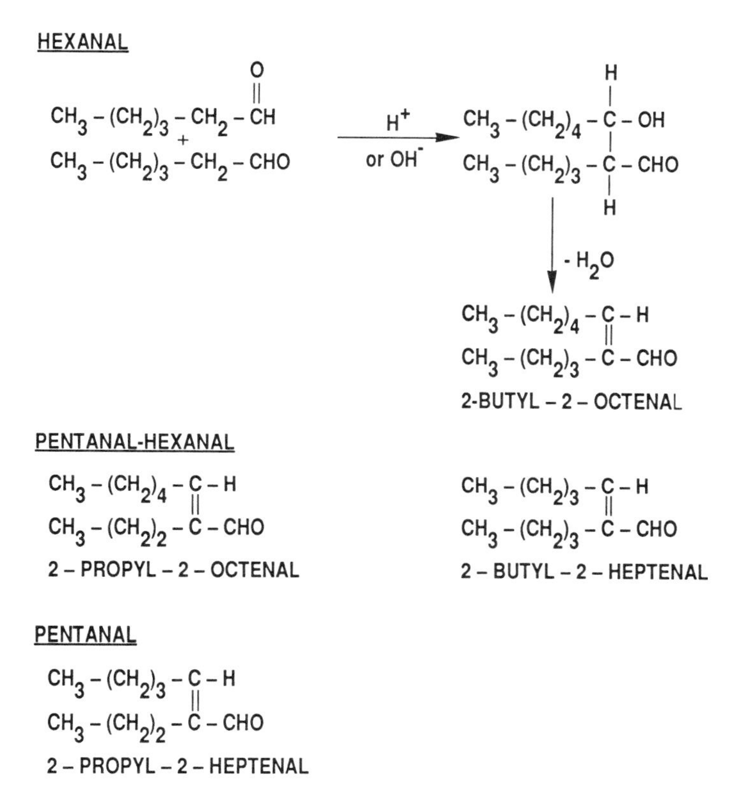

Figure 9.7 Aldehyde self-condensation products identified in rancid dry-heated oat groats [2].

French types of bread benefit in flavour from a certain amount of volatile lipid degradation products. Oats is a cereal that is subjected to lipid hydrolysis and oxidation during various processing leading to distinct rancidity. A new group of volatile compounds have been identified in rancid oat groats [2]. They consist of products formed by condensation of simple aldehydes. Examples are shown in Figure 9.7.

Fermentation by yeast or lactic acid bacteria and yeast (sour dough) gives rise to typical flavour compounds in bread. In the common breakdown of starch, sugars such as glucose, fructose, galactose and maltose are produced; this leads to the further production of ethanol and an array of organic acids such as acetic, propionic, butyric, lactic and malic acid, as well as citric and succinic acid. These acids, dominated by lactic and acetic acid, give a typical sour taste to some breads, especially rye bread. Other typical fermentation-derived flavour compounds are acetoin and diacetyl. Depending on the flour ingredients and micro-organisms used as well as the applied proofing time and temperature, the flavour profile can differ from one type of bread to another.

The Maillard and caramelization reactions are overwhelmingly responsible for flavour formation in cereal products upon heat treatment. Hundreds of different volatile compounds are produced in the Maillard reaction. This reaction involves simple precursors such as amino compounds and reducing compounds, which are represented mostly by amino acids or peptides and simple aldose or ketose sugars. The chemistry of the initial steps of this reaction is well documented; however, the further reactions develop into an extremely complex situation. The reaction consists of a series of condensations, dehydrations and decarboxylations and the most obvious results are colour (melanoidins) and aroma. It is beyond the scope of this presentation to review the many steps in the Maillard reaction. Reference is given to the series of symposia on the Maillard reaction [3–6] as well as to a recent symposium on thermal generation of aromas [7].

One important group of carbonyl-sugar derived volatile products is the alkylpyrazines. They are formed along with pyrrols, pyridines, furans and so on. Alkylpyrazines typically develop, for instance, in bread crust during baking [8], in beer malt upon extrusion cooking [9] and also occur in oils extracted from roasted oats [10]. Evidence is presented by the authors that pyrazines are important for the typical aroma of these materials. Bread crust flavour decreased along with the alkylpyrazines during storage of French-type bread [11].

In the work on extruded malt it was found that premilling and extrusion temperature were important parameters for the formation of alkylpyrazines. At extrusion temperatures of 130°C or below, or higher than 190°C, very low concentrations of alkylpyrazines were found in the extruded malt, while extrusion at 160°C was found to yield much higher concentrations. Premilling resulted in higher yields of alkylpyrazines – especially at 160°C.

The most common types of alkylpyrazines identified in malt extruded at 160°C are listed in Table 9.1 together with their concentrations. Table 9.2 shows their odour-detection thresholds. It can be noted that alkylpyrazines differ in thresholds depending on the number and type of ring substituents as well as on their ring position. The descriptors used by panellists for some of the alkylpyrazines found in extruded malt are found in Table 9.3.

Microwave cooking is a procedure of increasing use in food preparation – especially in homes. It is a fast and easy way to reheat preprepared food. However, consumers complain about lack of typical flavour when applying microwave cooking, for example, to the baking of bread and cakes. Owing to water evaporation, the surface is kept colder than the interior and this is why the Maillard reaction simply does not occur to a sufficient degree on the surface. This fact has led to the formulation of special flavour systems

Table 9.1 Mean concentration of pyrazines in malt (milled kernels) extruded at 160°C[a]

Pyrazine	Concentration (ng/g)[b]
Unsubstituted	1120±37
Methyl-	9500±660
2,5/2,6-Dimethyl-	8230±240
Ethyl-	1440±82
2,3-Dimethyl	1230±1
Vinyl-	257±20
2-Ethyl-6-methyl-	749±16
2-Ethyl-5-methyl-	393±29
Trimethyl-	696±23
2-Ethyl-3-methyl-	383±35
6-Methyl-2-vinyl-	342±33
5-Methyl-2-vinyl-	404±40
Isopropenyl-	88±21
2,5-Dimethyl-3-ethyl-	228±36
Quinoxaline	80±25

[a] Source: Fors and Eriksson [9].
[b] ng/g refers to the amount of pyrazine per gram of malt sample.

Table 9.2 Odour detection thresholds for 13 alkylpyrazines[a]

Pyrazine derivative	Panel estimate (mg/m^3)[b]	Panel estimate (ppb(v/v))[b]
Methyl-	1.9	490
Ethyl-	0.25	57
2,3-Dimethyl-	0.90	200
2,5-Dimethyl-	0.17	38
2,6-Dimethyl-	0.25	57
2,3-Diethyl-	0.050	8
2-Ethyl-5(6)-methyl-	0.036	7
2-Ethyl-3-methyl-	0.15	30
Trimethyl-	0.19	38
2,5(6)-Dimethyl-3-ethyl-	0.020	4
2,3-Diethyl-5-methyl-	0.023	4
Tetramethyl-	0.69	120
Vinyl-	0.33	76

[a] Source: Fors and Eriksson [9].
[b] Confidence limits at the 95% level.

being introduced in, for instance, bread or cake doughs for microwave baking. However, microwave cooking also leads to flavour loss owing to distillation as well as to sorption of flavour compounds to macromolecules as starch and proteins. The latter observations are quite recent and the studies have, so far, been applied primarily to simple model systems. The

Table 9.3 Odour descriptions for five alkylpyrazines

Pyrazine derivative	Concentration range (ppm)	Description
2,3-Dimethyl-	9–1200	Roasted (peanut), mousy, mouse cage, rye crispbread, pungent, solvent-like, musty
2,6-Dimethyl-	6–1100	Sweet, arrack-like, solvent-like, fruity, pungent
2,3-Diethyl	1–150	Earthy, vegetable-like, musty, green, grassy, pungent
Trimethyl-	10–850	Sweet, like ashtray, pungent, solvent-like
Tetramethyl-	6–560	Butterscotch, sweet, flowery (rose), arrack-like

[a] Source: Fors and Eriksson [9].

Table 9.4 The effect of microwave versus conventional heating on the loss of acids in an oil/water (90/10) mixture

	Acid (Percentage loss)				
Type of heating	Acetic	Propionic	Butyric	Valeric	Caproic
120°F					
Microwave heating	33	20	9	0	0
Conventional heating	1	0	0	0	0
140°F					
Microwave heating	66	44	22	8	1
Conventional heating	12	10	3	1	0
150°F					
Microwave heating	80	62	44	26	17
Conventional heating	20	12	5	2	0

Source: Steinke *et al.* [13].

effect of microwave and conventional heating at three different temperatures on the evaporation of short-chain organic acids is shown in Table 9.4 [13]. It is evident that the specific characters of microwave heating lead to significantly higher losses of acids than conventional heating at the same bulk temperature. A special theory has been launched, called the delta-T theory, by which phenomena like different evaporation kinetics of flavour compounds may be explained [14]. By proper application of this theory to a specific food and its typical flavour chemicals, it is claimed one can achieve better flavour of, for instance, microwave-baked goods. This theory is, however, disputed.

As already pointed out, investigations on the flavour chemistry and flavour development in raw grain are surprisingly scarce considering the

Table 9.5 Volatile compounds isolated and identified in raw wheat and wheat flour

	Intact raw wheat (number of compounds isolated)	Present in grain but not in flour	Flour (number of compounds isolated)	Present in flour but not in grain
Alcohols + phenols	21	Phenylethylalcohol	22	3-Pentan-1-ol Penten-3-ol 2-Heptan-1-ol
Carbonyls	7		12	4-Methyl-hex-2-enal Nonanal 2-Methyl-2-hepten-6-one 3-Octen-2-one
Esters + lactones	4	Ethyl benzoate	5	Butylacetate Octylacetate
Heterocyclic compounds	5		6	Alkylfuran
Aliphatic + alicyclic hydrocarbons	10		10	
Aromatic hydrocarbons	14		15	C_5-substituted benzene

[a] Source: Seltman [18].

immense importance of cereals in the human diet. The studies of volatile compounds in cereals seem to have had the main objective to study them as insect attractants or repellants [15] or as improvers of rat baits [16]. Maga [17] reviewed the literature on volatile compounds in cereals. With the techniques available before 1978, few compounds were identified compared with today's methods. The volatile fraction was found to be dominated by common aliphatic alcohols, aldehydes and lactones in rye, wheat, triticale, rice and corn. In general, it can be said that presence and quantities of these compounds in the cereals does not to any extent make up their positive flavour, rather they indicate lipid oxidation.

The most comprehensive modern work on volatile products in wheat is found in a dissertation thesis [18]. Wheat or wheat flour or wheat dough were stripped of their volatile fraction by vacuum condensation. The condensate was extracted with methylene chloride and the organic phase reduced by evaporation in a stream of nitrogen. The concentrate was then subjected to GC-MS analysis. Table 9.5 is a summary of the number of different chemical compounds found in intact raw wheat and wheat flour, respectively; it also indicates the differences in individual volatiles in wheat and flour, respectively. As can be seen, the milling of the flour introduces a few new compounds, especially among the carbonyls presumably caused by increased exposure to oxygen in the flour. The author also studied the influence of storage of the flour. Generally, storage in an open system (paper bags) resulted in higher concentrations, especially of alcohols and

Table 9.6 Volatiles in headspace gases (in ng/l of air) produced by different fungi during 6 days of cultivation with continuous air flow[a]

Fungus/volatile compound	Days					
	1	2	3	4	5	6
Aspergillus flavus						
2-Methylfuran	20	11	12	42	89	119
2-Methyl-1-propanol	4	24	12	48	84	76
3-Methyl-1-butanol	16	86	29	8	5	2
Fusarium culmorum						
Ethyl acetate	20	51	33	72	132	143
2-Methyl-1-propanol	3	30	89	145	201	202
Monoterpenes	6	2	7	8	15	17
Sesquiterpenes	2	4	24	25	36	34

[a] Source: Börjesson *et al.* [20].

carbonyl compounds compared with storage in a closed system (glass flasks), again suggesting that oxidative reactions cause these changes.

9.3 Bio off-flavour

During storage of cereal grain off-flavour can develop depending on variety and conditions of harvesting and storage. As already indicated there are few investigations of the volatile compounds in the raw grain. However, in many countries the estimation of grain standards involves odours like musty, sour, earthy, mouldy and so on. Off-odours can originate from mechanical damage to the grain, which can introduce advanced lipid oxidation and the development of typical rancid smelling carbonyls. However, one predominant cause of off-odour is the parasitic and saprophytic microflora of grain. The microflora consists, to a major part, of bacteria as *Pseudomonas, Flavobacterium, Xanthomonas* and *Erwinia* species. A minor yet important part of the flora consists of fungi as *Aspergillus, Penicillium, Absidia, Mucor* and *Rhizopus* species [19]. Cultivation of bacteria on wheat grain yielded volatile compounds of the classes alcohols, fatty acids, hydrocarbons, carbonyls, esters and amines. The corresponding trials with moulds gave rise to a lesser number of compounds within the classes alcohols, hydrocarbons, carbonyls, esters and heterocyclic nitrogenous compounds such as pyridine and pyrazines. 2-methylfuran, branched-chain lower alcohols and terpenes were identified in the headspace of wheat grain used as a growth medium of different fungi isolated from raw wheat [20]. Table 9.6 shows two examples from these investigations.

A comparison between wheat grains classified as normal and musty showed about the same composition of major volatile compounds. However, substances like 2-methylisoborneol and geosmine occurred in musty grain but were not present in normal wheat [21].

References

1. Frankel, E.N., Autoxidation of polyunsaturated lipids, in *New Aspects of Dietary Lipids. Benefits, Hazards, and Use*, Proceedings of the IUFOST Symposium, September 17–20, 1989, Göteborg, Sweden, 1990, pp. 103–22.
2. Haydanek, M.G., Jr. and McGorrin, R.J., Oat flavor chemistry: principles and prospects, in *Oats: Chemistry and Technology*, (ed. F.H. Webster), American Association of Cereal Chemists, St. Paul, 1986, pp. 335–68.
3. Eriksson, C. (ed.), *Maillard Reactions in Food. Chemical, Physical and Technological Aspects*. Prog. Fd. Nutr. Sci., 1981, **5**, 1–6.
4. Waller, G.R. and Feather, M.S. (eds), *The Maillard Reaction in Food and Nutrition, ACS Symposium Series No. 215*, Washington D.C., 1983.
5. Fujimaki, M., Namiki, M. and Kato, H. (eds), *Amino-Carbonyl Reactions in Food and Biological Systems. Developments in Food Science 13*, Elsevier, Tokyo, 1986.
6. Finot, P.A., Aeschbacher, H.U., Hurrell, R.F. and Liardon, R. (eds), *The Maillard Reaction in Food Processing. Human Nutrition and Physiology*, Birkhäuser, Basel, 1990.
7. Parliment, T.H., McGorrin, R.J. and Ho, C-T. (eds), *Thermal Generation of Aromas, ACS Symposium Series No. 409*, Washington DC, 1989.
8. Schieberle, P. and Grosch, W., Identifizierung von Aromastoffen aus der Kruste von Roggenbrot. *Z. Lebensm. Unters. Forsch.*, 1983, **177**, 173–80.
9. Fors, S.M. and Eriksson, C.E., Pyrazines in extruded malt. *J. Sci. Food Agric.*, 1986, **37**, 991–1000.
10. Fors, S.M. and Schlich, P., Flavor composition of oil obtained from crude and roasted oats, in *Thermal Generation of Aromas*, (eds T.H. Parliment, R.J. McGorrin, and C-T. Ho), *ACS Symposium Series No 409*, Washington DC, 1989, pp. 121–31.
11. Stöllman, U., Flavour changes in white bread during storage, in *The Shelf Life of Foods and Beverages*, (ed. G. Charalambous), Elsevier, Amsterdam, 1986, pp. 293–301.
12. Fors, S.M., Alkylpyrazines, flavour compounds in food. Thesis, Chalmers University of Technology, Göteborg, Sweden, 1987.
13. Steinke, J.A., Frick, C.M., Gallagher, J.A. and Strassburger, K.J., Influence of microwave heating on flavor, in *Thermal Generation of Aromas*, (eds T.H. Parliment, R.J. McGorrin and C-T. Ho), *ACS Symposium Series No 409*, Washington DC, 1989, pp. 519–525.
14. Shaath, N.A. and Azzo, N.R., Design of flavours for the microwave oven. Delta T theory, in *Thermal Generation of Aromas*, (eds. T.H. Parliment, R.J. McGorrin and C-T. Ho), *ACS Symposium Series No 409*, Washington DC, 1989, pp. 512–518.
15. Buttery, R.G., Xu, C. and Ling, L.C., Volatile components of wheat leaves and stems: possible insect attractants. *J. Agric. Food Chem.* 1985, **33**, 115–117.
16. Bullard, R.W. and Holguin, G., Volatile components of unprocessed rice (*Oryza sativa* L.), *J. Agric. Food Chem.*, 1977, **25**, 99–103.
17. Maga, J.A., Cereal volatiles, a review. *J. Agric. Food Chem.*, 1978, **26**, 175–178.
18. Seltman, A., Untersuchungen über flüchtige Komponenten von Weizen und Weizenmehlen. Thesis, Eidgenössische Technische Hochschule, Zürich, 1979.
19. Kaminski, E. and Wasowicz, E., Chemical indices of microbial activity in stored grain, in *Wheat End-Use Properties*. Proceedings of the ICC'89 Symposium, Lahti, Finland, 1989, pp. 47–58.
20. Börjesson, T., Stöllman, U., Adamek, P. and Kaspersson, A., Analysis of volatile compounds for detection of molds in stored cereals. *Cereal Chem.*, 1989, **66**, 300–304.
21. Wasowicz, E., Kaminski, E., Kollmansberger, H., Nitz, S., Berger, R.G. and Drawert, F., Volatile components of sound and musty wheat grain. *Chem. Microbiol. Technol. Lebensm.*, 1988, **11**, 161–168.

10 Meat flavour

D. MOTTRAM

Abstract

Flavour is a very important component of the eating quality of meat and there has been much research aimed at understanding the chemistry of meat flavour. This has resulted in the identification of over 1000 volatile compounds from cooked meats. Meat flavour develops during cooking from the complex interaction of precursors derived from both the fat and lean components of meat. The Maillard reaction and lipid degradation are the most important routes to flavour compounds in meat but the initial products of these reactions may undergo further interactions. Among the complex array of volatiles found in cooked meat, classes of compound that contribute to certain sensory characteristics can be identified. Lipid degradation products appear to be important in providing fatty aromas and in determining some of the aroma differences between species. Heterocyclic compounds, such as pyrazines and thiazoles derived from the Maillard reaction contribute to roast and grilled aromas, while certain aliphatic and heterocyclic sulphur compounds provide some of the characters of boiled meat. Sulphur compounds are extremely important in meat flavour and certain furanthiols and furan disulphides possess that characteristic meaty aroma that is associated with all cooked meats.

10.1 Introduction

Meat is a major component of our diet and good eating quality is of paramount importance to consumer acceptability. Flavour is a very important component of the eating quality of meat and there has been much research aimed at understanding the chemistry of meat flavour and at determining those factors during the production and processing of meat that influence flavour quality. The desirable characteristics of meat flavour have also been sought in the production of simulated meat flavourings, which are of considerable importance in convenience and processed foods.

Although meat flavour depends on the stimulation of the senses of taste and smell, most of the research over the past three decades has concentrated on the volatile compounds that contribute to the aroma, while investigations of the contribution of nonvolatile compounds to the taste

Table 10.1 Numbers of volatile compounds of different chemical classes reported in cooked meat

	Beef	Pork	Lamb/mutton	Chicken
Hydrocarbons	193	37	43	84
Alcohols and phenols	82	25	20	53
Aldehydes	65	41	39	83
Ketones	76	31	20	53
Carboxylic acids	24	30	51	22
Esters	59	33	11	16
Lactones	38	12	14	24
Furans and pyrans	47	28	5	16
Pyrroles and pyridines	39	16	19	24
Pyrazines	51	44	16	22
Other nitrogen compounds	28	9	2	7
Oxazoles and oxazolines	13	1	4	5
Non-heterocyclic sulphur	72	17	7	17
Thiophenes	35	15	2	7
Thiazoles and thiazolines	29	17	13	18
Other heterocyclic sulphur	13	1	4	6
Miscellaneous compounds	16	4	1	11
Total	880	361	271	468
Number of publications	70	11	12	20

sensations has been much more limited. The aroma of cooked meat is clearly recognizable; however it is a complex sensation with numerous different attributes that arise from a complex mixture of aroma compounds. Over 1000 different volatile compounds have been reported in over 120 publications on meat flavour (Table 10.1) [1,2]. Although several of these compounds are known to possess some of the characteristics of meat aroma, the desirable quality attribute that is perceived as meat flavour cannot be attributed to a single compound or group of compounds.

Meat flavour develops during cooking from the complex interaction of precursors derived from both the fat and lean components of meat. Several different flavour characteristics are clearly recognizable in cooked meat: all meats possess fatty flavours and the different species (e.g. beef, pork, lamb, chicken, etc.) have their own distinct flavours. Grilled or roast meats have characteristic roast aroma notes, which differ from the aromas associated with boiled or stewed meat. In addition, all meat, regardless of species or cooking method, has a characteristic 'meaty' aroma, which is a major component of the eating quality of meat.

In this chapter some of the reactions that occur between the flavour precursors will be examined, also how different classes of volatile compounds may be formed. The possible contribution of some of these compounds to the aroma characteristics of cooked meat will be discussed by examining the aroma properties and odour threshold values of selected volatile compounds. In this short review it is not possible to cover all the reported research in this area or all the types of volatile compounds found; more details can be found in reviews by Mottram [1], MacLeod and Seyyedain-Ardebelli [3], Ohloff and Flament [4].

10.2 Precursors of meat flavour

Raw meat has little aroma and only a blood-like taste, thus cooking is necessary to develop the characteristic flavour. During cooking, a complex series of thermally induced reactions occur between nonvolatile components of lean and fatty tissues, resulting in a host of volatile compounds that contribute to the aroma of the cooked meat. In addition, some of these nonvolatile components will contribute to the taste attributes of flavour, while other thermally derived nonvolatile compounds will add to the taste sensation.

Research during the 1950s and early 1960s demonstrated that the low molecular weight water-soluble fraction of meat contains meat flavour precursors and that the high molecular weight fibrillar and sarcoplasmic proteins are unimportant [5–8]. Water-soluble components of meat that could be flavour precursors include free amino acids and peptides, reducing sugars and sugar phosphates, nucleotides and other nitrogenous compounds such as thiamine.

In several studies the changes that occurred in the quantities of these water-soluble compounds on heating were examined. Depletions in the quantities of carbohydrates and amino acids were observed, the most significant losses occurring for cysteine and ribose. Subsequent studies of the aromas produced on heating mixtures of amino acids and sugars confirmed the important role played by cysteine in meat flavour formation and led to the classic patent of Morton *et al.* in 1960 [9], which involved the formation of meat-like flavour by heating a mixture of cysteine and ribose. Most subsequent patent proposals for 'reaction product' meat flavourings have involved sulphur, usually as cysteine or other sulphur-containing amino acids, or hydrogen sulphide [3,10].

The role of the lipid fractions of meat, both adipose tissues and fat contained within the lean, has been the subject of a continuing debate. Hornstein and Crowe [6,11] found that aqueous extracts of beef, pork and lamb had similar aromas when heated, while heating the fats yielded species-characteristic aromas. It was concluded that lipid provides volatile compounds that give the characteristic flavours of the different species, and the lean is responsible for a basic meaty flavour common to all species. It was also recognized that autoxidation of fat could produce undesirable flavour compounds and lead to rancidity [12]. Fat can also serve as a solvent for aroma compounds, obtained either from extraneous sources or as part of the flavour-forming reactions; it can therefore influence the release of flavour from the meat [13,14]. However, it is now realized that this is an over-simplification of the role of fat in meat flavour and that lipid-derived volatiles have an important part to play in desirable meat aroma both as aroma compounds and as intermediates to other compounds. Recently, it has been shown that phospholipids, the essential membrane lipids of all tissue, play an important part in the development of desirable flavours in meat during cooking [15].

10.3 Lipid-derived volatiles in meat

10.3.1 Lipid autoxidation

An important route to aroma volatiles during meat cooking is the thermally induced oxidation of the acyl chains of the lipids. Autoxidation of unsaturated fatty acid chains is also responsible for the undesirable flavours associated with the rancidity that develops during the storage of fatty foods. The reactions by which volatile aroma compounds are formed from lipids follow the same general routes for both thermal oxidation and rancid oxidation, although subtle changes in the mechanism give rise to different profiles of volatiles in the two systems.

The oxidative breakdown of the unsaturated alkyl chains of lipids involves a free radical mechanism and the formation of intermediate hydroperoxides [16]. Decomposition of these hydroperoxides involves further free radical mechanisms and the formation of nonradical products, including volatile aroma compounds [17,18].

The degradation of hydroperoxides (Figure 10.1) initially involves homolysis to give an alkoxy radical and a hydroxy radical; this is followed by cleavage of the fatty acid chain adjacent to the alkoxy radical. The nature of the volatile product for a particular hydroperoxide depends on the composition of the alkyl chain and the position where cleavage of the chain takes place (A or B). If the alkyl group is saturated and cleavage takes place at A a saturated aldehyde results, while cleavage at B gives an alkyl radical that can give an alkane or, alternatively, can react with oxygen to give a hydroperoxide. The latter, just like the hydroperoxide of the lipid, breaks down to give an alkoxy radical, which then gives a stable non-radical product much as an alcohol or aldehyde. One or more double bonds in the alkyl chain will give analogous compounds because further oxidation of the unsaturated chain can occur.

10.3.2 Contribution of lipid-derived volatiles to meat aroma

Several hundred volatile compounds derived from lipid degradation have been found in cooked meat, including aliphatic hydrocarbons, aldehydes,

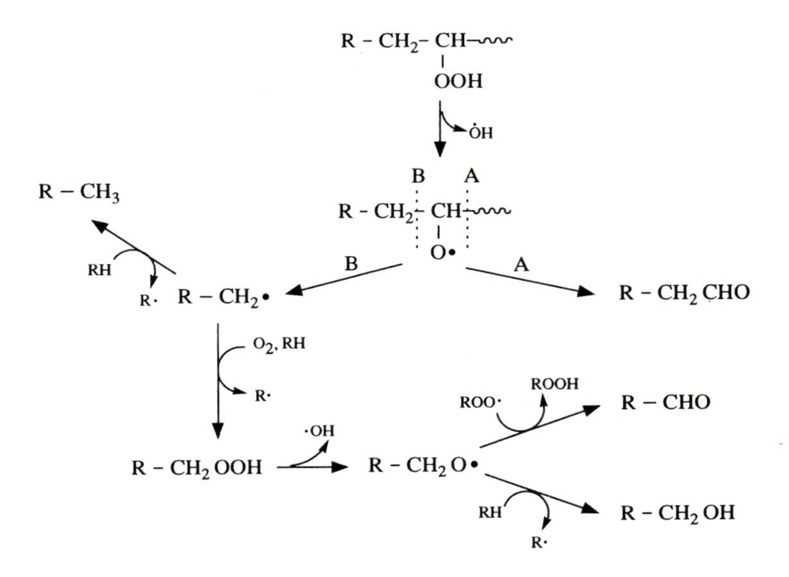

Figure 10.1 Simplified scheme showing breakdown of lipid hydroperoxides to give volatile products.

ketones, alcohols, carboxylic acids and esters. Some aromatic compounds, especially hydrocarbons, have also been reported as well as oxygenated heterocyclic compounds such as lactones and alkylfurans. In boiled and lightly grilled or roasted meat, lipid degradation products have been found to dominate the volatile extracts [15,19]. However, the contribution of many of these compounds to the overall flavour of the cooked meat may be very small because their odour threshold values are relatively high. In general, the major volatile components of cooked meat are present at concentrations in the part per million (ppm) range and the less abundant components may occur only at the part per billion (ppb) level or lower. Consequently, only those compounds with low odour threshold values are likely to contribute to meat flavour.

An examination of the odour threshold values of different lipid derived volatiles (Table 10.2) shows that the classes of compounds with the low odour threshold values are aldehydes and unsaturated alcohols and ketones. Saturated and unsaturated aldehydes with six to ten carbons are major volatile components of all cooked meats and therefore they probably play an important part in meat aroma. The aroma of these aldehydes is described as green, fatty, tallowy and 2,4-decadienal is reported to have an aroma of fat-fried food [17,20]. It seems likely that aliphatic aldehydes contribute to the fatty flavours of cooked meat.

Table 10.2 Odour threshold values of some classes of volatiles obtained from lipid degradation

Compound	Threshold (ppm in water)
Aliphatic hydrocarbons (saturated)	>2000
Aliphatic hydrocarbons (unsaturated)	10–20
Aromatic hydrocarbons	0.01–10
Aliphatic alcohols (saturated)	>20
1-Alken-3-ols	0.01–5
Aliphatic ketones	>10
1-Alken-3-ones	<0.003
Alkylfurans	1–10
Aldehydes	
Alkanals	<0.4
Alkenals	<0.6
Alkadienals	<0.1

Source: Badings [20]; van Gemert and Nettenbreijer [21].

10.3.3 Species-characteristic aromas

The characteristic flavour of the different meat species is generally believed to be derived from the lipid sources, although it has been suggested that interaction of lipid with other meat components may also be involved [13]. Aldehydes, as major lipid degradation products, are probably involved in certain species characteristics. More unsaturated aldehydes have been reported in the volatiles of chicken than from the other species (Table 10.3), and are believed to be important to the aroma of heated chicken fat [22]. These include some specific stereoisomers, such as *E*,*Z*-2,4-decadienal, *Z*-3-nonenal, *E*,*Z*,*Z*-2,4,7-decatrienal, *E*,*Z*-2,5-undecadienal, *E*,*Z* and *Z*,*Z*-2,6-dodecadienal, *E*,*Z*-2,4-tridecadienal, *E*,*Z*,*Z*-2,4,7-tridecatrienal, *E*,*Z*-2,4-tetradecadienal [23,24]. Several of these are typical breakdown products of arachidonic acid. Certain aldehydes also have been reported to contribute to the characteristics of beef and mutton fat flavour [25,26].

Sheep meat (lamb and mutton) shows a major difference from the other meats in the nature of the volatile free fatty acids. Sheep meat has been

Table 10.3 Numbers of aldehydes in cooked meat volatiles

	Beef	Pork	Lamb/mutton	Chicken
Alkanals	25	21	19	20
Alkenals	19	9	12	32
Alkadienals	8	3	6	20
Alkatrienals	–	–	–	2
Other	13	8	2	9
Total	65	41	39	83

found to contain 20 methyl-branched saturated fatty acids that have not been reported in other meats. These acids have been associated with the characteristic flavour of mutton, which results in low consumer acceptance of sheep meat in many countries, and which the Chinese describe as 'soo' flavour [27]. Two acids, 4-methyloctanoic (**1**) and 4-methylnonanoic (**2**), are considered to be primarily responsible for this flavour. The lipids of sheep, unlike other species, contain significant quantities of methyl-branched fatty acids, and these are known to arise from the metabolic process occurring in the rumen of sheep. Branched acids with methyl substituents at even-numbered carbon atoms result from fatty acid synthesis using methyl malonate (arising from propionate metabolism) instead of malonate in the chain lengthening [28].

Pork from uncastrated male pigs can have a characteristic and objectionable odour (boar taint) described as sweaty and urinous, although many people do not detect the odour. A steroid 5-α-androst-16-en-3-one (**3**) has been shown to be responsible. The compound is a pheromone produced by the male pig as a sexual stimulant for the female, and accumulates in the adipose tissue of older male pigs [29,30].

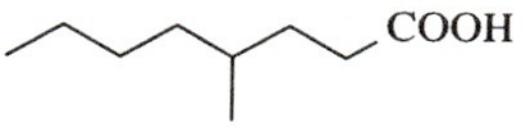

COOH

1

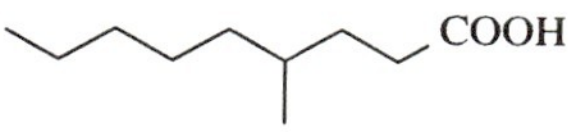

COOH

2

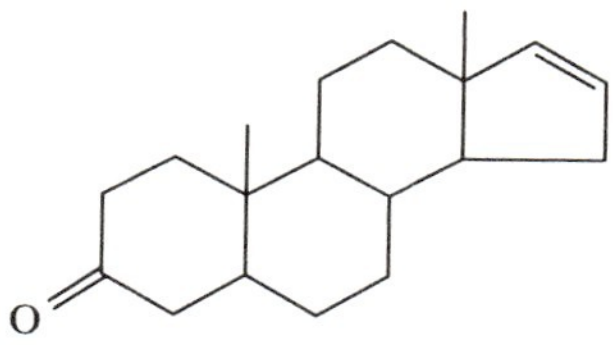

O

3

10.4 Volatiles from water-soluble precursors

The primary reactions of the water-soluble precursors occurring on heating, which can lead to meat flavour, include pyrolysis of amino acids and peptides, carbohydrate degradation, interaction of sugars with amino acids or peptides, also degradation of ribonucleotides and thiamine. This complicated array of reactions is made even more complex by a whole host of secondary reactions that can occur between the products of the initial reactions giving rise to numerous volatile aroma compounds.

The thermal decomposition of amino acids and peptides requires temperatures higher than those that are normally encountered during cooking. Temperatures approaching 200°C are required for pyrolysis of amino acids, when decarboxylation and deamination may occur with the formation of aldehydes, hydrocarbons, nitriles and amines [31,32].

When sugars are heated, caramelized flavours are formed but, in general, relatively high temperatures are required. At temperatures between 150°C and 180°C dehydration reactions occur, with the formation of furfural from pentoses or 5-hydroxymethylfurfural from hexose sugars. Further heating results in the formation of many other highly odoriferous compounds, including many furan derivatives, carbonyl compounds, alcohols and both aliphatic and aromatic hydrocarbons [33,34]. Since high temperatures are required for these pyrolysis and caramelization reactions, they are relatively unimportant in meat flavour and only occur when dehydration allows the temperature to rise significantly above the boiling point of water.

10.4.1 Maillard reaction

The Maillard reaction between reducing sugars and amino compounds is one of the most important routes to flavour compounds in cooked foods. The reaction does not require the very high temperatures associated with sugar caramelization and amino acid pyrolysis. Mixtures of sugars and amino acids stored at refrigerated temperatures can show signs of Maillard browning on storage [35,36]. The reaction rate increases markedly with temperature, and browning, which is accompanied by the formation of flavour compounds, generally occurs at the elevated temperatures associated with cooking. Although the Maillard reaction proceeds in aqueous solution, it occurs much more readily at low moisture levels, hence flavour compounds produced by the Maillard reaction tend to be associated with parts of grilled, roasted or baked foods that have been dehydrated by the heat source.

The first step of the reaction involves the addition of the carbonyl group of the open chain from a reducing sugar and the amino group of an amino acid, peptide or other compound with a primary amino group (Figure 10.2) [37]. The subsequent elimination of water and molecular rearrangement gives a 1-amino-1-deoxy-2-ketose (Amadori product). These Amadori intermediates do not contribute themselves to flavour; however, they are important precursors of flavour compounds. They are thermally unstable and undergo dehydration and deamination to give furans, similar to those obtained in sugar caramelization, as well as a host of other degradation products [38].

At 100–110°C the Amadori product will undergo irreversible 2,3-enolization with the elimination of the amine to give a 2,3-dicarbonyl compound (Figure 10.3). Further dehydration can lead to 5-methyl-4-hydroxy-3(2*H*)-furanone from pentose sugars or the 2,5-dimethyl homologue from hexoses. Alternatively, 1,2-enolization of the Amadori product will result in a 1,2-dicarbonyl, which, on dehydration, yields furfural from a pentose, or 5-hydroxymethylfurfural and 5-methylfurfural from a hexose [39]. The fragmentation of the carbohydrate chains of the Amadori

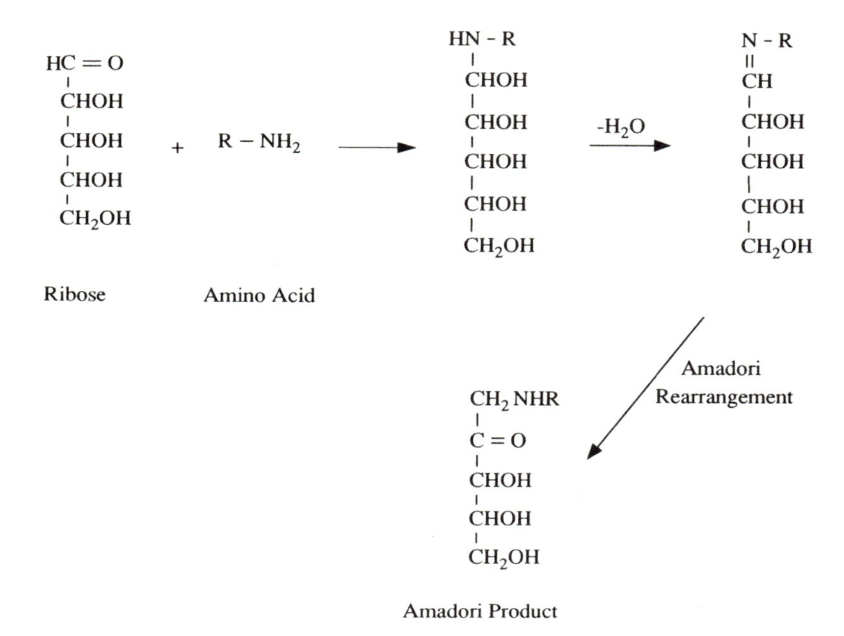

Figure 10.2 Formation of an Amadori compound in the Maillard reaction between ribose and an amino acid.

products can lead to a variety of α-dicarbonyl compounds such as pyruvaldehyde, 2,3-butanedione, hydroxyacetone, 3-hydroxy-2-butanone.

All these Maillard products are capable of further reaction, and the subsequent stages of the Maillard reaction involve the interaction of furfurals, furanones and dicarbonyls with other reactive compounds such as amines, amino acids, hydrogen sulphide, thiols, ammonia, acetaldehyde and other aldehydes. These additional reactions lead to many important classes of flavour compounds including heterocyclics such as pyrazines, oxazoles, thiophenes, thiazoles and other heterocyclic sulphur compounds.

10.4.2 Strecker degradation

Strecker degradation is one of the most important reactions associated with the Maillard reaction and involves the oxidative deamination and decarboxylation of an α-amino acid in the presence of a dicarbonyl compound (Figure 10.4). This leads to the formation of an aldehyde, containing one fewer carbon atom than the original amino acid, and an α-aminoketone. These aminoketones are important intermediates in the formation of several classes of heterocyclic compounds including pyrazines, oxazoles and thiazoles [38].

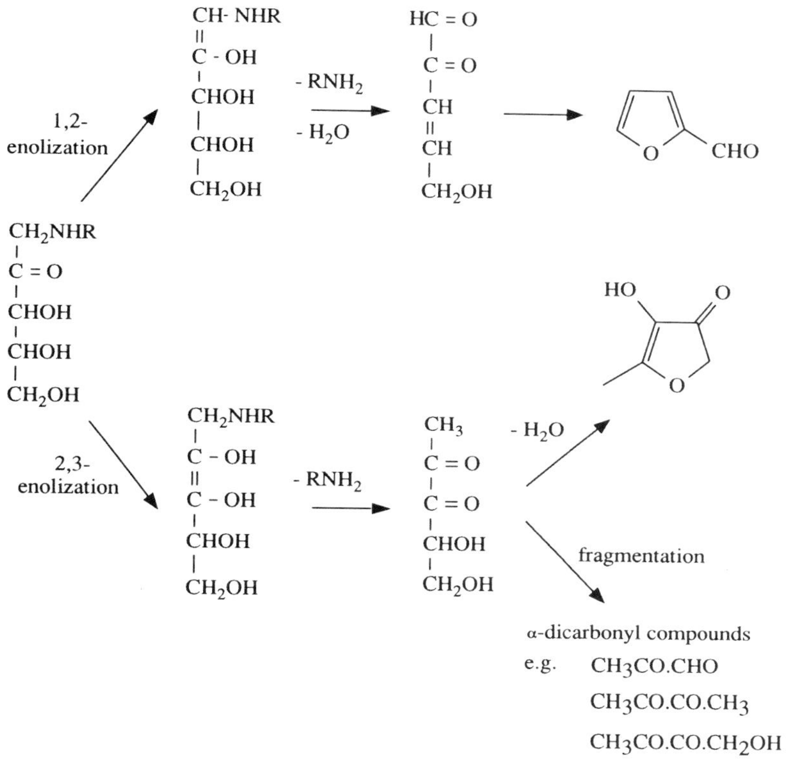

1,2-enolization

$CH-NHR$ / $C-OH$ / $CHOH$ / $CHOH$ / CH_2OH

$-RNH_2$, $-H_2O$

$HC=O$ / $C=O$ / CH / CH / CH_2OH

O CHO

CH_2NHR / $C=O$ / $CHOH$ / $CHOH$ / CH_2OH

HO O O

2,3-enolization

CH_2NHR / $C-OH$ / $C-OH$ / $CHOH$ / CH_2OH

$-RNH_2$

CH_3 / $C=O$ / $C=O$ / $CHOH$ / CH_2OH

$-H_2O$

fragmentation

α-dicarbonyl compounds

e.g. $CH_3CO.CHO$

$CH_3CO.CO.CH_3$

$CH_3CO.CO.CH_2OH$

Figure 10.3 Some routes for the decomposition of Amadori compounds in the Maillard reactions.

In the Strecker degradation of cysteine, hydrogen sulphide, ammonia and acetaldehyde are formed, as well as the expected Strecker aldehyde, mercaptoacetaldehyde and an aminoketone [40]. These compounds are important as reactive intermediates for the formation of many highly odoriferous compounds, which play important roles in meat flavour, and this emphasizes the importance of cysteine in the development of meat flavour.

10.4.3 Contribution of furfurals and furanones to aroma

Furans with oxygenated substituents (e.g. furfurals, furanones) occur in the volatiles of all heated foods and are generally formed from carbohydrates in the Maillard reaction (Figure 10.3).

Oxygenated furans, such as furfural (**4**), 5-methylfurfural (**5**) and 2-acetylfuran (**6**), generally impart caramel-like, sweet, fruity characteristics

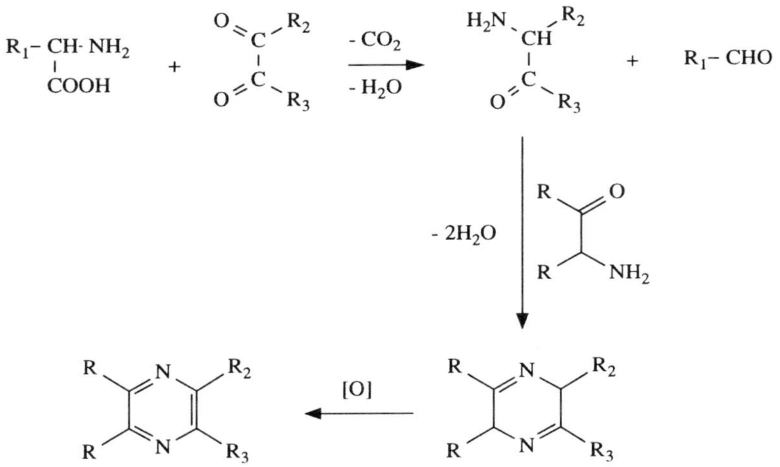

Figure 10.4 Strecker degradation of α-amino acids and the formation of alkylpyrazines.

to foods [4]. 2,5-dimethyl-4-hydroxy-3(2*H*)-furanone (**7**), found in beef broth [41] and several other foods, has an aroma described as caramel-like, burnt pineapple-like, although at low concentrations it attains a strawberry-like note [4]. The odour threshold values of furfurals and furanones are generally at the ppm level [42]. In their own right, oxygenated furans probably make only minor contributions to meat flavour; however, they are important intermediates to other flavour compounds, including thiophenes, furanthiols and other sulphur-containing compounds.

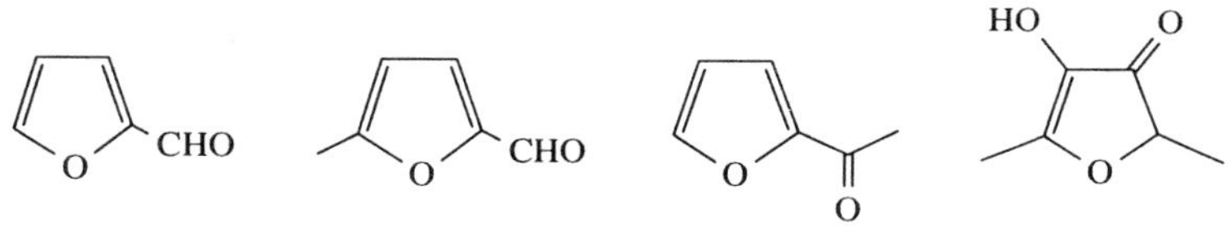

4 **5** **6** **7**

10.4.4 Nitrogen compounds in meat aroma

10.4.4.1 Pyrazines These important aroma compounds are believed to contribute to the pleasant and desirable flavour of many different foods, including meat. Mono-, di-, tri- and tetra-alkyl substituted pyrazines (**8**) are all found in meat volatiles. Substituents are most commonly methyl and ethyl, indeed all of the possible alkylpyrazine isomers containing methyl and ethyl groups have been found with the exception of 2,3-

diethylpyrazine and some tetra-substituted derivatives. The alkylpyrazines generally have nutty, roast aromas with some eliciting earthy or potato-like comments [42]. The odour threshold values of the mono-, di-, tri- and tetra-methylpyrazines are all relatively high (>1 ppm) and these pyrazines are unlikely to play a significant role in meat aroma. However, replacing one or more of the methyl groups with ethyl can give a marked decrease in threshold value [43], and some ethyl-substituted pyrazines have sufficiently low threshold values for them to be important in the roast aroma of cooked meat. Apart from acetylpyrazine (**9**), relatively few pyrazines with oxygen-containing substituents have been reported in meat. Acetylpyrazines have some importance as flavourings, having popcorn-like aromas and odour threshold values in the low ppb range [42].

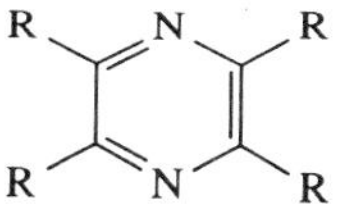

R = H or Alkyl

8 **9**

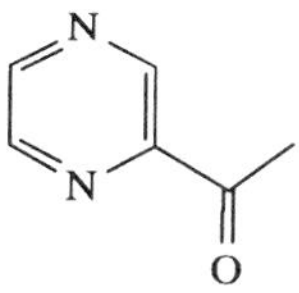

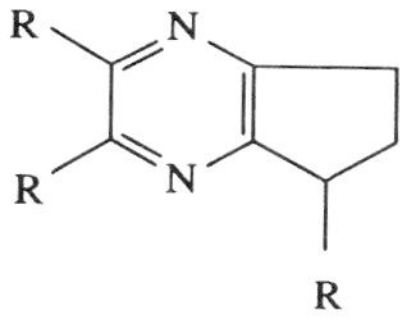

R = H or Me

10 **11**

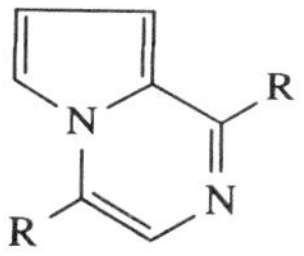

A particularly interesting group of pyrazines found in roast, grilled, fried and pressure-cooked meat are the 6,7-dihydro-5(*H*)-cyclopentapyrazines (**10**) [44]. These bicyclic compounds have roasted, grilled, burnt and animal notes and so could be important contributors to the roast aromas associated with grilled and roast meat [4]. A related group of bicyclic compounds, the pyrrolo [1,2a]pyrazines (**11**), were isolated from a heated extract of defatted beef by Flament *et al.* [45]. Seven of these alkyl substituted pyrrolopyrazines were isolated and compounds with this bicyclic structure have not been reported in any other food.

Pyrazines are one of the major classes of compound formed in the Maillard reaction, and their formation has been studied in many different model systems involving amino acids and reducing sugars, as well as simpler systems in which ammonia was the source of nitrogen or carbonyls replaced the sugar [38]. It has been shown that factors such as time and temperature of reaction, pH, reactant concentration and water activity are important variables in determining the nature and quantity of the products from such reactions [38,46,47], and this is a particularly important consideration when comparing volatile pyrazines, as well as other Maillard products, obtained from different meat preparations. Although pyrazines were found in a model system kept at −5°C for 30 days, significant pyrazine formation does not begin until 70°C and they are favoured by low moisture levels; thus in roast meat few pyrazines would be expected in the bulk of the interior where the temperature does not normally exceed 70–80°C, and most would occur in the browned surface [48]. In pork liver that was pressure-cooked at 160°C, 35 pyrazines, which quantitatively accounted for over 40% of the total volatiles, were reported [49]. However, the volatiles of boiled beef and pork were shown to comprise mainly aliphatic aldehydes and ketones with only very small quantities of pyrazines and other heterocyclics [19]. In a comparison of the heterocyclic and aliphatic compounds formed in pork that was cooked under different conditions, well-done, grilled meat contained 27 pyrazines (which accounted for 77% of the total amount of volatiles analysed by headspace concentration), while in lightly grilled, roast or boiled meat only methyl and dimethylpyrazines were found (comprising less than 1% of the total amount of volatiles), the main products being aliphatic aldehydes and alcohols [50].

Several mechanisms have been proposed for the formation of pyrazines in food flavours [38,46]. An important route to the alkylpyrazines is from the α-aminoketones, products of the condensation of dicarbonyl with an amino compound via Strecker degradation (Figure 10.4). Self-condensation of the aminoketones, or condensation with other aminoketones, affords a dihydropyrazine that is oxidized to the pyrazine.

10.4.4.2 Oxazoles and oxazolines. 2,4,5-Trimethyl-3-oxazoline (**12**) was first found in the volatiles of boiled beef by Chang *et al.* in 1968 [51] and, since then, has been reported on many other occasions. Several other oxazoles (**13**) and 3-oxazolines have been identified in meat volatiles but not so regularly as the trimethyloxazoline. Recent reports by Ho *et al.* on volatiles in roast beef [52], fried chicken [53] and fried bacon [54] extended the number of oxazoles found in meat; the compounds isolated included di- and tri- substituted derivatives containing 2-isobutyl, butyl or hexyl substitution. Oxazoles have been found in several foods with over 30 derivatives in cocoa and coffee; however, 3-oxazolines have so far only been detected in meat volatiles, with the exception of the mono-, di-, and tri-methyl derivatives, which have been reported in peanuts [2].

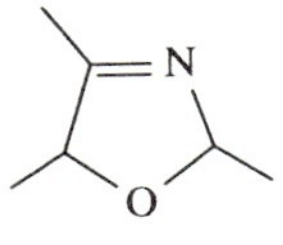

N O R N R O R

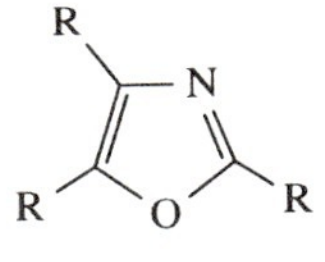

R = H or Alkyl

12 **13**

When 2,4,5-trimethyl-3-oxazoline was first identified in boiled beef it was hoped that the compound would possess the characteristic meat aroma; however, on synthesis it was shown to have a woody, musty, green flavour with a threshold of 1 ppm [55,56]. Other 3-oxazolines have nutty, sweet or vegetable-like aromas and the oxazoles also appear to be green and vegetable-like [56]; in contributing to the overall meat aroma, however, they are probably not so important as the closely related thiazoles and thiazolines.

10.4.5 Sulphur compounds in meat aroma

The importance of sulphur compounds, both aliphatic and heterocyclic, to the aroma of meat has been recognized by many groups of researchers and, in the search for meat-like aroma chemicals, much attention has been directed towards the organic chemistry of sulphur. In a review, MacLeod [10] listed 78 chemical compounds that have been reported in the literature as possessing meat-like flavours; seven are aliphatic sulphur compounds, 65 heterocyclic sulphur compounds and the remaining six nonsulphur heterocyclic compounds. Many of these compounds arise from the prolific patent literature on this subject and only 25 of the compounds have actually been identified in meat. Hydrogen sulphide is the essential sulphur compound that has been identified repeatedly in the volatiles of all meats. It is formed in the Strecker degradation of cysteine, together with ammonia, acetaldehyde and mercaptoacetaldehyde, all of which are reactive compounds providing an important source of reactants for a wide range of flavour compounds.

10.4.5.1 Nonheterocyclic sulphur compounds. The volatiles of meats have been shown to contain many aliphatic thiols, sulphides and polysulphides, several thiophenols and other aromatic thiols and sulphides. In general, the aliphatic sulphur compounds and other nonheterocyclic sulphur compounds are associated with boiled meat; 70 of the 72 compounds of this type reported in beef have been found in meat cooked by boiling. In contrast, only 22 out of the 72 compounds have been associated with beef cooked by grilling, roasting or frying [1,2]. The other meats do not show the same high numbers of aliphatic sulphur compounds. The reason may be that with beef the researchers specifically analysed for volatile thiols and sulphides, while in pork, lamb or chicken these compounds have not received any particular attention. Indeed, these compounds are reactive and could be lost in many of the extraction techniques normally used for volatile analysis.

The majority of the thiols and methyl sulphides result from the reaction of hydrogen sulphide or methanethiol with aliphatic aldehydes or alcohols. The reactions do not necessarily terminate with these compounds and further reaction can give a range of different acyclic and heterocyclic compounds, including thiophenes, thiazoles, trithiolanes and other heterocyclic sulphur-containing aroma compounds [57].

Aliphatic sulphur compounds generally have low odour threshold values [21,42] and they have attracted some interest in their possible contribution to meat aroma, especially that of boiled meat. Alkanethiols and sulphides have powerful, very objectionable odours described as sulphurous, cabbage- or onion- or garlic-like [42,58]. Nevertheless, there are some exceptions reported in the literature: 1-(methylthio)-1-ethanethiol (**14**), reported on several occasions in beef volatiles, has a meaty, onion-like character, which becomes more meaty in very dilute aqueous solution at levels approaching the threshold value of 5 ppb [59,60]. Without doubt many nonheterocyclic sulphur compounds play an important role in meat aroma. With low odour threshold values, relatively small amounts could have a significant effect. Some compounds may contribute directly to the meaty characteristics, while others will provide an overall sulphurous note, which is a part of meat aroma. For many of these compounds their reactivity and role as intermediates to other aroma substances is extremely important.

10.4.5.2 Thiazoles and thiazolines. The thiazoles and thiazolines are closely related in structure to the oxazoles and oxazolines but, in general, have lower threshold values and are more prolific in food volatiles than their oxygenated analogues. The first thiazoles were isolated from food volatiles in 1966 and they are now recognized as important constituents of food aromas [61]. In meat 41 different thiazoles (**15**) and five thiazolines (**16**) have been found, nearly all with alkyl substituents and usually in roast, grilled or fried products. The only thiazole containing another functional group that has been regularly reported in meat volatiles is 2-acetylthiazole (**17**).

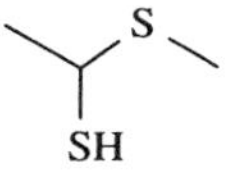

SH S **14**

R R N R S R R R N R S R

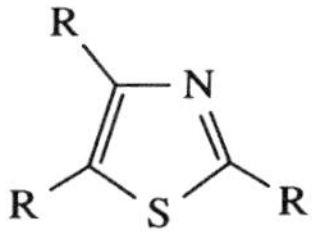

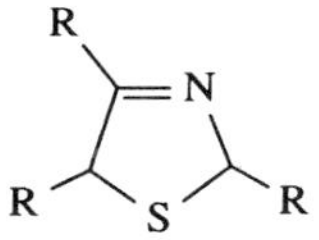

R = H or alkyl

15 **16**

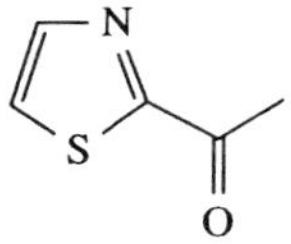

N S O **17**

The thiazoles and thiazolines are important aroma compounds in meat and probably contribute to both the meaty and the roast characteristics. Petit and Hruza [62] examined the sensory properties of 26 synthesized thiazoles. In general, 2-alkyl-thiazoles possessed green vegetable-like properties, while increasing the substitution added nutty, roasted and sometimes meaty characteristics. The nature and number of alkyl substituents appears to be important in determining the aroma character and certain di- and tri-alkyl derivatives have roast, meaty notes. A particularly

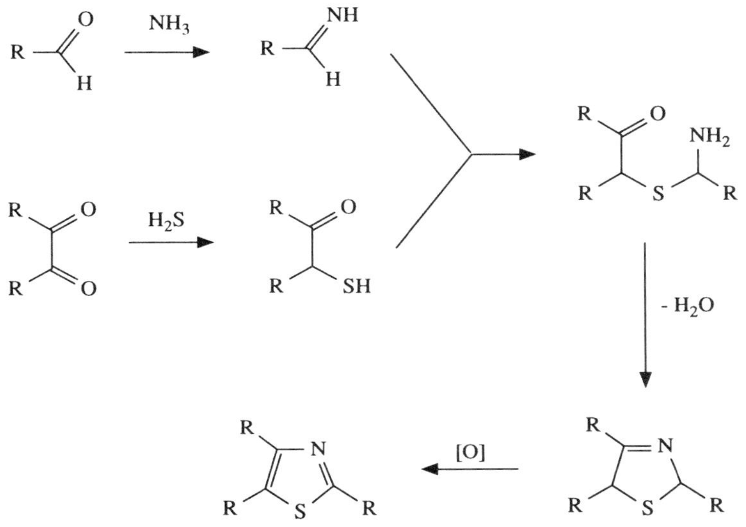

Figure 10.5 Formation of thiazoles and thiazolines from intermediates of the Maillard reaction.

interesting compound is 2,4-dimethyl-5-ethylthiazole, which has a nutty, roast, meaty, liver-like flavour and a low odour threshold value of 2 ppb [63]. This compound has been found in roast beef [64], grilled pork [50] and fried chicken [53].

Several mechanisms have been proposed for the formation of thiazoles. An important route involves the action of hydrogen sulphide and ammonia on mixtures of aliphatic aldehydes and 1,2-dicarbonyl compounds (Figure 10.5). In heated foods, the main source of the aldehydes for these reactions is via Strecker degradation of amino acids. However, in fat-containing foods an additional source of aldehydes is from lipid oxidation, and the trialkylthiazoles containing long-chain 2-alkyl substituents recently reported in chicken, deep fried in beef fat [53] and in roast beef [52], could only result from the participation of lipid-derived aldehydes in the above reactions.

10.4.5.3 Thiophenes. At least 35 thiophenes have been reported in meat volatiles, although so far many more have been found in beef than in other meats. In contrast to aliphatic sulphur compounds, they are found in both grilled and boiled beef. Most of the thiophenes found in meat volatiles are substituted in the 2-position and include all n-alkyl substituents **(18)** up to n-octyl and the corresponding acyl derivatives from formyl to octanoyl. The thiophenes with the long alkyl chains must arise from lipid sources, probably by the interaction of hydrogen sulphide with unsaturated fatty acids or their oxidation products. Long-chain 2-alkylthiophenes have recently been shown to be formed when a phospholipid was heated in a Maillard system of cysteine and ribose [65,66], and a pathway has been proposed involving the reaction of hydrogen sulphide with a 2,4-dienal derived from lipid degradation [67]. There are several possible routes to the other thiophenes found in meat volatiles, in general involving the reaction of hydrogen sulphide, or some other sulphur compound also derived from the sulphur amino acids, with sugar-degradation products from the Maillard reaction [38].

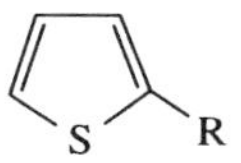

S R **18** S CHO **19** S R O **20**

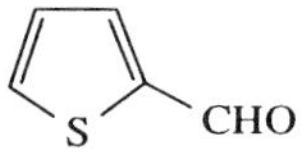

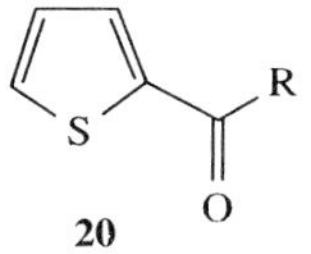

The majority of thiophenes that have been examined for their sensory properties show odour threshold values in the ppb range and therefore they should be considered as potential contributors to the aroma of meat [42]. The alkyl thiophenes are reported to have aromas reminiscent of roast onions [68,69]; 2-formylthiophene (**19**) is reported as benzaldehyde-like, while 5-methyl-2-formylthiophene gives a cherry-like odour at a level of 0.5 ppm [4]. 2-Acetylthiophene (**20**) has an onion- or mustard-like aroma although in coffee it is reported to develop a malty, roast note [4]. Its odour threshold value in water was found to be 0.1 ppb, which was some 2000 times lower than 2-formylthiophene [70]. Although none of the thiophenes so far identified in meat appears to be directly responsible for the meaty characteristics, thiophenes with a thiol group substituted in the 3-position, which have been found in model systems but not yet found in meat itself, are known to possess meaty properties [10,71].

10.4.5.4 Other heterocyclic sulphur compounds. Several non-aromatic heterocyclic compounds with two or three sulphurs in five- and six-membered rings have been reported in meat volatiles and have attracted some interest as possible contributors to meat aroma. They are associated mainly with boiled meat rather than the roast or grilled product and most have been found in beef. The most frequently found is 3,5-dimethyl-1,2,4-trithiolane (**21**), which was first isolated from the volatiles of boiled beef by Chang *et al.* in 1968 [51] and has subsequently been found in pork, lamb and chicken. When this trithiolane was first identified in beef volatiles, it was isolated from a fraction that had an aroma of boiled beef. It was therefore thought to be responsible for this characteristic odour; however, subsequent synthesis showed that the compound did not have a boiled-beef aroma [55,56]. Nixon *et al.* [72] observed that 3,5-dimethyl-1,2,4-trithiolane comprised between 10% and 20% of the total volatile extract of boiled mutton.

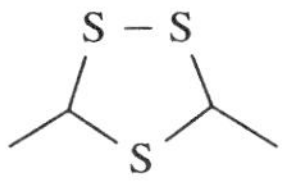

21 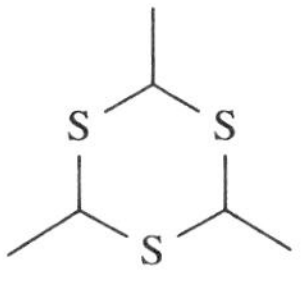 **22** 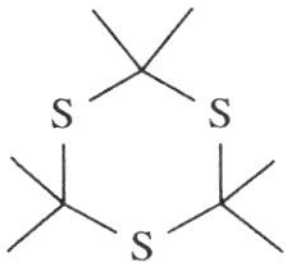 **23** 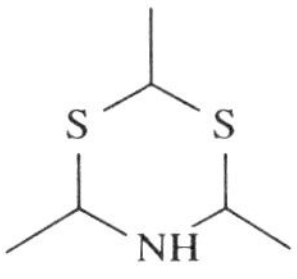 **24**

The sulphur heterocyclics with six-membered rings found in beef include trithioacetaldehyde (2,4,6-trimethyl-1,3,5-trithiane) **(22)** and trithioacetone (hexamethyl-1,3,5-trithiane) **(23)**. The trithioacetaldehyde has also been found in the volatiles of fried chicken [53]. Thialdine (5,6-dihydro-2,4,6-trimethyl-1,3,5-dithiazine) **(24)** has been reported in the volatiles of all the meat species and was the main volatile component obtained from a sample of boiled mutton [72]. Thialdine was first described in 1847 as a product from the reaction of hydrogen sulphide, ammonia and acetaldehyde [60] and, in recent years, has been reported in several heated foods. Although the reaction occurs readily, the compound is thermally labile and therefore larger amounts might be expected when cooking and volatile extraction are carried out at lower temperatures [57]. The compound has also been shown to be formed during volatile extraction and this, together with its thermal lability, may explain why large amounts have been obtained in some systems but not in others. Thialdine has been reported to have a roast-beef-like aroma [73,74]. Trithioacetaldehyde and trithioacetone were also reported in the patent literature to have meaty characters [75], although Boelens *et al.* [57] described trithioacetaldehyde as dusty, earthy and nutty.

One of the most important routes to sulphur-containing volatiles in meat involves the reaction of carbonyls and/or dicarbonyls with ammonia, hydrogen sulphide or thiols, which are products of the Strecker degradation of cysteine or methionine. An example of this has already been discussed in relation to the formation of thiazoles. Boelens *et al.* [57] examined the products from the reaction of several aldehydes with hydrogen sulphide and thiols. Acetaldehyde reacted with gaseous hydrogen sulphide with the formation of trimethyl-substituted dioxathianes, oxadithianes and trithianes (Figure 10.6). When the reaction was carried out with liquid hydrogen sulphide, under pressure, bis-(1-mercaptoethyl) sulphide was formed, which readily oxidized to 3,5-dimethyl-1,2,4-trithiolane, while the inclusion of ammonia in the reaction produced thialdine.

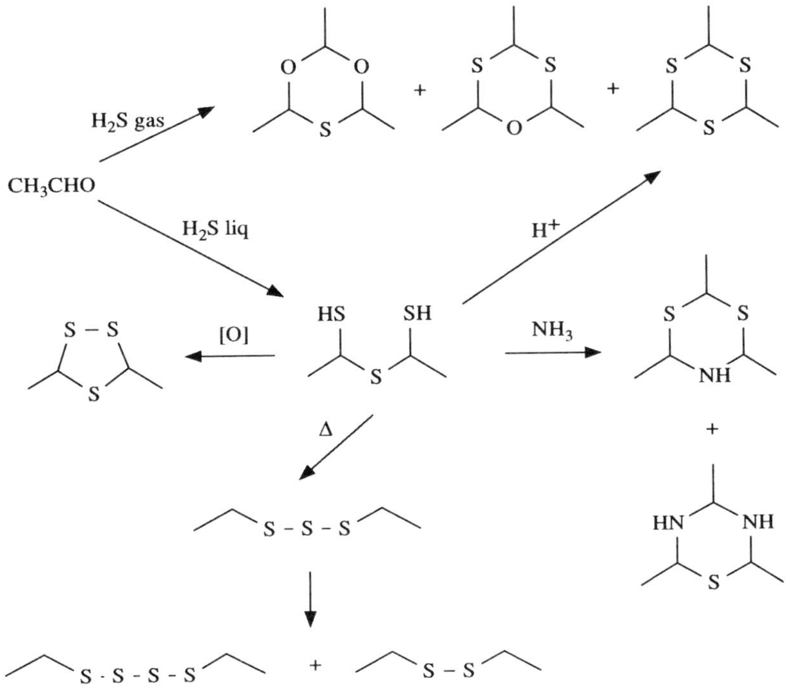

Figure 10.6 Formation of some sulphur-containing aroma compounds from the reaction of acetaldehyde, hydrogen sulphide and ammonia 1571.

10.4.5.5 Furan thiols, sulphides and disulphides. The odour properties of furans with thiol and methlthio substituents have been examined along with corresponding furyl disulphides; some have been found to have distinct meaty characteristics [71,76]. Furans (and also thiophenes) with a thiol group in the 2-position appear to have burnt or sulphurous aroma characteristics, while the isomers with 3-thiol substitution have meat-like aromas. Several furanthiols, thiophenethiols and disulphides have been found as the volatiles from heated model systems containing hydrogen sulphide or cysteine and pentoses or other sources of carbonyl compounds, while some are reported to have meaty aromas. However, until recently, none of these compounds had been found in meat. MacLeod and Ames [77] identified 2-methyl-3(methylthio)furan **(25)** as a character-impact compound in cooked beef. It has been reported to have a low odour threshold value (0.05 ppb) and a meaty aroma at levels below 1 ppb, becoming thiamine-like at higher concentrations [78]. Subsequently Glasser and Grosch [79] used the technique of flavour-dilution analysis to investigate the beef volatiles with high aroma values. Two compounds that showed the highest aroma values had meaty aromas and were identified as 2-methyl-3-furanthiol (**26**) and the corresponding disulphide, bis-(2-methyl-3-furyl) disulphide (**27**). The odour threshold value of the disulphide has been reported as 0.00002 ppb, one of the lowest known threshold values [80]. These two compounds were considered to be very important in determining the characteristic flavour of meat.

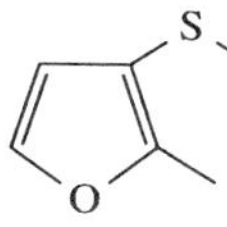

25 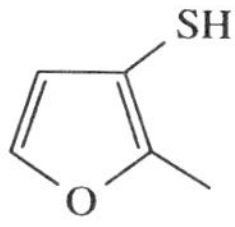 **26** 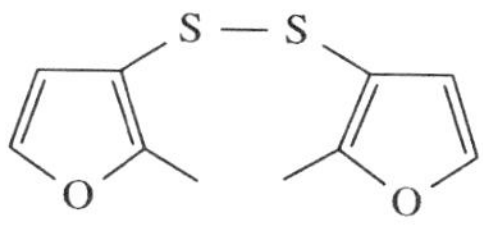 **27**

2-Methyl-3-furanthiol was shown to be one of the products formed in the reaction of hydrogen sulphide with 4-hydroxy-5-methyl-3(2*H*)-furanone, which is a dehydration product of ribose [71]. The meaty characteristics of the reaction mixture led to patents dealing with several related compounds with potential as meat flavourings [3,10]. 2-Methyl-3-furanthiol, and the analogous thiophenethiol, have been found in the volatiles from the reaction of cysteine and ribose [66,81], from heated thiamine [82] and from a model meat-flavour system containing cysteine, thiamine, glutamate and ascorbic acid [83]. These thiols are readily oxidized to the corresponding disulphides, which have also been found in these model systems. Several mixed sulphides and disulphides from furan- and thiophenethiols and furylmethanethiol were found in the cysteine-ribose [81] and the thiamine-cysteine-glutamate-ascorbate reaction mixtures [83], and several of these compounds were reported to have meaty aromas. Farmer and Patterson [84] reported 2-methyl-3-furyl-2-methyl-3-thienyl disulphide (**28**), 2-furfuryl-2-methyl-3-furyl disulphide (**29**) as well as bis-(2-methyl-3-furyl) disulphide (**27**), in the volatiles of heated beef muscle, and evaluation of their aroma on elution from the GC column indicated meaty characteristics.

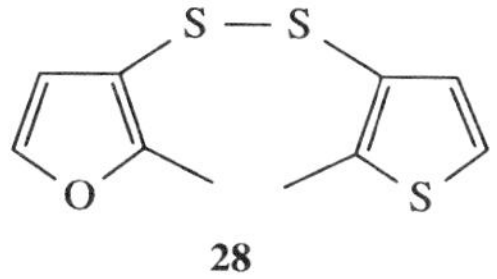

28 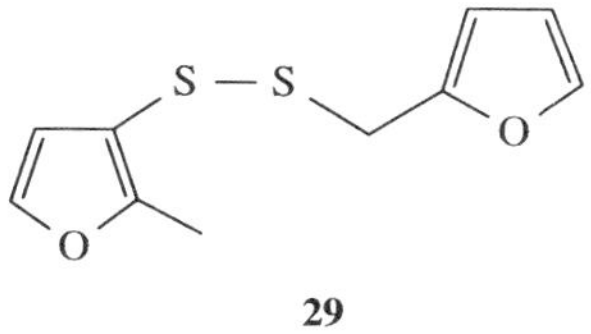 **29**

From the work on the model systems and from attempts to synthesize structures with meat-like aromas, it has become clear that the furanthiols and thiophenethiols and their disulphides have particularly meaty aroma characteristics. The importance of this type of compound in meat flavour has now been confirmed by their detection in meat itself.

References

1. Mottram, D.S., Meat, in *Volatile Compounds in Foods and Beverages*, (ed. H. Maarse), Marcel Dekker, New York, 1991, pp. 107–117.
2. Maarse, H. and C.A. Visscher, *Volatile Compounds in Food – Qualitative and Quantitative data*, 6th edn., TNO-CIVO Food Analysis Institute, Zeist, The Netherlands, 1989.
3. MacLeod, G. and Seyyedain-Ardebili, M., Natural and simulated meat flavours (with particular reference to beef). *CRC Crit. Rev. Food Sci. Nutr.*, 1981, **14**, 309–437.
4. Ohloff, G. and Flament, I., Heterocyclic constituents of meat aroma. *Heterocycles*, 1978, **11**, 663–695.
5. Kramlich, W.E. and Pearson, A.M., Separation and identification of cooked beef flavour components. *Food Res.*, 1960, **25**, 712–719.
6. Hornstein, I. and Crowe, P.F., Flavour studies on beef and pork, *J. Agric. Food Chem.*, 1960, **8**, 494–498.
7. Macey, R.L. Jr., Naumann N.D. and Bailey, M.E., Water-soluble flavour and odour precursors of meat. *J. Food Sci.*, 1964, **29**, 136–148.
8. Wasserman, A.E. and Gray, N., Meat flavour. I. Fractionation of water-soluble flavour precursors of beef. *J. Food Sci.*, 1965, **30**, 801–807.
9. Morton, I.D., Akroyd, P. and May, C.G., Flavouring substances and their preparation. British Patent No. 836 694, 1960.
10. MacLeod, G., The scientific and technological basis of meat flavours, in *Developments in Food Flavours*, (eds G.C. Birch and M.G. Lindley), Elsevier, London, 1986, pp. 191–223.
11. Hornstein, I. and Crowe, P.F., Meat flavour: lamb. *J. Agric. Food Chem.*, 1963, **11**, 147–149.
12. Pippen, E.L., Mecchi, E.P. and Nonaka, M., Origin and nature of aroma in fat of cooked poultry. *J. Food Sci.*, 1969, **34**, 436–442.
13. Wasserman, A.E. and Spinelli, A.M., Effect of some water-soluble components on aroma of heated adipose tissue. *J. Agric. Food Chem.*, 1972, **20**, 171–174.
14. Pippen, E.L. and Mecchi, E.P., Hydrogen sulfide, a direct and potentially indirect contributor to cooked chicken aroma. *J. Food Sci.*, 1969, **34**, 443–446.
15. Mottram, D.S. and Edwards. R.A. The role of triglycerides and phospholipids in the aroma of cooked beef. *J. Sci. Food Agric.*, 1983, **34**, 517–522.
16. Frankel, E.N., Lipid oxidation. *Prog. Lipid Res.*, 1980, **19**, 1–22.
17. Forss, D.A., Odour and flavour compounds from lipids. *Prog. Chem. Fats Other Lipids*, 1972, **13**, 181–258.
18. Grosch, W., Lipid degradation products and flavours, in *Food Flavours*, (eds I.D. Morton and A.J. MacLeod), Elsevier, Amsterdam, 1982, pp. 325–398.
19. Mottram, D.S., Edwards, R.A. and MacFie, H.J.H., A comparison of the flavour volatiles from cooked beef and pork meat systems. *J. Sci. Food Agric.*, 1982, **33**, 934–944.
20. Badings, H.T., Cold storage defects in butter and their relation to the autoxidation of unsaturated fatty acids. *Neth. Milk Dairy J.*, 1970, **24**, 147–256.
21. van Gemert. L.J. and Nettenbreijer, A.H., *Compilation of Odour Threshold Values in Air and Water*, TNO-CIVO Food Analysis Institute, Zeist, The Netherlands, 1977.
22. Noleau, I. and Toulemonde, B., Volatile components of roast chicken fat. *Lebensm. Wiss. Technol.*, 1987, **20**, 37–41.
23. Pippen, E.L., Nonaka, M., Jones, F.T. and Stitt, F., Volatile carbonyl compounds of cooked chicken. I. Compounds obtained by air entrainment. *Food Res.*, 1958, **23**, 103–113.
24. Harkes, P.D. and Begemann, W.J., Identification of some previously unknown aldehydes in cooked chicken. *J. Am. Oil Chem. Soc.*, 1974, **51**, 356–359.

25. Caporaso, F., Sink, J.D., Dimick, P.S., Mussinan, C.J. and Sanderson, A., Volatile flavour constituents of ovine adipose tissue, *J. Agric. Food Chem.*, 1977, **25**, 1230–1234.
26. Hoffman, G. and Meijboom, P.W., Isolation of two isomeric 2,6-nonadienals and two isomeric 4-heptenals from beef and mutton tallow. *J. Am. Oil Chem. Soc.*, 1968, **45**, 468–470.
27. Wong, E., Nixon, L.N., and Johnson, C.B., Volatile medium chain fatty acids and mutton fat. *J. Agric. Food Chem.*, 1975, **23**, 495–498.
28. Christie, W.W., The composition, structure and function of lipids in the tissues of ruminant animals. *Prog. Lipid Res.*, 1978, **17**, 111–205.
29. Patterson, R.L.S., 5-α-Androst-16-en-3-one: a compound responsible for taint in boar fat. *J. Sci. Food Agric.*, 1968, **19**, 31–38.
30. Patterson, R.L.S., Identification of 3-hydroxy-5-α-androst-16-ene as the musk odour component of boar submaxillary salivary gland and its relation to the sex odour taint in pork meat. *J. Sci. Food Agric.*, 1968, **19**, 434–438.
31. Kato, S., Kurata, T. and Fujimaki, M., Thermal degradation of aromatic amino acids. *Agric. Biol. Chem.*, 1971, **35**, 2106–2112.
32. Lien, Y.C. and Nawar, W.W., Thermal decomposition of some amino acids: valine, leucine and isoleucine. *J. Food Sci.*, 1974, **39**, 911–913.
33. Fagerson, I.S., Thermal degradation of carbohydrates: a review. *J. Agric. Food Chem.*, 1970, **17**, 747–750.
34. Feather, M.S. and Harris, J.F., Dehydration reactions of carbohydrates. *Adv. Car bohyd. Chem. Biochem.*, 1973, **28**, 161–224.
35. Hurrell, R.F., Maillard reaction in flavour, in *Food Flavours*, (eds I.D. Morton and A.J. MacLeod), Elsevier, Amsterdam, 1982, pp. 399–437.
36. Mauron, J., The Maillard reaction in food: a critical review from the nutritional standpoint, in *Maillard Reactions in Food*, (ed. C. Eriksson), Pergamon Press, Oxford, 1981, pp. 3–35.
37. Hodge, J.E., Chemistry of browning reactions in model systems. *J. Agric. Food Chem.*, 1953, **1**, 928–943.
38. Vernin, G. and Parkanyi, C., Mechanisms of formation of heterocyclic compounds in Maillard and pyrolysis reactions, in *Chemistry of Heterocyclic Compounds in Flavours and Aromas*, (ed. G. Vernin), Ellis Horwood, Chichester, 1982, pp. 151–207.
39. Tressl, R., Grunewald, K.G., Silwar, R. and Bahri, D., Chemical formation of flavour substances, in *Progress in Flavour Research*, (ed. D.G. Land and H.E. Nursten), Applied Science, London, 1979, pp. 197–214.
40. Kobayashi, N. and Fujimaki, M., Formation of mercaptoacetaldehyde, hydrogen sulfide and acetaldehyde by boiling cysteine and carbonyl compounds. *Agric. Biol. Chem.*, 1965, **29**, 698–699.
41. Tonsbeek, C.H.T., Plancken, A.J. and Weerdhof, T.v.d., Components contributing to beef flavour. Isolation of 4-hydroxy-5-methyl-3-(2H)-furanone and its 2,5-dimethyl homolog from beef broth. *J. Agric. Food Chem.*, 1968, **16**, 1016–1021.
42. Fors, S., Sensory properties of volatile Maillard reaction products and related compounds, in *The Maillard Reaction in Foods and Nutrition*, (eds G.R. Waller and M.S. Feather), American Chemical Society, Washington, 1983, pp. 185–286.
43. Guadagni, D.G., Buttery R.G. and Turnbaugh, J.G., Odour threshold and similarity ratings of some potato chip components. *J. Sci. Food Agric.*, 1972, **23**, 1435–1444.
44. Flament, I., Kohler, M. and Aschiero, R., Sur l'arome de viande de boeuf grillée. II. Dihydro-6,7-5H-cyclopenta-[b]pyrazines, identification et mode de formation. *Helv. Chim. Acta*, 1976, **59**, 2308–2313.
45. Flament, I., Sonnay P., and Ohloff, G., Sur l'arome de viande de boeuf grillée. III. Pyrrolo[1,2a]pyrazines, identification et synthèse. *Helv. Chim. Acta,* 1977, **60**, 1872–1883.
46. Maga, J.A., Pyrazines in flavour, in *Food Flavours*, (eds I.D. Morton and A.J. MacLeod), Elsevier, Amsterdam, 1982, pp. 283–323.
47. Shibamoto, T. and Bernhard, R.A., Effect of time, temperature and reactant ratio on pyrazine formation in model systems. *J. Agric. Food Chem.*, 1976, **24**, 847–852.
48. Wasserman, A.E., Chemical basis for meat flavour: a review. *J. Food Sci.*, 1979, **44**, 6–11.

49. Mussinan, C.J. and Walradt, J.P., Volatile constituents of pressure cooked pork liver. *J. Agric. Food Chem.*, 1974, **22**, 827–831.
50. Mottram, D.S., The effect of cooking conditions on the formation of volatile heterocyclic compounds in pork, *J. Sci. Food Agric.*, 1985, **36**, 377–382.
51. Chang, S.S., Hirai, C., Reddy, B.R., Herz K.O. and Kato, A. Isolation and identification of 2,4,5-trimethyl-3-oxazoline and 3,5-dimethyl-1,2,4-trithiolane in the volatile flavor compounds of boiled beef. *Chem. Ind. (London)*, 1968, 1639–1641.
52. Hartman, G.J., Jin, Q.Z., Collins, G.J., Lee, K.N., Ho, C.T. and Chang, S.S., Nitrogen-containing heterocyclic compounds identified in the volatile flavour constituents of roast beef. *J. Agric. Food Chem.*, 1983, **31**, 1030–1033.
53. Tang, J., Jin, Q.Z., Shen, G.H., Ho, C.T. and Chang, S.S., Isolation and identification of volatile compounds from fried chicken. *J. Agric. Food Chem.*, 1983, **31**, 1287–1292.
54. Ho, C.T., Lee, K.N. and Jin, Q.Z., Isolation and identification of flavour compounds in fried bacon. *J. Agric. Food Chem.*, 1983, **31**, 336–342.
55. Hirai, C., Herz, K.O., Pokorny J. and Chang, S.S., Isolation and identification of volatile flavour compounds in boiled beef. *J. Food Sci.*, 1973, **38**, 393–397.
56. Mussinan, C.J., Wilson, R.A., Katz, I., Hruza A. and Vock, M.H., Identification and flavour properties of some 3-oxazolines and 3-thiazolines isolated from cooked beef, in *Phenolic, Sulphur and Nitrogen Compounds in Food Flavours*, (eds G. Charalambous and I. Katz), American Chemical Society, Washington, 1976, pp. 133–145.
57. Boelens, M., van der Linde, L.M., de Valois, P.J., van Dort, H.M. and Takken, H.J. Organic sulphur compounds from fatty aldehydes, hydrogen sulfide, thiols and ammonia as flavour constituents. *J. Agric. Food Chem.*, 1974, **22**, 1071–1076.
58. Maga, J.A. The role of sulphur compounds in food flavour. Part III. Thiols. *CRC Crit. Rev. Food Sci. Nutr.*, 1976, **7**, 147–192.
59. Schutte, L. Precursors of sulphur-containing flavour compounds, *CRC Crit. Rev. Food Technol.*, 1974, **4**, 457–505.
60. Brinkman, H.W., Copier, H., de Leuw, J.J.M. and Tjan, S.B. Components contributing to beef flavour. Analysis of the headspace volatiles of beef broth. *J. Agric. Food Chem.*, 1972, **20**, 177–181.
61. Maga, J.A. The role of sulphur compounds in food flavour. Part I. Thiazoles. *CRC Crit. Rev. Food Sci. Nutr.*, 1975, **6**, 153–176.
62. Petit, A.O. and Hruza, D.A. Comparative study of flavour properties of thiazole derivatives. *J. Agric. Food Chem.*, 1974, **22**, 264–269.
63. Vernin, G. Recent progress in food flavours: the role of heterocyclic compounds. *Ind. Aliment. Agric.*, 1980, **97**, 433–449.
64. Galt, A.M. and MacLeod, G. Headspace sampling of cooked beef aroma using Tenax GC. *J. Agric. Food Chem.*, 1984, **32**, 59–64.
65. Whitfield, F.B., Mottram, D.S., Brock, S., Puckey, D.J. and Salter, L.J. The effect of phospholipid on the formation of volatile heterocyclic compounds in heated aqueous solutions of amino acids and ribose. *J. Sci. Food Agric.*, 1988, **42**, 261–272.
66. Farmer, L.J., Mottram, D.S. and Whitfield, F.B. Volatile compounds produced in Maillard reactions involving cysteine, ribose and phospholipid. *J. Sci. Food Agric.*, 1989, **49**, 347–368.
67. Mottram, D.S. and Salter, L.J. Flavour formation in meat-related Maillard systems containing phospholipids, in *Thermal Generation of Aroma*, (eds T.H. Parliment, C.T. Ho and R.J. McCorrin), American Chemical Society, Washington, 1989, pp. 442–451.
68. Boelens, M., de Valois, P.J., Wobben, H.J. and van der Gen, A. Volatile flavour compounds from onions. *J. Agric. Food Chem.*, 1971, **19**, 984–991.
69. Sakaguchi, M. and Shibamoto, T. Formation of sulphur-containing compounds from the reaction of D-glucose and hydrogen sulfide. *J. Agric. Food Chem.*, 1978, **26**, 1260–1262.
70. Golovnja, R.V. and Rothe, M. Sulphur-containing compounds in the volatile constituents of boiled meat. *Nahrung*, 1980, **24**, 141–154.
71. van den Ouweland, G.A.M. and Peer, H.G. Components contributing to beef flavour. Volatile compounds produced by the reaction of 4-hydroxy-5-methyl-3(2H)-furanone and its thio analog with hydrogen sulfide. *J. Agric. Food Chem.*, 1975, **23**, 501–505.

72. Nixon, L.N., Wong, E., Johnson, C.B. and Birch, E.J. Nonacidic constituents of volatiles from cooked mutton. *J. Agric. Food Chem.*, 1979, **27**, 355–359.
73. Kubota, K., Kobayashi, A. and Yamanishi, T. Some sulphur-containing compounds in cooked odour concentrate from boiled Antarctic krills. *Agric. Biol. Chem.*, 1980, **44**, 2677–2682.
74. Ledl, F. Analysis of a synthetic onion aroma. *Z. Lebensm. Unters. Forsch.*, 1975, **157**, 28–33.
75. Wilson, R.A., Vock, M.H., Katz, I. and Shuster, E.J. Aroma mixtures. British Patent No. 1 364 747, 1974.
76. Evers, W.J., Heinsohn, H.H., Mayers, B.J. and Sanderson, A. Furans substituted in the three position with sulphur, in *Phenolic, Sulphur and Nitrogen Compounds in Food Flavours*, (eds G. Charalambous and I. Katz), American Chemical Society, Washington, 1976, pp. 184–93.
77. MacLeod, G. and Ames, J.M. 2-Methyl-3-(methylthio)furan: a meaty character impact aroma compound identified from cooked beef. *Chem. Ind. (London)*, 1986, 175–177.
78. Tressl, R. and Wilwar, R. Investigation of sulphur-containing components in roast coffee. *J. Agric. Food Chem.*, 1981, **29**, 1078–1082.
79. Gasser, U. and Grosch, W. Identification of volatile flavour compounds with high aroma values from cooked beef. *Z. Lebensm. Unters Forsch.*, 1988, **186**, 489–494.
80. Buttery, R.G., Haddon, W.F., Seifert, R.M. and Turnbaugh, J.G., Thiamin odor and bis(2-methyl-3-furyl)disulphide. *J. Agric. Food Chem.*, 1984, **32**, 674–676.
81. Farmer, L.J. and Mottram, D.S. Recent studies on the formation of meat aroma compounds, in *Flavour Science and Technology*, (eds Y. Bessiere and A.F. Thomas), Wiley, Chichester, 1990, pp. 113–116.
82. van der Linde, L.M., van Dort, J.M., de Valois, P., Boelens, H. and de Rijke, D., Volatile compounds from thermally degraded thiamin, in *Progress in Flavor Research*, (eds D.G. Land and H.E. Nursten), Applied Science, London, 1979, pp. 219–224.
83. Werkhoff, P., Emberger, R., Guntert, M. and Kopsel, M. Isolation and characterisation of volatile sulphur-containing meat flavour components in model systems. *J. Agric. Food Chem.*, 1990, **38**, 777–791.
84. Farmer, L.J. and Patterson, R.L.S., Compounds contributing to meat flavour. *Food Chem.*, 1991, **40**, 201–205.

11 Consumer perceptions of natural foods

K. DREW

Abstract

With particular reference to flavours, this chapter considers the concept of 'natural foods', consumer perceptions of and attitudes to 'natural foods', and the acceptability of new technological food treatments.

11.1 Introduction

Consumer perceptions of foods are of interest to commercial organizations insofar as they affect food purchasing and consumption. They are also of concern from the welfare and regulatory points of view: if they differ significantly from scientific truth they could alter food consumption in a way that compromises consumers' safety or nutritional status or they could invite fraud.

An understanding of consumers' views of foods allows:

- consumers to be more satisfied with the range of food products available to them;
- businesses to gain commercial advantage by offering products more sensitively tuned to consumer tastes;
- food professionals and consumers to communicate with each other more effectively; and
- research effort to be invested more productively.

Written by a food scientist interested in product development and involved in investigations into consumer attitudes to food, this chapter considers consumer attitudes to natural and processed foods with particular reference to flavours and flavouring substances.

Flavours are an issue because much processed food is of itself insipid. This is the result of some of the processes to which it has been subjected, and to selective breeding programmes aimed at maximizing yield and disregarding sensory quality. Until recently, the main answer to the problem came in the form of flavouring additives. Now genetic engineering can be used, for example, to boost and improve food commodities' inherent flavours.

11.2 Factors influencing perception of food

When considering consumers' views of natural foods, one should take account of consumers' *perceptions* [1] of food, which in turn form part of *attitudes* [2], which contribute significantly to *behavioural intention*, which importantly influence *behaviour* [3,4]. The extensive work done to elucidate factors affecting food preferences and choices has been reviewed by Shepherd [5].

The factors influencing perception that will be considered are: (i) myths that help to form the consumer's world view; (ii) information that the consumer has about the food and the naturalness of the food.

This author has suggested [6] a series of myths that seem to underly perceptions of and attitudes to our experience of the world. These are:

- increasingly there should be personal responsibility for informed decision-taking;
- increasingly the world should be viewed holistically and one should consider the wider effects of a local decision;
- 'big business' is out to make a 'fast buck', if necessary at the expense of the consumer;
- the Government is on the side of 'big business';
- technology and technologists should not be trusted;
- consumers have a right to good, quality products, to information and to be heard; and
- 'natural' implies 'safe'.

By myths are meant wide-ranging beliefs held by a significant proportion of the population. They represent many people's expectations and, in the absence of clear information or experience, myths help us to order and control a potentially chaotic world. Such myths may or may not be 'true' but they are reflected in people's attitudes to many aspects of life.

It seems that *information* about a food can significantly influence perception of and attitudes towards that food. Consumers' reported sensory perceptions of foods are affected by accompanying information. For example, responses were more favourable toward beef labelled according to percentage lean compared with reactions to the same beef labelled according to its percentage fat [7]. Brand also affects reported sensory perception [8].

The way that 'the story is told' has its effect on attitude and marketing copywriters are masters of the reassuring 'blurb'. Quorn, a mycoprotein product is described in marketing literature as 'vegetable in origin' and 'a distant relative of the mushroom'.

The source of the information affects attitudes (i.e trust or approval) to that information. In a poll [9] the question 'who do you trust most to tell the truth?' produced the following rank order: (i) doctor; (ii) priest;

(iii) solicitor; (iv) policeman; (v) teacher, (vi) 'none of these'; (vii) scientist; (viii) trade union leader; (ix) politician. Journalists and 'military top brass' are apparently not trusted at all. Industrialists were not mentioned in the question.

If *natural foods* are those produced without human intervention then people can have eaten little that is natural since the demise of hunter-gathering some 10,000 years ago. For example, it is unlikely that, left to itself, nature would produce a field of wheat. As Bender points out [10], the only food tailored to human needs by nature is mother's milk, and that only suits babies. The composition of breast milk can vary with the mother's diet, especially in its vitamin content. There is even a report in the *New England Journal of Medicine* [11] of anaemia in a breast-fed infant because the mother was vegetarian who consumed no vitamin B12.

The difficulties of defining 'natural' for the purposes of controlling food claims have been described by Bender [12]. He poses such questions as 'are heat processes natural?', 'is smoking natural?' and 'are nature-identical ingredients natural?'

Current interpretations in law, including that in the voluntary guidelines developed by the Food Advisory Committee of the UK Ministry of Agriculture, Fisheries and Food, are broader and more pragmatic than the strict approach taken in the previous paragraph.

The definition of a natural flavour ingredient has been considered by Moyler [13] and Heath [14].

11.3 Consumer perceptions and attitudes

Most of the published information about consumer perceptions of, and attitudes to, foods is gained by standard market research methods. The attitudes of any one person may appear contradictory and surveys planned to elicit information about the whole population may appear to contradict market data.

Consumers rank ingredients after smoking and pollution as causes for (safety) concern. Food ingredients top the list for women aged 25–34 years [15]. One survey reveals 'an extensive loss of confidence in food quality'. It reports people's main food concerns as safety, guilt and confusion [16]. Only about 20% said that they prefer to eat food that they like without worrying about any consequences to health. Processed foods are not considered as nourishing as fresh, unprocessed foods [17].

Atkinson [18] suggests that the term 'natural' carries overtones of 'purity' and 'health' and field research supports this. Fenwick [19] reports a study by the Good Housekeeping Institute in the USA. When consumers were asked to rate the importance of specific characteristics in relation to shopping for food, safety, taste and purity were all rated higher than price

– 'the criterion against which all other attributes are usually judged'. Other terms considered important, 'without preservatives' and 'natural' are positively correlated with purity. The authors paraphrased the consumers' concern as:

- artificial food ingredients are chemicals;
- chemicals cause cancer; and
- artificial food ingredients can cause cancer.

In a survey by Wright [20]:

- 59% strongly agree that additives in food can be dangerous;
- 77% strongly agree that there are too many additives in food; and
- 61% strongly agree that they preferred to buy natural foods whenever possible:

While, in another survey by the KMS Partnership [21]:

- 67% of the sample believe that 'natural' foods taste better than other foods; and
- 92% of the sample believe that 'natural' foods yield greater benefits to health.

Another survey by Yeomans [22] reports that a majority have the following expectations of a '100% natural' food:

- it would be 'healthier';
- it would be 'good for you';
- indicates 'good living' which has 'snob value';
- it would be more expensive; and
- it would contain no chemicals.

In a survey into attitudes to additives [23], 35% of respondents claim to have stopped buying a product because of additives and 43% claim to have started buying a product because it is additive-free or low in additives. Damage to health is given as the main reason for switching to additive-free formulations. Additives are seen as less damaging than fat but more dangerous than sugar or salt. One third of women respondents aged 25–34 years perceive food additives to be the most dangerous [15]. The deliberate purchase of additive-free food increases dramatically with responsibility for feeding a partner and/or children [23].

Of respondents, 55% claim to be 'very' or 'fairly' concerned about the use of additives in general and 20% consider that no additives are 'necessary' in our food and drink [23]. However, most women surveyed consider natural additives to be safe [15].

Added flavourings are second only to colours in the disapproval they engender in consumers [23]. They are considered to be unnecessary by 38% (colours by 64%).

Organically grown foods are considered to have health benefits, and to be less harmful than those produced with the help of artificial fertilizers and pesticides. These benefits are deemed to warrant higher prices [24]. They are presumably seen as more natural than conventional crops. Respondents to a survey in the USA cited safety as the main reason for using organic foods [25].

Traditional foods are acceptable despite their processing. Anecdotal evidence suggests that familiarity brings acceptance:

1. There was considerable resistance to frozen foods when they were introduced. They are now well accepted.
2. Most consumers do not welcome the prospect of irradiated food [26]; however, the group of people conspicuous in their professed acceptance of irradiated food are those employed in the medical field and familiar with irradiated items [27].
3. Younger people are generally less opposed to additives than are older people. Those claiming to be 'not at all' concerned about the use of additives are most likely to be under 24 years [23] and, therefore, presumably, well used to them.
4. Well established products such as cured meats, although produced with sophisticated technology and of questionable safety, are generally uncontroversial. Brown bread (containing the usual array of additives) may be described as 'natural' by consumers [21].

Ingredients lists on food labels are scrutinized by 50% of shoppers, mostly to avoid artificial flavourings and colourings [16]. Among those who look at ingredient labels, 44% say that they avoid E numbers in general and 40% claim to avoid artificial flavourings specifically.

Consumers are reported to be doubtful about claims such as 'no additives' on labels [24]. When asked:

- about one-half said that they would not, necessarily, believe claims;
- two-fifths admitted that they were confused by claims;
- seven out of ten stated that they would be more likely to believe claims on the pack if they trusted the manufacturer or shop from which they were buying.

The confusion admitted by respondents was borne out by their views on eight specific claims about which they were questioned. The general findings were:

1. Purchase of products with the different claims varied between the claims and between different groups of people. Of the claims, respondents most commonly bought food with 'no artificial additives' and 'low fat' or 'reduced fat' on the packaging. For most of the claims, purchases were common among women, younger people, and those in higher socio-economic groups.

2. Consumers are not particularly aware of whether claims are regulated.
3. Consumers showed uncertainty over the definition of the claims. For example, 13% said that 'low fat' and 'reduced fat' were the same.
4. Consumers are misled by the claims. For example when asked about a product which had a 'high fibre' claim, nearly half the sample wrongly thought that the product would have more fibre than a similar product without such a claim. And nearly half wrongly assumed that the colour in a product labelled 'cherries with natural colour' would come from cherries.
5. Consumers infer from the use of the word 'natural' in food claims that these products are 'healthy'. Respondents were asked if a '100% natural' product claim would make them think that the food was likely to be 'healthy'; four out of five thought that it would. Almost three quarters of the sample would not expect a product with a 'natural' claim to contain additives, which, of course, it could.

The Government is not trusted to make clear to consumers food-associated hazards and to protect the consumer from them. Nine out of ten say they would like to see more controls on how food is produced and manufactured, and nearly eight out of ten think the Government should publicize the possible health hazards associated with different kinds of food. Nearly seven out of ten feel that the Government is doing too much to protect the food industry, and too little to protect consumers. A majority say they are either not very confident or not at all confident, that the Government ensures that all food additives are safe [16].

Consumers, when describing their concern about unnatural/artificial ingredients and additives, focus primarily on issues of *risk to health*. Perception of risk may bear little relationship to statistical ratings and has been discussed elsewhere [6]. The gravity of concern relates to: (i) how well the risk is understood; (ii) the severity of the consequences, and (iii) the number of people exposed to the risk.

These conclusions are congruent with, for example, the results of a survey in the USA [28]. When two sets of respondents were asked to rank the hazards from eating food, the lists in Table 11.1 were obtained.

A pilot scale survey in Sheffield [29] produced opinions about the safety of different food treatments (Table 11.2).

11.3.1 Discussion

When evaluating market research results, the weaknesses of the investigative methods must be acknowledged. These have been rehearsed elsewhere [6]. Consumers seem to see foods as either 'good for you' or 'bad for you'. It is not evident that consumers know that a food may contain nutrients *and* toxins.

Table 11.1 Ordering of food-associated hazards[a]

The experts	The public
1. Microbial safety	1. Pesticides
2. Over-nutrition	2. New food chemicals
3. Nonmicrobial safety:	3. Chemical additives
(a) contaminants;	4. Familiar hazards
(b) natural toxicants	(a) fat and cholesterol
(c) agricultural chemicals	(b) microbial spoilage
(d) food additives	(c) junk foods

[a] 1 = greatest hazard.
Source: Lee [28].

Table 11.2 Perceived safety of food treatments[a]

	Percentage of sample		
Food treatment	(Very) Safe	Uncertain	(Very) Unsafe
Canning	80	10	10
Irradiation	6	40	54
Drying	74	23	3
Genetic engineering	13	63	23
Freezing	80	13	7
Refrigeration	77	17	6
Additives	13	40	47
Vacuum-packing	33	37	30
Pesticides	0	3	97
Hormones	0	17	83

[a] Figures in the first column represent the percentage (to the nearest whole number) of respondents who considered the treatment to be safe or very safe, and so on.
Source: Kilner [29].

Consumers believe that processing lowers the nutritional value of foods and may introduce dangers into natural and therefore safe raw materials. They are wary of, and many try to avoid, additives in general and colourings and flavourings in particular. Natural is good, additives and flavourings are bad, even apparently if they are natural (they are in the food due to human intervention).

Traditional foods are acceptable even if they are very 'unnatural'. This is reasonable and understandable to the extent that any dangers associated with a well established food should by now be well known. It is possible that another complaint against 'new' products is an aesthetic one. For example, margarine was, by all accounts, very unappetizing when it was introduced. It is now acceptable not only because it is familiar and apparently safe but also because the technology is more sophisticated and the product more convenient and very acceptable from a sensory point of view. There is little confidence in Government to give consumers enough information or to protect them from hazard. The attitudes reported here tally with the myths listed earlier.

11.3.2 The safety of natural foods

The natural world and its products often hold hazards for humans. Natural toxicants have been listed [30,31] and discussed elsewhere [10,19]. Consumer concern is out of line with expert estimates of risk, for example:

Table 11.3 Relative risk of food-associated hazards[a]

Food safety hazard	Relative risk
Food-borne disease	100 000
Poor nutrition	100 000
Environmental pollution	100
Natural toxicants	100
Pesticides	1
Additives	1

[a] Source: Gormley *et al.* [32].

Conning [33] has discussed the extent to which concern about the risks posed by the 'uncontrolled use of flavourings' is justified. He concludes that the concern expressed is disproportionate.

11.4 Short-term future

Can foods produced using new technology ever be instantly acceptable? The situation with regard to additives is far from hopeless. One crucial element in getting an additive accepted is that of perceived consumer benefit. Many consumers see additives as being used for the producer's commercial benefit and this fuels their distaste for such ingredients. However, the marketing of aspartame under the brand name Nutrasweet was a notable success. It is unusual for such ingredients to be branded at all and to brand a 'high-tech' one could be considered risky. However, there are clear advantages for the consumer, namely, few calories and a good taste and, in advertising, this product is presented as: 'made from parts of protein like those found in Nature. Clever Nature. Clever Nutrasweet'.

The immediate question is whether foods improved using the gene technologies will be regarded as natural and therefore safe. If risks that are poorly understood and with potentially severe consequences for many people cause the greatest concern, the gene technologies could be in for a rough ride. It looks currently as if opinion could go either way. A question about the safety of genetic engineering produced the largest number of 'uncertain' responses (Table 11.2). There is clearly an information gap to be filled somehow. Different views about this are already emerging, for example [34,35]:

> The enzyme technology of the '90s is set to enhance the green revolution by continuing to contribute to the development of new 'green technologies' displacing the more traditional chemical processes which are currently relied on
>
> *(Hanson, 1990)*

and

> There are those who, on religious and ethical grounds, regard genetic engineering as odious. Living things do not exist solely for the benefit of humans.
>
> *(Brunner, 1990)*

Some consumers already have clear views. For 2 years running, in The Netherlands, activists have destroyed the fields of genetically engineered potatoes, causing 'millions of guilders of damage' [36].

The opinions formed by the majority of consumers who are currently uncertain about genetic engineering could hinge on how the information is presented. Kilner [29] presents genetic engineering as a way of solving consumer problems with fresh tomatoes (e.g. lack of flavour, soon go soft, artificially ripened). On this basis, 73% of respondents agree or strongly agree with the statement: 'These (genetically engineered) tomatoes should be available and I would buy them'. It may help if a new technology is introduced incrementally so that there is time for review and adjustment after each step.

Consumers in general are confused about foods and claim to want more information [15,23]. If 'reader-friendly' information were easily available, explaining the consumer benefits of food treatments as well as the risks, interested consumers could make use of it. The next question to be asked is 'who should present the information?'. Not many professions have consumer confidence when it comes to supplying information; only doctors and priests have significant 'street credibility'.

11.5 Long-term future

For any but the simplest information to be understood, people need to be scientifically 'literate'. This is currently far from being the case: 68% describe themselves as only 'slightly knowledgeable about science' or 'baffled' [9].

Science education, then, is likely to be very important to the management of many aspects of technology by society. However, it is not the complete answer: one should also try to change some of the myths listed earlier. Such broadly based beliefs are likely to be slow to alter. In a society largely geared to short-term ends (e.g. winning the next election, maximizing this year's profit figures), attempts at change will entail major reorientations.

Food professionals are particularly well placed to tackle beliefs about technology, corporate business and government. To be effective, their credibility ratings must improve.

11.5.1 Technology

The myth is that technology and technologists should not be trusted. According to Professor Stephen Hawking [37]:

> The public has a rather ambivalent attitude to science at present. It has come to expect the steady increase in the standard of living that science and technology have brought but it distrusts science because it doesn't understand it. Even today, this distrust is caricatured in the picture of the mad scientist working in his laboratory to produce a Frankenstein. The public also has a great interest in science, particularly astronomy . . . What can be done to harness this interest and give the public the scientific background it needs to make informed decisions on subjects like acid rain, the greenhouse effect . . . or genetic engineering? It must be sold on science. *(Hawking, 1990)*

Making technology more acceptable to consumers is difficult in the aftermath of Chernobyl and Three-Mile Island disasters, and amid reports of acid rain and polluted waterways. Societies using advanced technology are playing for higher stakes than ever before: 'They can blow the world to bits or mess up our genes'. Suspicions of unacceptable risk-taking are exacerbated by areas of secrecy (e.g. military information, diplomatic activity); however justifiable, they fuel mistrust and alarm: 'what are they doing to us?', 'they obviously have something to hide'. The benefits of technology are often taken for granted. To improve the image of technology:

1. Maximum information must be available to anyone interested. If information is not available there should as far as possible be a full explanation of why it is not available.
2. Consumer interests must be seen to be a real (the major?) force in technology decision-making.
3. Scientists and technologists must share with lay people their enthusiasm for and information about their work.

This latter point is a situation well recognized by leading scientists, such as Sir Walter Bodmer [37]:

> The scientist has an absolute responsibility to explain to the public what he is doing. However, the scientist who does explain . . . to the public runs the risk of not being taken seriously by fellow scientists. *(Bodmer, 1990)*

The image of science and technology as an authoritative, exclusive and jargon-ridden club must be dismantled. There are hurdles to be cleared and much has already been done to this end; much more remains to be done.

11.5.2 Government

The myth is that the Government backs 'big business' against the consumer. Consumers are aware, for example, of the power of the agricultural lobby but are far less aware of the Government's regulatory activities on their behalf. To increase confidence, the Government must be seen:

- to be technically and administratively competent;
- to take decisions in an open and even-handed way; and
- to give consumers access to information whenever feasible.

11.5.3 Corporate business

The myth is that 'big business' is out to make a 'fast buck', if necessary at the expense of the consumer. Businesses operate to make a profit and newspaper stories confirm suspicions that ethics are sometimes sacrificed for financial gain. Ethical companies may be suspect because of secrecy or poor communications. To improve the image of corporate business, businesses must:

- be seen to be acting openly and ethically;
- communicate effectively; and
- be seen to respect the consumer.

This approach apparently paid off when used by Quaker companies such as Rowntrees, Twinings and Cadbury's in the last century.

11.5.4 Does natural food imply safe food?

This leaves us to consider the myth that *'natural' implies 'safe'* and what food professionals believe to be a reasonable view of the situation. The myth is that if food were not 'interfered with' there would be no cause for concern. The author believes that this notion, while understandable, is a side issue. Let us consider first what constitutes a natural diet for a human. *Homo sapiens* evolved some 3 000 000 years ago and for all but the last 10 000 years or so survived, so far as we can tell, by hunting and gathering.

This diet was varied, including hundreds or even thousands of food species, which were unimproved by humans.

It takes a much larger land area and more work to support one person in this way than current food production methods. It would be impossible for the world to sustain its current population with this sort of diet and it is likely that few people today would find such a diet attractive. However, a 'wild' diet had a built-in safety mechanism, namely that of spreading risk. By consuming only a small amount of any one food, the attendant risks would be correspondingly small.

By crop improvement and the use of intensive production systems for the few species that have been domesticated, the balance, refined over several millions of years is tested. Similarly, processing that involves extracting one component of a food crop may lead to an increase in risk as the physiological effects of the food are also potentiated.

For most people, aiming for a truly natural diet is a waste of effort. One can sensibly aim to have a more varied and less refined diet but not a natural one. Our efforts should go towards ensuring that our diet:

- provides only risks that are understood and are acceptable;
- provides the necessary nutrients;
- is aesthetically excellent (e.g. delicious);
- produced in a sustainable and non-exploitative way; and
- provides choice about how much an enjoyable and wholesome diet costs to provide (in money, time and expertise).

By making such a diet readily available we may even be in danger of extinguishing the opportunity for feeling guilty! The author believes that food professionals should take all opportunities to explain this to interested lay people, and to listen to their concerns and discuss them openly and fully.

11.6 Conclusion

The current vogue for 'natural' foods is a romantic one born of confusion and anxiety about the uses of science. It is conceivable that it always takes one generation before a benevolent technology is accepted. Such a passive reaction means a generation missing the benefits of new technology.

Improving the situation for all concerned involves education and wide-ranging communication. This way forward demands not the abandoning of technology, which some consumers would clearly favour, but rather renewed efforts to understand people and their physiological and aesthetic needs and to understand our food resources and how to use them wisely.

References

1. Krondl, M. and Lau, D., cited in [5].
2. Kretch D. and Crutchfield R.S., *Theory and Problems in Social Psychology*, McGraw-Hill, New York, 1948
3. Fishbein, M. and Ajzen, I., *Belief, Attitude, Intention and Behaviour: An Introduction to Theory and Research*, Addison-Wesley, Reading, Massachusetts, 1975.
4. Ajzen, I. and Fishbein, M., *Understanding Attitudes and Predicting Social Behaviour*, Prentice Hall, Eaglewood Cliffs, New Jersey, 1980.
5. Shepherd R., Factors influencing food preference and choice, in *Handbook of the Psychophysiology of Human Eating* (ed. R. Shepherd), Wiley & Sons, Chichester, 1989, pp. 3–24.
6. Drew K., Consumer attitudes to food quality, in *Processing and the Quality of Foods – 1*, (ed. P. Zeuthen, J.C. Cheftel, C. Eriksson, T.R. Gormley, P. Linko and K. Paulus), Elsevier Applied Science Publishers, London, 1990, pp. 12–23.
7. Levin I.P. and Gaeth, G.J., How consumers are affected by the framing of attribute information before and after consuming the product. *Journal of Consumer Research*, 1988 **15**(3), 374–378.
8. Martin, D., The impact of branding and marketing on perception of sensory qualities, *Food Science and Technology Today*, 1990, **4**(1), 44–49.
9. Blakemore, C., Baffled by the boffins – that's Britain, *The Daily Telegraph*, London, 1990, 20 August, p. 10.
10. Bender, A., *Health or Hoax*, Elvendon Press, Goring-on-Thames, 1985, pp. 33–36
11. Higginbottom, M.C., Sweetman, L. and Nyman, W.L., A syndrome of methylmalonic aciduria, homocystinuria, megoblastic anaemia and neurologic abnormalities in a Vit. B12 deficient breast-fed infant of a strict vegetarian. *New England Journal of Medicine*, 1978, **229**, 317–323.
12. Bender, A., What is natural? *Food Chemistry*, 1989, **33**, 43–51.
13. Molyer, D.A., Definition of a natural flavour ingredient, in *Food Technology International Europe*, (ed., A Turner), Sterling Publications, London, 1989.
14. Heath, H., Flavouring materials – just what is natural? *Food Flavourings Ingredients Packaging and Processing*, 1980, **1**(8), 19–20.
15. Ministry of Agriculture, Fisheries and Food, *Survey of Consumer Attitudes to Food Additives*, HMSO, London, 1987.
16. Hodgkinson, N., Britain's hunger for good health, *Sunday Times*, London, 1989, 1 October pp. B8–B9.
17. British Nutrition Foundation, *Eating in the Early 1980s*, British Nutrition Foundation, London, 1985.
18. Atkinson P., Eating Virtue, cited in *The Sociology of Food and Eating*, (ed. A. Murcott), Gower, Aldershot, 1983, p. 17.
19. Fenwick, R., Natural Toxicants, *Food Science and Technology Today*, 1987, **1**(2) 93.
20. Wright, G., *Milk and the Consumer*, Food Policy Research Unit, University of Bradford, 1987.
21. KMS Partnership, *Eating What Comes Naturally*, Presto Stores, UK, 1986.
22. Yeomans, L., How Natural is 'Natural'? – The Consumers' View, *Food Science and Technology Today* 1987, **1**(2) p. 84.
23. Drew, K. and Lyons, H.M., *Special Report: Additives and the Consumer*, Mintel, London, 1987.
24. Survey Research Group, *Consumers and Food Claims*, Consumers' Association, London, 1989.
25. Jolly, D.A., Schutz, H.G., Diaz-Knauf, K.V. and Johal, J., Organic foods: consumer attitudes and use. *Food Technology*, November 1989, 60, 62, 64, 66
26. *Food Irradiation – the Consumers' View*, Consumers' Association, London, 1989
27. Daly, L., Irradiated foods and the consumer, *Food Marketing*, 1986, **2**(3), 67–76.
28. Lee, K., Food neophobia: the major causes and treatments, *Food Technology*, December 1989, p. 63.
29. Kilner, D.R., *Consumer attitudes to biotechnology*. BSc Thesis, Food Marketing Sciences, Sheffield City Polytechnic, 1990.

30. Roberts H.R., Food safety in perspective, in *Food Safety*, (ed. H.R. Roberts), Wiley-Interscience, New York, p. 6.
31. Rechcigl, M., Jr, *Handbook of Naturally Occurring Toxicants*, CRC Press, Boca Raton, 1983.
32. Gormley, T.R., Downey, G., O'Beirne, D., *Technological Change in Agriculture and the Food Industry, and Public Policy in Relation to Food Production, Nutrition and Food Safety*, FAST Occasional Paper No. 107, Commission of the European Communities, Brussels, 1986.
33. Conning, D., Flavours – little but large, *Nutrition Bulletin 55*, **14**(1), 1989 63–70.
34. Hanson, M., Green reactions, *Food Production*, London May 1990, 27.
35. Brunner, E., Food genes the case against, *The Food Magazine*, Jan/March 1990, 25.
36. Biedermann F., Holland's hot potatoes, *Chemistry and Industry*, 20 August 1990, p. 506
37. Hawking, S., Awakening the scientist in all of us. *The Daily Telegraph*, August 1990, p. 1
38. Bodmer W., Science friction. *Radio 4*, 3 September 1990, 9.05pm.

12 Biotechnical production of flavours – current status

R.G. BERGER

Abstract

Volatile flavours are considered suitable target compounds of novel biotechnologies. As most of the industrially used flavours are directly or indirectly derived from plant metabolism, this chapter will focus on the application of *in vitro* plant cells. Topics under discussion are exogenous substrates and cell vitality, artificial sites for the accumulation of products, nutritional improvements, the metabolism of monoterpenes and photomixotrophic cells.

12.1 Introduction

Biotechnology is often associated with the industrial production of antibiotics and other pharmaceuticals; the oldest and commercially most important sector of biotechnology, however, is traditional food biotechnology. Generating or refining food by this process may comprise alcoholic, acetic acid, propionic acid, and lactic acid fermentations, which, on complex substrates, frequently occur as mixed forms. As the starter culture or the endogenous microflora develop, precursors and components of the volatile and nonvolatile flavour fraction accumulate.

Based on the historical background and also on the expanded experience with advanced methods of bioengineering and genetic engineering of industrial-scale fermentations of nonvolatile flavours such as organic acids or amino acids, bioprocesses for the generation of volatile flavours have recently received increased attention. Among the origins of this trend are:

1. An increasing demand of the food industry for volatile flavours, related to an increasing market share of thermally processed food and to the consumer's preference for intensely flavoured products.
2. A decreasing availability of certain natural sources due to agricultural or ecological problems in the producing countries.
3. The lack of convincing chemosynthetic alternatives. The physiological properties of many aroma chemicals depend on a precise regiospecificity or stereospecifity [1–3] that may be achieved by using highly substrate

and reaction specific (bio)catalysts but not easily using a multi-step chemosynthetic approach. Moreover, the environmental compatibility of bioprocesses, usually proceeding in an aqueous phase, is regarded as superior.
4. In many countries legal regulations distinguish 'natural' flavours from others; products obtained from bioprocesses are classified as 'natural'. This results in a higher market value.
5. Many consumers tend to reject food containing non-natural additives. The high esteem of the attribute 'natural' in the food industry is illustrated by the fact that although products of chemosynthesis account for 90% of all flavour and fragrance chemicals worldwide [4]; inversely, about two-thirds of the aromas processed by the German food industry are labelled 'natural' [5].

In this chapter it is hoped to demonstrate that some fundamental characteristics of plant cell cultures were underestimated in previous work but also that suspended plant cells may represent a future source of volatile and nonvolatile flavours.

12.2 What are bioflavours?

Although the title of an international symposium [6], there exists no clear-cut definition of the term. Schreier equated 'bioflavours' with 'biologically derived aroma chemicals generated by microbial fermentation, by the action of endogenous or processing enzymes, and by plant metabolism' [7]. Drawert linked naturally produced flavours with biotechnology, bioprocess technique, bioactivity and bioanalytical methods [8].

The key elements include a selected biocatalyst that is able to perform a single-step transformation of a substrate, or a multiple-step conversion starting with an intermediate metabolite, or a *de novo* synthesis from the basic nutrients of a fermentation broth, and a controlled and optimized technical process. Target compounds of operating or suggested bioprocesses belong to the group of chiral character impact components, as well as to structurally simple volatiles such as acetaldehyde, benzaldehyde or ethyl acetate. Classical or genetically modified food micro-organisms (or parts thereof) and model mixtures that imitated food were the starting points for the development of new processes. Several chapters in this book are devoted to classical fermentation flavours and fermentation-based processed flavours. However, the overwhelming majority of the aromas used in industrial food processing depends ultimately on the biosynthetic potential of living plant cells. This chapter shall, therefore, concentrate exclusively on the description and discussion of the current status of plant cell biotechnology in terms of flavour biogeneration.

12.3 Plant cells as biocatalysts

Shortly after the first successful submerged cultivation of plant cells in the late 1950s, reviews on their commercial exploitation appeared [9]. A continuous biotechnological use of plant cells necessitates breaking down the highly structured cellular arrangement of a whole plant, and transferring the cells to a state of unlimited division in a sterile environment. This involves the treatment of wounding-induced plant tissue with phytoeffectors, and the conversion of the growing callus into a finely dispersed cell suspension for subcultivation and scaling-up in bioreactors.

The uniform chemical and physical surroundings, the absence of the intercellular communication that the cell was acquainted with, and the synthetic growth stimuli induce a shift of the plant cell's metabolism towards primary routes. As a result, it is common that *in vitro* cells of flavour-producing plants do not contain any volatile flavours at all [10]. Conversely, two reviews cited over 120 and over 170 references, respectively, describing the formation of volatiles. This clearly indicates that *in vitro* cells still retain the required genetic information and may produce volatile flavours under suitable conditions [11,12]. The pathways of flavour formation may only be expressed at low levels, for limited periods, or in quantitative ratios differing from the whole plant. Despite these drawbacks the tremendous diversity of secondary plant products, unequalled by any other biosystem, has continued to keep the scientific interest alive.

12.4 Exogenous substrates and cell vitality

Organic solvents were discussed as a means to increase product yields by preventing substrate or product inhibition or degradation in a two-phase system [13]. Supplementing lipophilic aroma precursors in transformation studies may also require the use of organic solvents. A modified 'airlift' loop reactor was introduced for pilot-scale applications that is agitated by a stream of a water-immiscible solvent instead of air [14]. Little systematic research, however, has been published on the physiological effects of solvents on *in vitro* plant cells [15].

During studies on the transformation of exogenous β-carotene by suspended plant cells, various solvents proved useful and were examined further for their biocompatibility [16]. The solvents chosen all affected the vitality of stationary phase cells of lemon after a 5 h exposure, with methanol, dimethyl formamide (DMF), dioxane (DIOX) and ethanol being the most compatible ones (Figure 12.1). At higher concentrations ($> 2\%$ v/v), methanol exerted less toxic effects than ethanol but stabilized cell vitality. Tetrahydrofurane represented a group of more toxic solvents. The unexpected behaviour of the methanol-treated cells and the missing

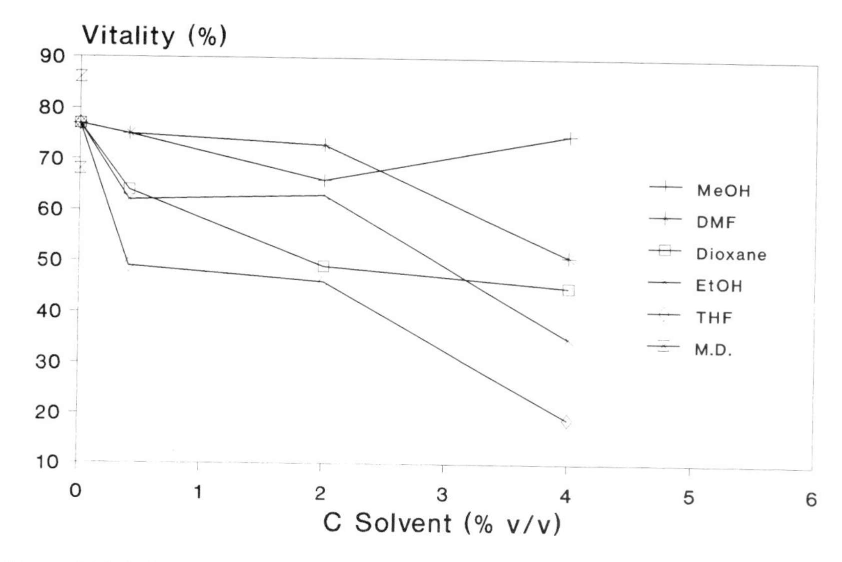

Figure 12.1 Solvent concentration and vitality of suspended cells of *Citrus limon* cv Ponderosa (late log phase, LS medium, 5 h incubation period).

correlation between these results and the dielectric constant or partition coefficient data [17] of the solvents necessitated a closer look.

The bioconversion of exogenous geraniol to related monoterpene alcohols, a feature present in many cell-suspension cultures, was selected as a metabolic trait to study the solvent sensitivity of the plant enzymes involved. Using the same species and cultivation conditions as for the light microscopic vitality tests, the adverse effect of ethanol was confirmed, while, by contrast, methanol was found not to be superior to DMF or DIOX (Figure 12.2). Using the same cultivation conditions but suspended cells of apple (Figure 12.3), a different picture was obtained: ethanol was now superior to methanol at lower concentrations and DMF and DIOX started to exhibit noticeable inhibitory effects with increasing concentrations. In conclusion, only very low concentrations of DMF and DIOX and, to a lesser extent, ethanol were tolerated without significantly reducing cellular activities.

In order to eliminate a hidden toxicity of the applied geraniol, suspended lemon cells were treated with various concentrations in 1% v/v DIOX (Figure 12.4). Far beyond the concentrations used in the above experiments (8.7 μM) the critical threshold, determined as vitality of stationary phase cells, appeared to be in the range of 2–3 mM geraniol. Similar threshold values were found for suspensions of *Pelargonium fragrans* supplied with limonene [18] or terpenols [19]. It was concluded that geraniol should not have interfered in the above solvent experiments.

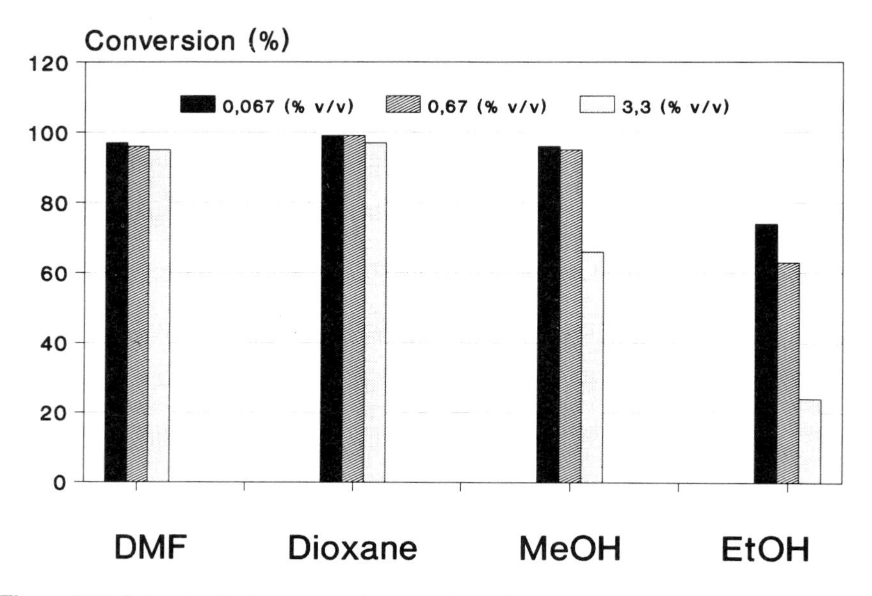

Figure 12.2 Solvent effect on rate of conversion of exogenous geraniol to terpenols by suspended cells of *Citrus limon* (stationary phase, LS medium, 5 h incubation period).

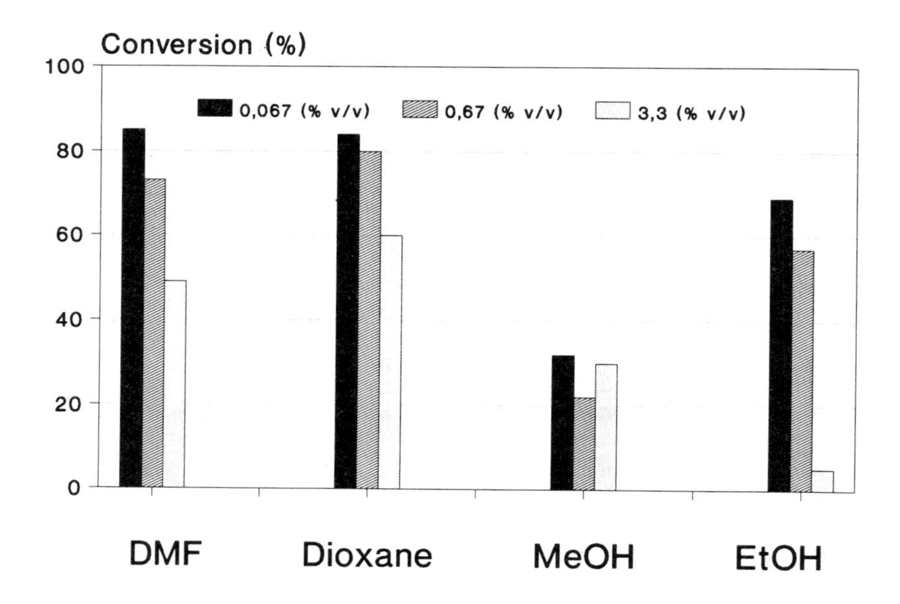

Figure 12.3 As for Figure 12.2 and *Malus sylvestris* cell suspension (MS medium).

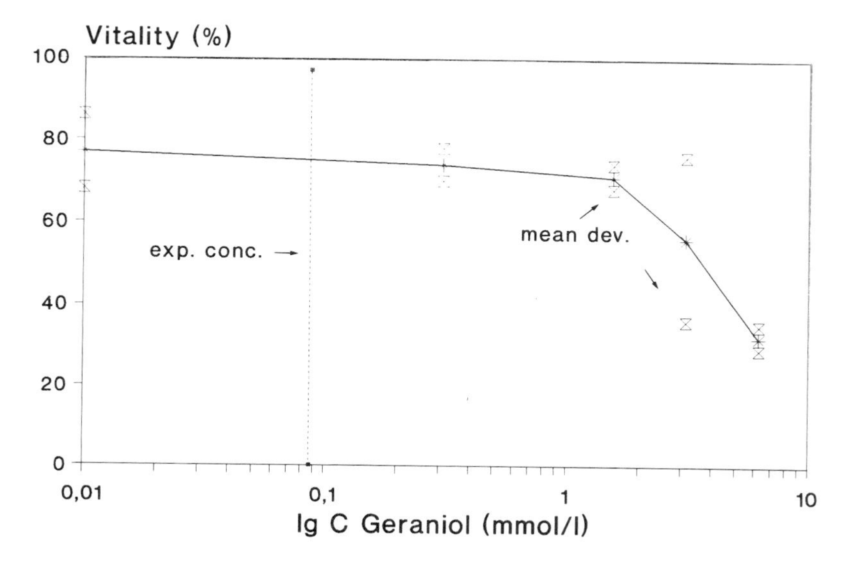

Figure 12.4 Geraniol effect on vitality of suspended cells of *Citrus limon* (for experimental conditions see Figure 12.2).

On the other hand, as millimolar concentrations of essential oil constituents are harmful even after a short-term exposure, the accumulation of significant amounts in the nutrient medium of growing cultures, as expected earlier, must clearly be excluded.

Plant metabolites of commercial interest are often stored intracellularly. To overcome the problem, particularly in immobilized plant cells [20], dimethyl sulphoxide (DMSO) was introduced as a permeabilizing agent: cells of *Catharanthus roseus* released product upon treatment, and remained viable under the growth conditions used [21]. However, investigation of suspensions of *Cinchona ledgeriana* yielded the opposite results [22]. Increased DMSO concentrations correlated with increased release of product and with increased damage of cells; many of the permeabilized membranes did not recover their integrity. This is in agreement with results of studies on suspended cells of apple. The extracellular activity of peroxidase (POD), a highly abundant activity in many cell suspensions [23], was measured as a metabolic marker (Figure 12.5). After 4 h of incubation in the presence of 0.1% v/v DMSO, total POD activity was remarkably reduced; at increased concentrations of the solvent even lower activities and a reduced cell vitality were observed. The relative sensitivities of log-phase and stationary-phase cells were quite similar. Other harvesting techniques, such as electroporation or ultrasonication, were claimed to be less destructive [24]. At this time the question whether permeabilization is inevitably associated with cell lysis is still a matter of discussion [25].

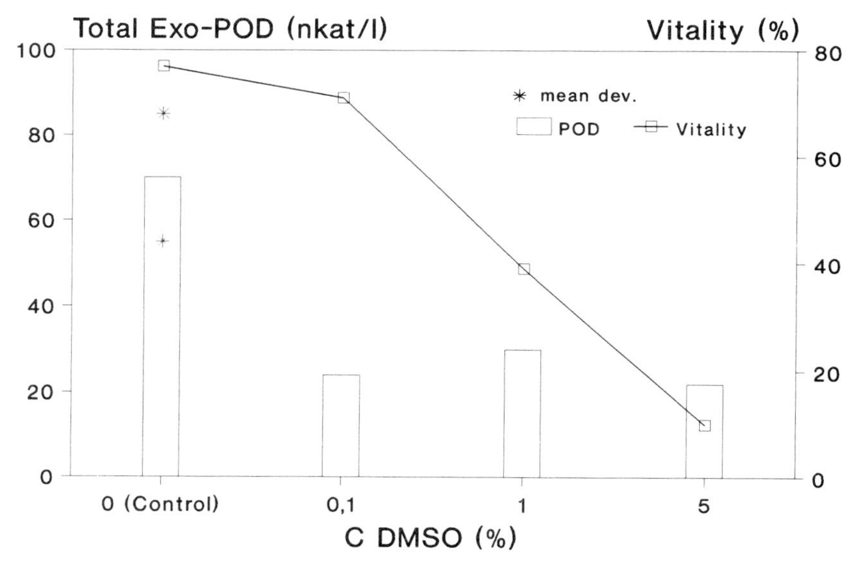

Figure 12.5 Effect of DMSO on Exo-POD and vitality of suspended cells of *Malus sylvestris* (log phase, MS medium, 6 h incubation period).

12.5 Artificial sites for the accumulation of product

The toxicology and, on a larger scale, the safety problems of organic solvents have induced the search for alternative lipophilic phases as synthetic accumulation sites. The beneficial effects on product yields of charcoal, defined triglycerides, or of RP-8 silica gel [26–29] may be interpreted by an efficient protection of traces of secreted metabolites from subsequent catabolism or evaporation losses. Cytotoxic effects of accumulating terpenes (Figure 12.4), for example which the whole plant rectifies by secretion into specialized storage structures, are thus prevented. Miglyol™, a synthetic triglyceride, was successfully added to suspended camomile and grape cells [27,28]. Using apple and lemon cell cultures the necessary metabolic inertness could not be confirmed, again demonstrating (Figures 12.2 and 12.3) the risks of an unseen generalization of results. When log-phase apple cells were incubated in the presence of Miglyol™ a broad spectrum of volatile acyl metabolites was isolated from the nutrient medium (Table 12.1). Methyl and ethyl esters of the acyl moieties of Miglyol™, odd- and even-numbered aliphatic alcohols, alkanals and alkenals were identified. Structure of products, time course of formation, and the parallel experiment using the corresponding free fatty acids strongly point to a direct precursor activity of the exogenous triglyceride upon lipolysis. Lemon cells produced a similar spectrum of esters but no alcohols and lactones and only traces of aldehydes [30]. The metabolic

Table 12.1 Formation of volatiles in cell suspensions of *Malus sylvestris* by exogenous triglyceride (Miglyol™) and free fatty acids (log phase, 4-day incubation period)

Compound (μg/l)	Miglyol™ (5% v/v)	C6C8C10 FA (4 mM)
Ethyl hexanoate	15	23
Methyl octanoate	171	131
Ethyl octanoate	81	94
Methyl decanoate	92	40
Ethyl decanoate	276	441
1-Hexanol	128	58
1-Heptanol	163	139
1-Octanol	214	173
1-Nonanol	59	40
1-Decanol	22	24
n-Hexanal	167	285
n-Decanal	18	30
2(*E*)-Hexenal	69	110
2(*E*)-Heptenal	19	34
2(*E*)-Octenal	17	24
3(*E*)-Nonenal	7	7
2(*E*)-Decenal	46	59
4-Hexanolide	–	240
5-Hexanolide	–	28
4-Octanolide	–	70
5-Octanolide	–	trace
4-Nonanolide	–	trace
5-Nonanolide	–	trace
4-Decanolide	–	trace
5-Decanolide	–	trace

overflow upon addition of excess C-source appears to give rise to this formation of volatiles. Even metabolites not found in the whole plant, such as the lactones, may occur in this unbalanced metabolic situation. These results also shed some light on the genesis of lactones in ageing fruit tissues.

More recently, solid lipophilic polymers were introduced into plant cell systems [25]. Problems of cytotoxic monomer and oligomer contaminants can be solved by thermal or chemical purification procedures. A comparative study using a cell-free model system showed that some of the materials efficiently removed volatile reference compounds from the nutrient medium (Table 12.2). The overall recovery of some of the compounds was unsatisfactory. *In praxi* the situation is complicated by difficulties in separating the polymer particles from the suspended cell aggregates. More promising from a technical point of view would be switching a fixed bed of polymer into an external loop of a bioreactor. Another novel approach is the use of an aqueous two-phase polymer system comprising polyethylene glycol and dextran [31].

Table 12.2 Accumulation of volatiles on porous polymers (48 h, 130 rpm, ambient light)

	Percentage recovery of added compound			
Compound	Control	Chromosorb 102	XAD 4	Silicone oil
Methyl octanoate	29	98	95	69
Linalool	48	81	87	78
α-Humulene	23	72	78	42

12.6 Nutritional improvement

While improved techniques for product accumulation will be available in the future, improved net production rates of target molecules, the key to commercial success, will continue to be the scientific challenge. The conventional strategy has been to vary and to modify the chemical environment of the cell. Although time-consuming and somewhat random, these efforts have resulted in some high-yielding cell cultures [32]. Recent reports deal, for example with inexpensive C-substrates such as whey permeate [33], or corn syrup [34], with more-adequately buffered media [35,36], gelling agents for callus culture [37], conditioning factors [38], and elicitors [39].

Biotic and abiotic elicitors induce plant defence reactions and may be grouped among external stress factors. Stress on cultivated plant cells, by definition, restricts growth but may induce enzymes of secondary pathways. For example, in immobilized chilli cells, the increased accumulation of capsaicin at low oxygen concentrations was traced back to a stress phenomenon [40].

Freely suspended cells of pear, cultivated in a 5 l airlift reactor, were used to demonstrate the stress imposed on the cells by elevated concentrations of oxygen (Figure 12.6, [41]). Log-phase cells stopped growing when a certain threshold concentration was exceeded. Between days 4 and 6 the hydrolysis of sucrose proceeded but growth remained inhibited. After this long phase of adaptation, called ‘secondary lag phase’, growth and glucose consumption commenced again. Exhaust port and capillary gas chromatographic sniffing of solvent extracts revealed the presence of some odorants such as hexanal in the waste air [41]. The dissolved oxygen concentration is not a crucial factor in supporting the respiration of airlift cultured cells but a carefully controlled oxygen profile might help to establish conditions of enhanced production of volatiles.

Basic data on the use of nutrient components by bioreactor-cultured plant cells are still lacking. Strains of pear and peppermint grew very similar with eight- to ten-fold increases of dry weight within 1 week.

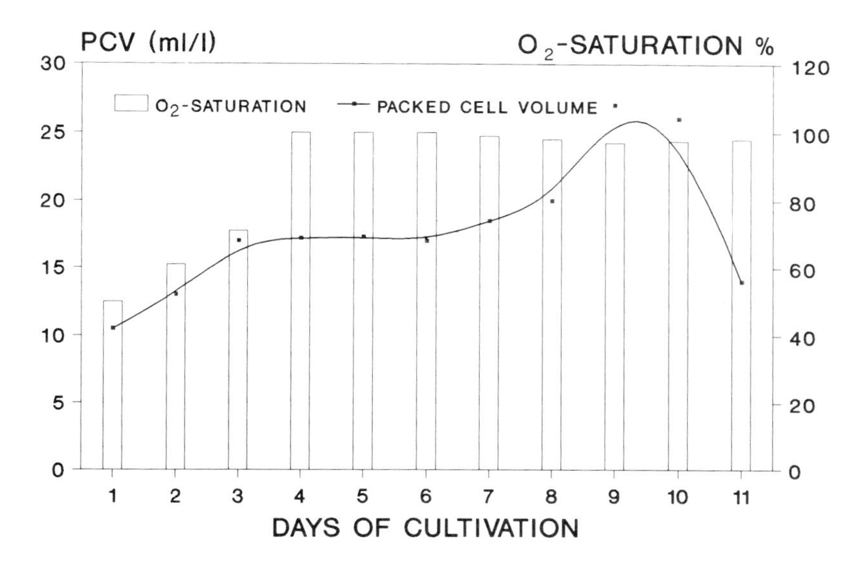

Figure 12.6 Airlift cultivated cells of *Pyrus communis* cv Alexander Lucas and the secondary lag phase effect (MS medium).

Although sucrose was completely hydrolysed after 3 days the β-fructosidase activity remained high until day 7 (Figure 12.7, [41]). Parallel cultures of suspended lemon cells grew very slowly. Based on a recently developed leaky cell model [42] a low initial activity of β-fructosidase was suggested as the cause. The measured activities were, however, twice as high, on the average, compared with pear and peppermint, and they were maintained beyond day 10 without an apparent physiological reason.

The concentrations of ionic components of the medium were monitored in the same pear suspension. With the exception of P_i, the concentrations of all ions quantified remained almost unchanged during the entire growth cycle. Less than 20 % of the initial iron concentration, and less than 10% of the calcium and magnesium was taken up, probably even less was metabolized. Nitrate and ammonia values exhibited weak increases during the log phase, and the nitrate concentration dropped, only in the stationary phase, partly resulting in a weak increase of ammonia (Figure 12.8). The sometimes slower growth of the cells at bioreactor conditions and pre-culture effects may account for a reduced consumption of nutrients; nevertheless, the fact remains that the bulk ingredients of this widely used nutrient medium were by far overdosed [41]. The adverse effect of a high substrate concentration on secondary metabolite formation was substantiated, keeping a cell line of tobacco at low levels of internal P_i [43].

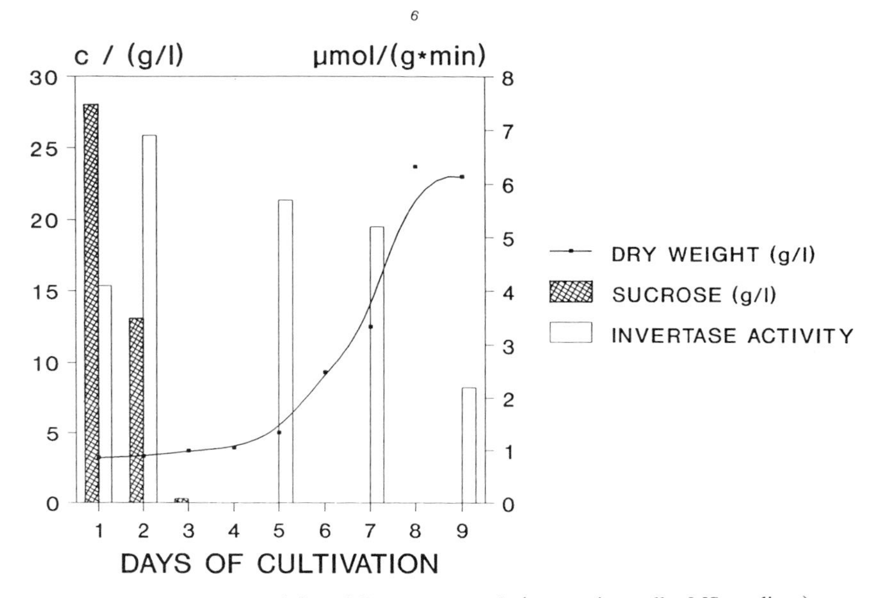

Figure 12.7 Invertase activity of *Pyrus communis* (suspension cells, MS medium).

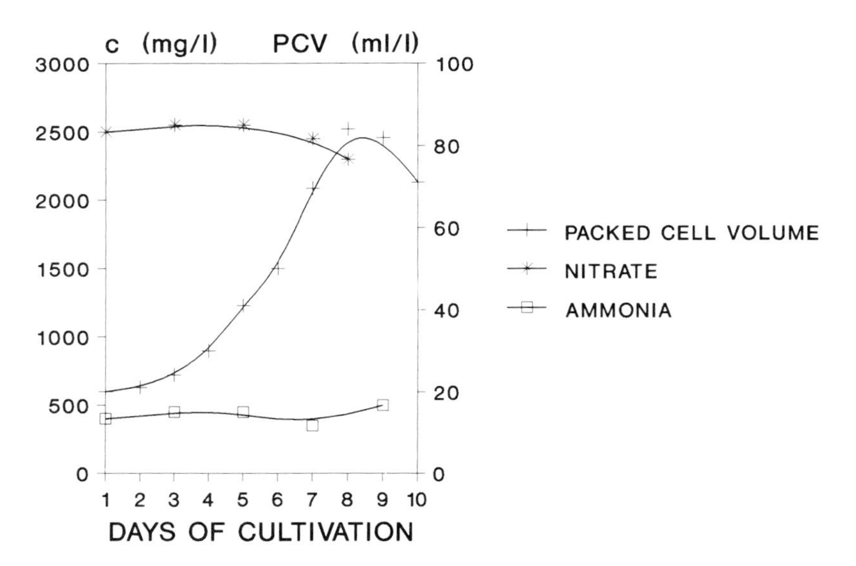

Figure 12.8 Airlift cultivated cells of *Pyrus communis* and time course of nitrate and ammonia concentrations (MS medium).

Similar effects on secondary product formation of other regulatory ionic constituents of the nutrient media, such as calcium, may be predicted.

12.7 Metabolism of monoterpenes

Since the metabolic instability and the biological roles of volatile terpenes have more widely been recognized [44], isoprenoids can no longer be considered inert end-products of plant metabolism [45]. This has led to a new look at the reasons why most suspension-cultured cells do not accumulate significant amounts of essential oil constituents. More important than the existence of active key enzymes [46] and specialized storage sites could be an insufficient removal of products from the catabolic activities inside the cell. This raises the question of how to attain a status of cytodifferentiation favouring intracellular product stability; since any kind of artificial accumulation site can only store the extracellular proportion of product and may only indirectly contribute to increased yields by shifting intracellular equilibria towards a desired product [47]. This view is supported by the often observed rapid decrease of exogenous monoterpenes upon application to regular cell cultures. In some cases, volatile conversion products were found; these undergo a transient accumulation phase and then, like the original substrate, also disappear – usually within hours [28,48–50]. The further fate of the terpenic substrates and metabolites in cell suspensions remained a matter of speculation.

To elucidate the structure of some of the intermediate degradation products, geraniol was added to suspended lemon cells [51]. A rapid catabolism was recorded and, at the same time, geranic and citronellic acid were formed, followed by a rapid accumulation of bound forms of branched and unbranched fatty acids. A concerted action of combined α/β-oxidation enzymes on the terpenic acids resulting in an enlarged pool of acetyl CoA would explain the time course of formation and the structure of the products. As the nutrient medium is a rich source of oxidases and hydrolases [23,52], this catabolism may already have begun in the extracellular 'lytic compartment' [53].

Another typical plant conversion of hydroxy compounds is glycosilation. Cyclic terpenols or terpenols devoid of a δ2 element appear to better resist an enzymic attack and are channelled into this alternative pathway of detoxification. Glycosides of terpenols are more stable and more water-soluble than the free aglyca, they may be technologically useful as flavour depot in food, and some of them are not easily amenable to chemosynthesis. The glycosylation efficiency and the distribution of added menthol was followed in peppermint suspensions at different points of the growth cycle (Figure 12.9, [54]). Conversion rates of menthol were greater

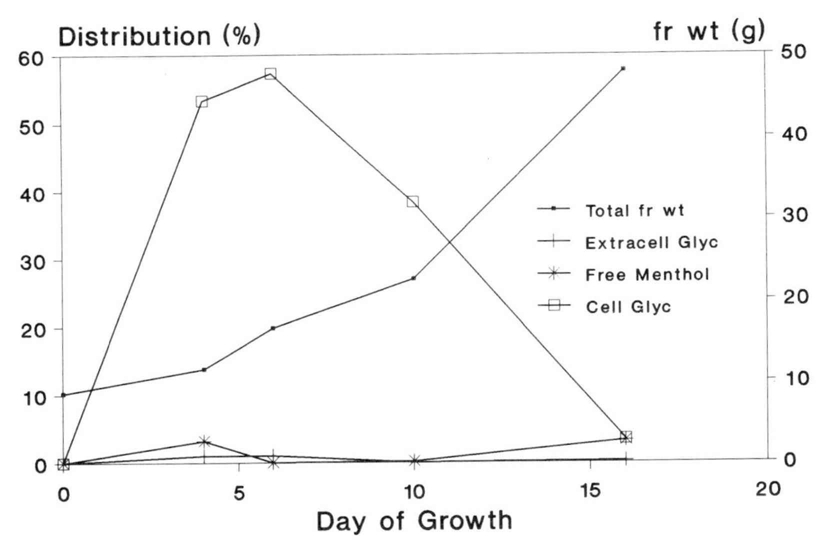

Figure 12.9 Glycosilation efficiency and distribution of added menthol in cell cultures of *Mentha piperita* (8.8 μmol menthol/gram fresh weight added at various phases of the growth cycle, 24 h each incubation period).

than 70% at optimized conditions, yielding concentrations of glycosides exceeding those reported for intact leaves of Japanese mint [55] by a factor of more than 20. Early log-phase cultures possessed the highest conversion activity. This may correlate with the activity of rather nonspecific glycosidases involved in the construction of cell wall materials at this point of growth. Catabolic pathways dominate over glycosilation as the culture ages [56].

12.8 Photomixotrophy

Quantitative determinations showed that the glycosilation of menthol and of some phenolics was stimulated by light. Evidence in the literature and physiological considerations led to the assumption that the formation of other metabolites such as volatiles could also be enhanced by light. Photoheterotrophic cells at high densities experience conditions resembling those existing in root cells. Indeed, many reports on successful secondary metabolite production in plant cell cultures refer to alkaloids, typically root-synthesized in whole plants. The synthesis of volatile terpenes, phenylpropanoids or aliphatic compounds, by contrast, is often associated with the green parts of the plant.

Pioneering work on tobacco [57] and rue [58] has established conditions

Table 12.3 Light-dependent formation of volatiles in plant cell culture (different light regimes)

Species	Year	Products
Ruta graveolens callus	1975	Ketones, esters, terpenes
Pinus radiata	1984	α-/β-Pinene
Mentha piperita	1985	Menthol
Jasminum officinalis	1986	Monoterpene alcohols
Ruta graveolens suspension	1986	Terpenes, esters, ketones
Malus sylvestris	1987	2(*E*)-Hexenal
Psidium guajava	1989	Hexanol, hexenols

for growing *in vitro* plant cells photoautotrophically. The genuine ability of a plant cell to synthesize complex organic materials using CO_2 and light is obviously extremely economic. Only few attempts have been made to exploit this potential for commercial applications. A closer look at reports on *in vitro* plant cells producing volatiles revealed that most of these 'heterotrophic' cultures were actually grown in light. More recently, quantitative comparisons of cultures grown in dark and light were published, indicating a significant effect of light on the formation of volatiles (Table 12.3, [45,59–64]). Although mechanisms other than phytochrome regulation may be involved, and although chlorophyll accumulation does not necessarily correlate with the accumulation of lower terpenoids, the essential role of light cannot be overlooked anymore.

Light-induced cytodifferentiation and related accumulation of mono- and sesquiterpenes was observed recently in mixotrophic cell cultures of *Coleonema album*, a Rutaceae [65]. In stems and leaves 41 volatiles, mainly terpenes and phenylpropanoids, were detected. Z-β-ocimene and myrcene were the major monoterpenes, while α-guajene was the most abundant sesquiterpene, and eugenol the most abundant phenolic compound.

Photoheterotrophic and mixotrophic cells on various media did not contain any volatiles. On a modified medium containing a weak auxin and a high concentration of kinetin, essential oil production was initiated in the green cells, while the dark cells remained nonproducing. β-Phellandrene and δ-selinene were the most prominent volatiles of the light culture; numerous terpenoids and unidentified volatiles contributed to the spectrum of volatiles. Thirteen volatiles were common to the intact leaf and the green cell culture. Indeed, the absolute concentrations of volatiles in the cell culture were much lower than in the differentiated organ; however, the calculated productivities of both sources differed by a factor of about five only, in favour of the plant. As a result, both the physical stimulus light and an appropriate selection and ratio of phytoeffectors were required for the generation of volatiles; a polymer trapping material provided for product accumulation. Similar light and phytoeffector conditions were reported to favour the production of cardiac glycoside in *Solanum* calli [66] and of volatile terpenes in *Mentha* plantlets [67]; again, one should avoid making immediate conclusions.

The *Coleonema* cultures did not exhibit any light-microscopically visible morphological differentiation. Electron microscopy showed that the cells contain a heterogeneous population of plastids. Chloroplasts and leucoplasts were never seen in the same cell; the osmiophilic droplets detected in some of the leucoplasts were, in an earlier study [68], assumed to represent terpenes. Heterotrophically grown parallel cultures were devoid of plastids. Indications that plastids may generally be the site of monoterpene biosynthesis in plant cells were reviewed by Kleinig [69].

12.9 Conclusion

At this time the industrial application of novel plant biotechnologies in the food industry is limited to clonal micropropagation and selection of plantlets with improved performance. First attempts to alter the sensory properties of food materials by genetic manipulations have become known. The few operating large-scale processes using submerged plant cells aim at the production of pharmaceutically interesting metabolites. A much better knowledge of the pathways of formation of plant volatiles and of the interrelated factors affecting and regulating them will be required for a more concerted development of high-yielding processes. Parameters that lead reproducibly to product accumulation and secretion need to be identified.

In vitro plant cells are up to six orders of magnitude larger than some bacteria, contain three different genomes per cell, and are derived from a highly complex macro-organism. Thus, their cultivation on the basis of data collected from bacterial bioprocesses should be discontinued. Considering the special physiology of plant cells in bioengineering, for example in developing improved gas and substrate supplies, and *in situ* product recovery will assist in establishing further bioflavour producing laboratory-scale cultures. This, in turn, could induce the degree of industrial research activity plant cell cultures ultimately deserve.

Acknowledgements

This work was in part funded by the Federal Minister of Research and Technology (BMFT, Bonn), Project No. 0318980A. F. Drawert, Z. Akkan, W. Brunner, G. Büch, L. Enders and R. Godelmann are thanked for their essential support and collaboration.

References

1. Mosandl A., Günther, C., Gessner, M., Deger, W., Singer, G. and Heusinger, G., in *Bioflavour '87*, (ed. P. Schreier), deGruyter, Berlin, 1988, p. 55.
2. Berger, R.G., Drawert, F., Kollmannsberger, H., Nitz, S., and Schraufstetter, B., *J. Agric. Food Chem.*, 1985, **33**, 232.
3. Pickenhagen, W., in *Flavour Chemistry*, (eds R. Teranishi, R.G. Buttery, F. Shahidi), *ACS Symposium Series No. 388*, ACS, Washington DC, 1989, p. 151.
4. Roels, J.A. and Thijssen H.A.C., in, *Food Biotechnology*, (ed. D. Knorr), Marcel Dekker, New York, 1987, p. 21.
5. Emberger, R., 43. Disk. Tagung FK der Ernährungsindustrie, Forschungskreis, Hannover, 1985, p. 41.
6. Schreier, P., (ed.) *Bioflavour '87*. Proceedings of an International Conference, deGruyter, Berlin, 1988.
7. Schreier, P., *Food Rev. Intern.*, 1989, **5**, 289.
8. Drawert, F., in *Bioflavour '87*, (ed. P. Schreier), deGruyter, Berlin 1988, 3.
9. Klein, R.M., *Econ. Bot.* 1960, **14**, 286.
10. Drawert, F. and Berger, R.G., *Chem. Mikrobiol. Technol. Lebensm.*, 1982, **7**, 143.
11. Koch-Heitzmann, I. and Schultze, W. in *Bioflavour '87*, (ed. P. Schreier), deGruyter, Berlin, 1988, 365.
12. Mulder-Krieger, T., Verpoorte, R., Baerheim Svendsen, A. and Scheffer, J.J.C., *Plant Cell Tissue Organ Culture*, 1988, **13**, 85.
13. Deno, H., Suga, C., Morimoto, T. and Fujita, Y., *Plant Cell Rep.*, 1987, **6**, 197.
14. Tramper, J., Wolters, I. and Verlaan, P., in *Biocatalysis in Organic Media*, (eds, C. Laane, J. Tramper, M.D. Lilly), Elsevier, Amsterdam, 1987, p. 311.
15. Davis, D.G., Frear, D.S. and Shimabukuro, R.H. *Pesticide Biochem. Physiol.*, 1978, **8**, 84.
16. Brunner, W., MS. Thesis, University of Tübingen, 1987.
17. Rekker, R.F. and deKort, H.M., *Eur. J. Med. Chem.-Chimica Therapeutica*, 1979, **14**, 479.
18. Charlwood, B.V. and Brown, J.T., *Biochem. Soc. Trans.*, 1988, **16**, 61.
19. Brown, J.T., Hegarty, P.K. and Charlwood, B.V., *Plant Sci.*, 1987, **48**, 195.
20. Hulst, A.C. and Tramper, J., *Enzyme Microb. Technol.*, 1989, **11**, 546.
21. Brodelius, P. and Nilsson, K., *Eur. J. Appl. Microbiol. Biotechnol.* 1983, **17**, 275.
22. Parr, A.J., Robins, R.J. and Rhodes, M.J.C., *Plant Cell Rep.*, 1984, **3**, 262.
23. Berger, R.G., Drawert, F., Kinzkofer, A., Kunz, C. and Radola, B.J., *Plant Physiol.* 1985, **77**, 211.
24. Hunter, C.S. and Kilby, N.J. in *Manipulating Secondary Metabolism in Culture*, (eds R.J. Robins, M.J.C. Rhodes), Cambridge University Press, 1988, p. 285.
25. Robins, R.J., Parr, A.J. and Rhodes, M.J.C., *Biochem. Soc. Trans.*, 1988, **16**, 67.
26. Knoop, B. and Beiderbeck, R., *Z. Naturforsch.*, 1983, **38c**, 484.
27. Bisson, W., Beiderbeck, R. and Reichling, J., *Planta Med.*, 1983, **47**, 164.
28. Cormier, F. and Ambid, C., *Plant Cell Rep.*, 1987, **6**, 427.
29. Becker, H. and Herold, S., *Planta Med.*, 1983, **49**, 191.
30. Godelmann, R., PhD Thesis, TU München 1985.
31. Hooker, B.S. and Lee, J.M., *Plant Cell Rep.*, 1990, **8**, 546.
32. Fowler, M.W. and Scragg, A.H., in *Plant Cell Biotechnology*, (eds M.S.S. Pais, F. Mavituna, J.M. Novais), NATO ASI Series, Ser H, Vol. 18, Springer, Berlin, 1988, p. 165.
33. Callebaut, A., Motte, J.-C., Hoenig, M., de Cat, W. and Baeten, H., in *Plant Cell Biotechnology*, (eds M.S.S. Pais, F. Mavituna, J.M. Novais), NATO ASI Series, Series H, Vol. 18, Springer, Berlin 1988, 285.
34. Kinnersley, A.M. and Henderson, W.E., *Plant Cell Tissue Organ Cult.*, 1988, **15**, 3.
35. Banthorpe, D.V. and Brown, G.D., *Plant Sci.*, 1990, **67**, 107.
36. Parfitt, D.E., Almehdi, A.A. and Bloksbeg, L.N., *Scientia Hortic.*, 1988, **36**, 157.
37. Morimoto, H. and Murai, F., *Plant Cell Rep.*, 1989, **8**, 210.
38. Schröder, R., Gärtner, F., Steinbrenner, B., Knoop, B. and Beiderbeck, R., *J. Plant Physiol.*, 1989, **135**, 422.

39. Barz, W., Daniel, S., Hinderer, W., Jacques, U., Kessmann, H., Köster, J., Otto, C. and Tiemann K., in *Application of Plant Cell and Tissue Culture*, Ciba Foundation Symposium 137, Wiley, Chichester 1988, p. 178.
40. Büch, G., MS Thesis, TU München 1989.
41. Wilkinson, A.K., Williams, P.D. and Mavituna, F. in *Plant Cell Biotechnology*, (eds M.S.S. Pais, F. Mavituna, J.M. Novais), NATO ASI Series, Series H, Vol. 18, Springer, Berlin, 1988, p. 373.
42. Frazier, G.C., *Biotechnol. Bioengin.*, 1988, **33**, 313.
43. Schiel, O., Jarchow-Redecker, K., Piehl, G.-W. and Lehmann, J. and Berlin, J., *Plant Cell Rep.*, 1984, **3**, 18.
44. Charlwood, B.V. and Banthorpe, D.V. *Progr. Phytochem.*, 1978, **5**, 65.
45. Croteau, R., *Dev. Food Sci.*, 1988, **18**, 65.
46. Banthorpe, D.V., Branch, S.A., Njar, V.C.O., Osborne, M.G. and Watson, D.G., *Phytochemistry*, 1986, **25**, 629.
47. Payne, G.F., Payne, N.N. Shuler M.L. and Asada, M., *Biotechnol. Lett.*, 1988, **10**, 187.
48. Carriere, F., Gil, G., Tapie, P. and Chagvardieff, P., *Phytochemistry*, 1989, **28**, 1987.
49. Gbolade, A.A. and Lockwood, G.B., *Z. Naturforsch.*, 1989, **44c**, 1066.
50. Berger, R.G. and Drawert, F., in *Flavour Science and Technology*, (eds M. Martens, G.A. Dalen, H. Russwurm Jr.), Wiley, Chichester 1987, 199.
51. Berger, R.G., Akkan, Z. and Drawert, T.F., *Biochim. Biophys. Acta*, 1990, **1055**, 234.
52. Berger, R.G., Drawert, F. and Kunz, C., *Plant Cell Tissue Organ Cult.*, 1988, **15**, 137.
53. Wink, M., *Naturwiss.*, 1984, **75**, 635.
54. Berger, R.G. and Drawert, F., *Z. Naturforsch.*, 1988, **43c**, 485.
55. Sakata, I. and Mitsui, T., *Agric.Biol.Chem.*, 1975, **39**, 1329.
56. Berger, R.G. and Drawert, F., Ger. Offen. DE 3718340 A1 1988.
57. Bergmann, L., *Planta*, 1967, **74**, 243.
58. Corduan, G., *Planta*, 1970, **91**, 291.
59. Nagel, M. and Reinhard, E., *Planta Med.*, 1975, **27**, pp. 151, 263.
60. Banthorpe, D.V. and Njar, V.C.O., *Phytochem.*, 1984, **23**, 295.
61. Bricout, J. and Paupardin, C., *C.R. Acad. Sci. Paris*, 1985, 383
62. Jordan, M., Rolfs, C.H., Barz, W., Berger, R.G., Kollmannsberger, H. and Drawert, F., *Z. Naturforsch*, 1986, **41c**, 809.
63. Berger, R.G., Kler, A. and Drawert, F., *Plant Cell Tissue Organ Cult.*, 1987, **8**, 147.
64. Prabha, T.N., Narayanan, M.S. and Patwardhan, M.V., *J. Sci. Food Agric.* 1989, **50**, 105.
65. Berger, R.G., Akkan, Z. and Drawert, F., *Z. Naturforsch*, 1990, **45c**, 187.
66. Nigra, H.M., Alvarez, M.A. and Giulietti, A.M., *Plant Cell Rep.*, 1989, **8**, 230.
67. Hirata, T., Murakami, S., Ogihara, K. and Suga. T., *Phytochem.* 1990, **29**, 493.
68. Carde, J.-P., Bernard-Dagan, C. and Gleizes, M., *Can. J. Bot.*, 1979, **57**, 255.
69. Kleinig, H., *Ann. Rev. Plant Physiol. Plant Mol. Biol.*, 1989, **40**, 39.

13 Natural flavours for alcoholic beverages

J. BRICOUT, P. BRUNERIE, J. DU MANOIR,
J. KOZIET and C. JAVELOT

Abstract

The production of alcoholic beverages is probably one of the oldest biotechnological processes. It has long been recognized that fermentation stabilizes and imparts new desirable organoleptical properties to sweet plant extracts, and efforts have been made to improve or modify these flavours in order to diversify the palette of the alcoholic beverages. For this reason a great variety of aromatic plants or their extracts have been used. Some spirits are flavoured with fruit extracts such as raspberries, peach, etc. Unfortunately fruit flavours generally lack intensity and it is necessary to boost the flavour with other plant extracts or natural substances produced by biotechnology. The development of the chemical industry has led to the production of nature identical flavours which are not allowed in some spirits, so it is necessary to develop analytical methods in order to detect the presence of synthetic flavouring substances. A complete understanding of the quality of the flavour needs further research with the collaboration of chemists, microbiologists and food technologists. In this way new improvements in the quality of the flavour of alcoholic drinks can be expected in the near future.

13.1 Introduction

The consumption of alcoholic beverages is a very old tradition in many countries. Wine and beer were the first products available but efforts were made in order to diversify the range of the alcoholic beverages. A new category of products was created: the aperitifs and spirits. The world market of this category was 45 467 $\times$ 10^6 litres in 1988. From 1980, it has declined by 1.4% each year. Approximately 30% of this category comprises wine or spirits, which are flavoured.

Table 13.1 indicates the market of the different flavoured alcoholic beverages. With more than 60% consumed in Europe, these products appear to reflect European taste.

Table 13.1 Market of flavoured alcoholic beverages in 1988 in 1×10^6

	Europe	North America	World
Vermouth aromatic wines	237	25	404
Gin genever	176	144	353
Liqueurs	175	200	400
Bitters/aperitifs	95	0.5	114
Anis/ouzo	182	0.4	184
Other flavoured alcoholic beverages	186	43	261

13.2 Aromatized wines

The flavouring of wine is a very old tradition, many recipes were developed, often characteristic of the origin areas. In Italy the use of aromatic herbs to flavour wine has yielded a famous category of products, namely vermouth, the name of which is derived from the name of one the aromatic herb used for flavouring the wine – *Artemisia absinthium*. There are many different types of vermouth. Traditionally they are flavoured by macerating a mixture of plants in the wine. However, some plants contain aromatic substances whose concentration must be limited for safety reasons. Table 13.2 indicates the list of the active principals limited in food and beverages [1].

Methods were developed to quantify these substances [2]. After solvent extraction, the concentrated extract is analysed by capillary gas chromatography. A typical chromatogram is represented in Figure 13.1 with the reference chromatogram of the reference substances. Since a vermouth chromatogram is complex, it is generally recommended that mass spectrometry in the selective ion monitoring mode is used to quantify these substances.

Biotechnology offers new ways for flavouring wines. Grape juice is generally fermented with strains of *Saccharomyces cerevisiae*, and it is recognized that this genus does not synthesize any volatile terpenoid compounds. Nevertheless, *Saccharomyces* spp. contain all the enzymes

Table 13.2 Provisional limits for active principals

	Beverages	Alcoholic beverages
β-Asarone	0.1 mg/kg	1 mg/kg
Safrole	<1 mg/kg	5 mg/kg
Thujones (α and β)	0.5 mg/kg	5 mg/kg in alcoholic beverages containing less than 25% of alcohol by volume; 10 mg/kg in alcoholic beverages containing more than 25% alcohol by volume; 35 mg/kg in bitters
Coumarin	<2 mg/kg	10 mg/kg in alcoholic beverages

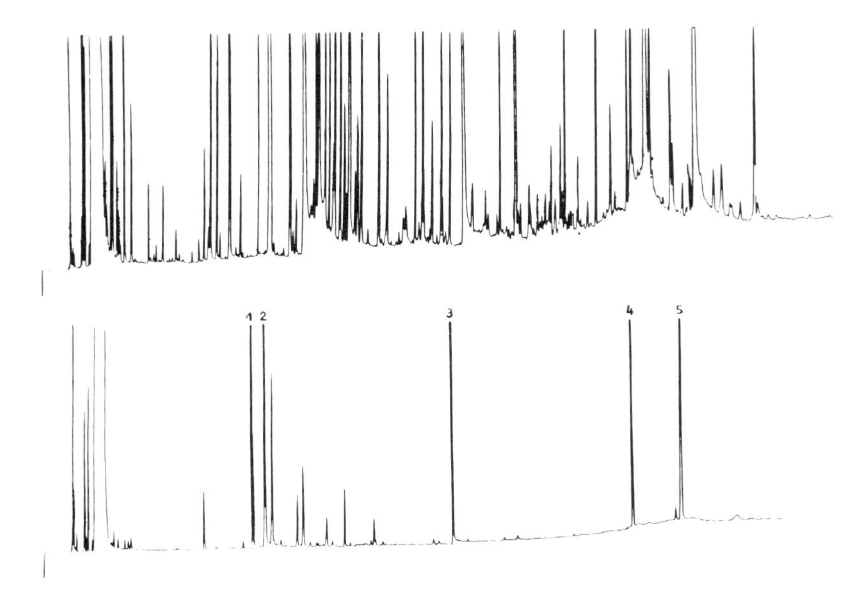

Figure 13.1 Gas chromatogram of a vermouth extract (top) and of reference substances (bottom); (1 = α-thujone; 2 = β-thujone; 3 = saffrole; 4 = β-asarone; 5 = coumarin).

catalysing the synthesis of geranyl pyrophosphate and farnesyl pyrophosphate, which are precursors of volatile terpenic compounds. Unlike in many plants these two pyrophosphates are used primarily for the biosynthesis of ergosterol and other polyprenols (Figure 13.2).

Different mutants of a strain of *S. cerevisiae* (FL 100) blocked in the sterol pathway, particularly in the prenyl transferase step (erg 20) and auxotroph for ergosterol were screened for the presence of volatile terpenes. One of them, VL 134, accumulates in the fermentation medium isopentenyl alcohol, dimethyl allyl alcohol, linalol and geraniol [3].

A spontaneous mutant of VL 134 called CM 592 was isolated and was shown to have an increased production of terpenic alcohols, particularly linalol. Figure 13.3 represents the CH_2Cl_2 extract of a synthetic medium fermented with the CM 592 strain. The presence of prenyl and terpenic alcohols appears clearly, together with normal constituents of *Saccharomyces* spp. fusel alcohols phenylethanol, whose biosynthesis is not affected by the mutation in the prenyl transferase step.

The concentration of the volatiles terpenoid alcohols obtained by fermentation of grape juice with CM 592 and the reported concentration of the same compounds in Muscat grape are shown in Table 13.3. The development of such new yeast strains opens a new field for the production of aromatic wines.

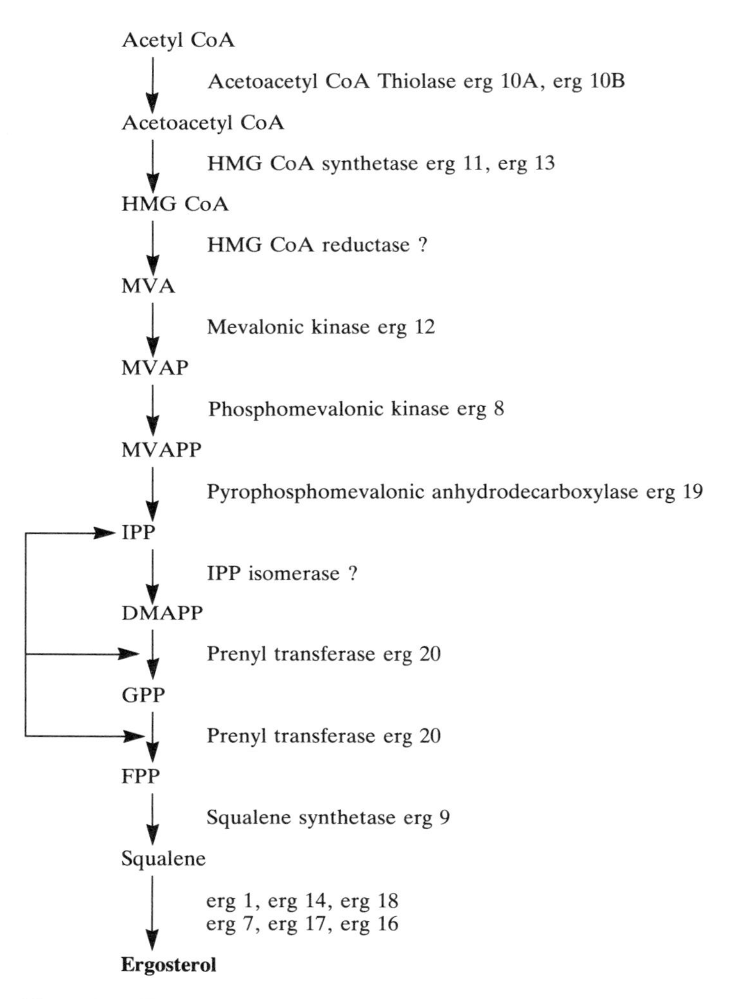

Figure 13.2 Early steps in ergosterol biosynthesis in *Saccharomyces cervisiae* and the gene–enzyme relationships.

13.3 Anise-flavoured spirits

This category of spirit was developed mainly in the south of Europe namely: Greece (Ouzo), Italy (Sambuca), south of France (Pastis), Spain (Anisette) and also in Turkey (Raki). The main aromatic component is *trans* anethole, which can be obtained from different plants: aniseed (*Pimpinella anisum* L.); star anise oil (*Illicum verum* Hooker); and fennel (*Foeniculum vulgare* Miller). However, the essential oils obtained by steam distillation of the seeds of these three plants are very different. Their chromatograms are reproduced in Figure 13.4.

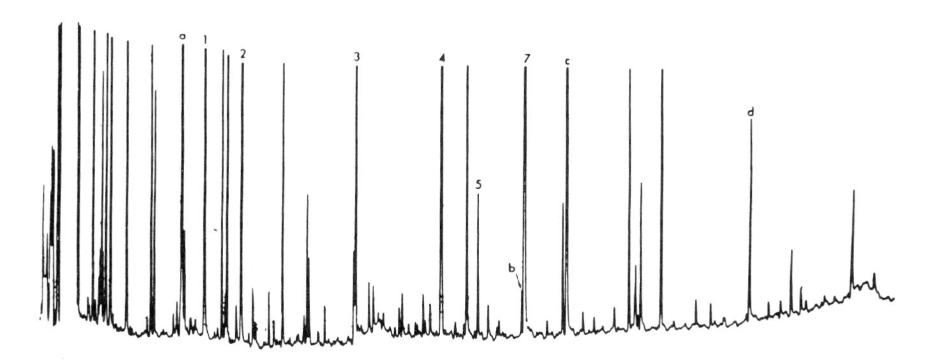

Figure 13.3 Gas chromatogram of an extract of a synthetic medium fermented with CM 592 strain (a = isoamyl alcohol; c = phenyl ethanol; 1 = isopentyl alcohol; 2 = dimethyl allyl alcohol; 3 = linalol; 4 = ISTD; 5 = citronellol; 7 = geraniol.

Table 13.3 Comparison of the amounts (mg/l) of linalool and geraniol in several Muscatel grape varieties with those found in the cultures of CM 592

	Perception threshold[a]	Muscat Alexandrie[a]	Muscat de Frontignan[a]	CM 592
Linalool	0.10	0.45	0.47	1.19
Geraniol	0.13	0.50	0.33	1.80

[a] Source: Terrier [4].

Fennel oil contains terpenic hydrocarbons and fenchone [5]. Star anise oil is characterized by linalol, sesquiterpene hydrocarbon and foeniculine [6], while aniseed contains a peculiar ester [4-methoxy-2-(transpropenyl)-phenyl-2-methyl-butanoate] [7] and sesquiterpenes hydrocarbon (γ-himachelene) [8].

These oils have very different organoleptic properties and, in some cases, they are rectified under vacuum to give a consistent determined quality.

As *Illicium verum* grows only in China and Vietnam, efforts were made to cultivate fennel as an alternative source of anise oil. Fields of fennel were therefore established mainly in France and Tasmania (Australia). However, fennel is an allogam plant, giving rise to a great variability in the fennel population. In order to improve the productivity, a programme of selection was established. Plants in the field showing interesting properties were selected, buds were picked and propagated by *in vitro* tissue culture.

This technique allowed the isolation of clones with the same genotype. Table 13.4 shows the performance of some fennel clones, which have better productivity than the original population [9], and can be used to produce new hybrids.

Pastis is the most famous anise-flavoured spirit in France. Its composition is defined by the EC regulation No. 1576/89.

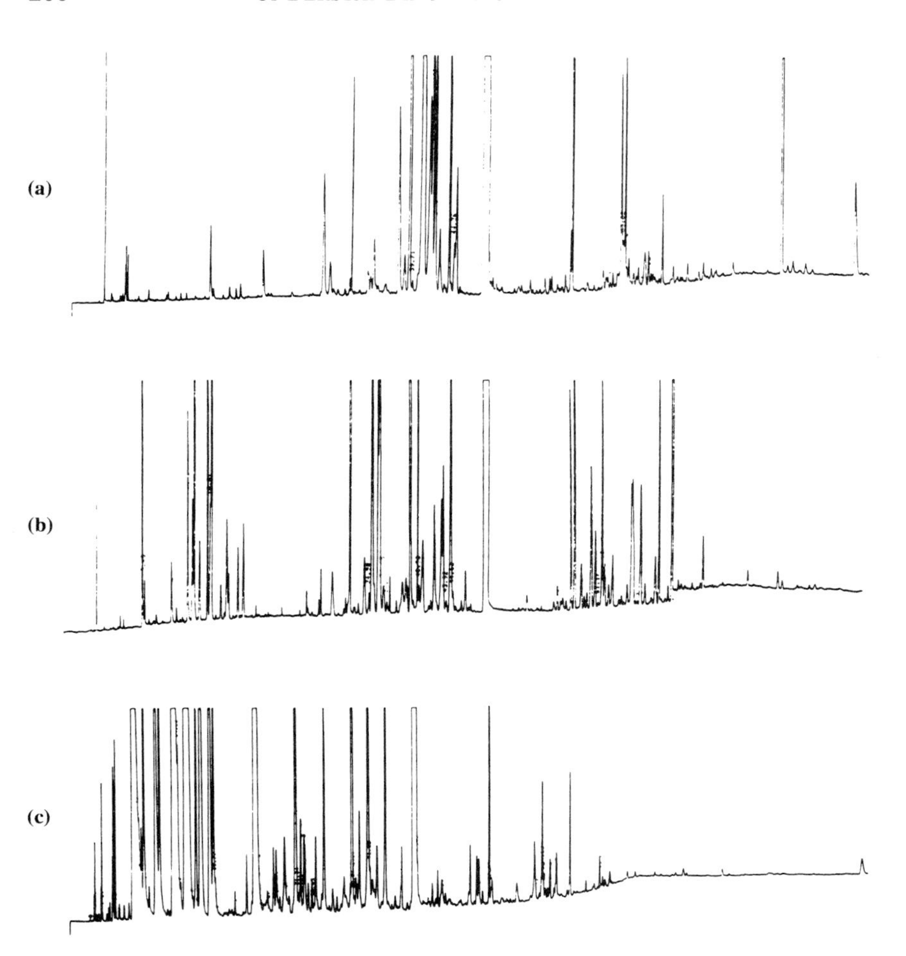

Figure 13.4 Gas chromatogram of: (a) *Pimpinella anisum* essential oil; (b) star anise oil; and (c) fennel oil.

Table 13.4 Yield features of several clones obtained through micropropagation of selected plants

Clone number	Fruit yield (g dry fruit/plant)	Anethole content (g/kg dry fruit)	Anethole yield (g/plant)
Control	83.0	43.8	3.5
34	114.0	55.4	5.9
6	97.8	63.7	6.1
16	96.6	64.7	6.3
28	94.7	68.9	6.5
12	100.2	68.2	6.5
20	142.0	58.1	6.9

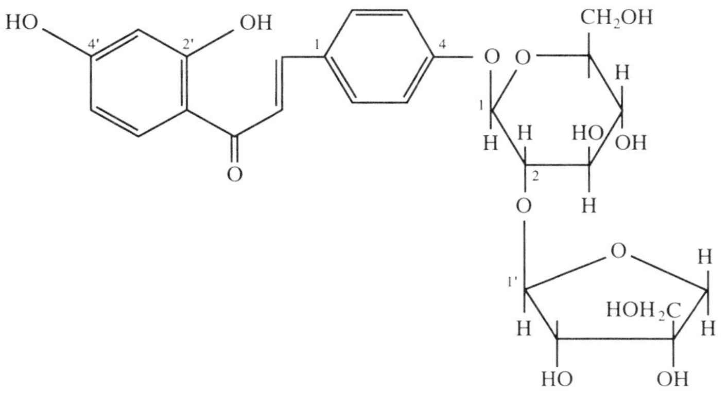

HO OH 4' 2' 1 4 O O CH_2OH H 1 H H HO OH 2 O H O H 1' HOH_2C H H HO OH

Figure 13.5 The structure of licurasid.

> For an aniseed-flavoured spirit drink to be called 'pastis' it must also contain natural extracts of liquorice root (*Glycyrrhiza glabra*), which implies the presence of the colorants known as 'chalcones' as well as glycyrrhizic acid, the minimum and maximum levels of which must be 0.05 and 0.5 grams per litre respectively. Pastis contains less than 100 grams of sugar per litre and has a minimum and maximum anethole level of 1.5 and 2 grams per litre respectively.
> (*EC, 1989*)

The particular feature of pastis is the presence of liquorice extract from *Glycyrrhiza glabra* roots. Liquorice imparts to the product a long-lasting sweet taste owing to the presence of glyccyrrhizin and a yellow colour owing to the presence of chalcones. Chalcones are glycosides of iso-liquiritegenine with glucose or glucoapiose. The structure of one of these chalcones, licurasid, is shown in Figure 13.5 [10].

The glycosyl pattern varies with the liquorice variety. For example, in the variety *typica* the glucosyl derivatives are predominant, whereas in the variety *glandulifera* the glucosylapioside derivative is more abundant. These chalcones are in equilibrium with the noncoloured flavanone.

A high performance liquid chromatography (HPLC) procedure was developed usine a diode array detector to monitor the concentration of glycyrrhizin at 254 nm and chalcones at 370 nm (Figure 13.6).

13.4 Fruit liqueurs

Many spirits are flavoured with fruit to produce liqueurs. However, fruit has a delicate flavour that is generally weak, thus it is necessary to boost the flavour with natural extracts or natural substances. For example, raspberry flavour can be enhanced by the addition of iris root extract,

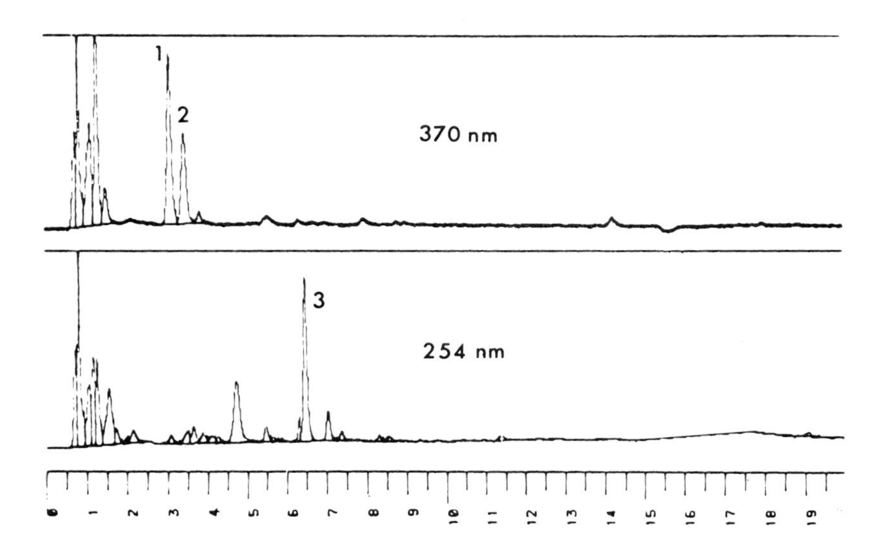

Figure 13.6 HPL chromatogram of a pastis (1, 2 = chalcones; 3 = glycyrrhizin).

which contains irone in particular. The flavour of irone is similar to the ionones naturally present in raspberries.

For blackcurrant liqueur a valuable extract is the 'bud extract'. Using CO_2 in a supercritical condition an 'essential oil' can be extracted from the bud, with an overall yield of about 0.1%. This oil has a typical blackcurrant flavour due particularly to sulphur compounds. Figure 13.7 represents the chromatogram of the supercritical CO_2 extract obtained with two detectors: FID and a chemiluminescent sulphur detector.

The high selectivity of the sulphur detector is clearly demonstrated. A sulphur compound can be detected at 22 min and may correspond to the substance identified in blackcurrant [11] (Figure 13.8).

One interesting source of blackcurrant flavour is buchu leaves (*Betulina crenulata*). This plant produces an essential oil whose chromatogram is represented in Figure 13.9 with two detection systems: an FID and a sulphur chemiluminescent detector. Two sulphur compounds are clearly detected, they are the *trans* and *cis* isomers of 8-mercapto-*p*-menthane-3-one (Figure 13.9) [12]. Buchu oil contains pulegone and their derivatives particularly diosphenols.

A new flavour, peach, introduced in spirits some years ago, has become successful. Peach flavour contains many volatile compounds. Figure 13.10 represents the chromatogram of a peach extract.

It has been demonstrated that, from a sensory point of view, lactones are character-impact compounds [13]. However, there is no other natural source of lactones than fruits. So the resources of biotechnology were used

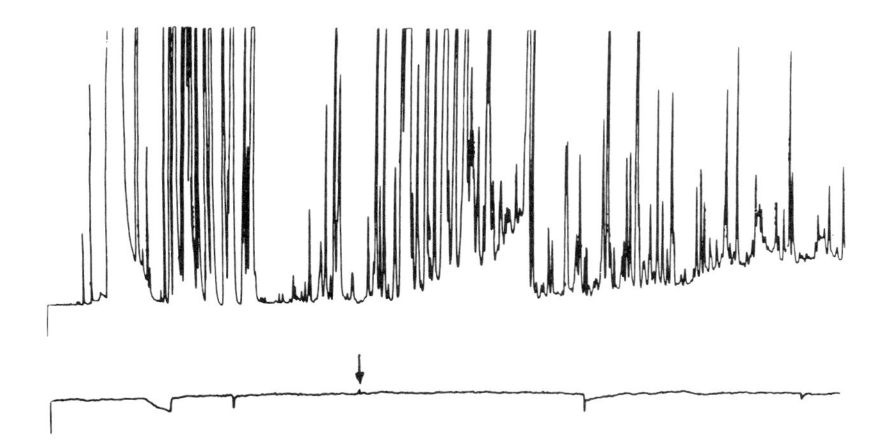

Figure 13.7 Gas chromatogram of blackcurrant bud extra (top = FID; bottom = sulphur detector).

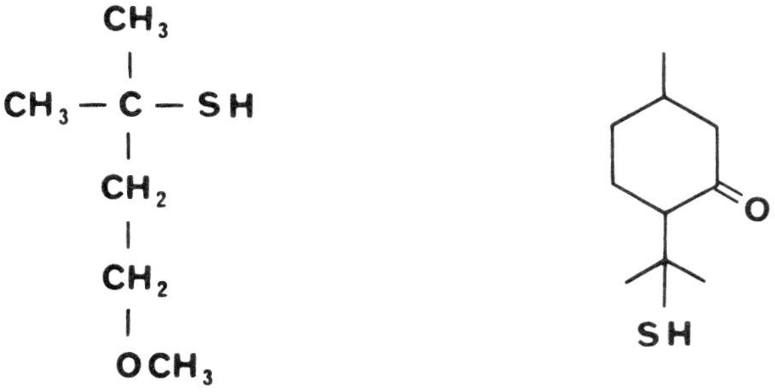

$CH_3-C(CH_3)(SH)-CH_2-CH_2-OCH_3$

SH O

Figure 13.8 Structure of the sulphur compounds of blackcurrant (left) and buchu (right).

for the production of one very important lactone, γ-decalactone. Different micro-organisms are able to β oxidize ricinoleic acid. After the elimination of 4-acetyl co-enzyme A units, 4-hydroxydecanoic acid remains in the fermented medium where it can be cyclized at acidic pH [14].

As in natural ricinoleic acid (from castor oil) the hydroxy group has the *R* configuration: γ-decalactone with an enantiomer excess of 90% *R* is produced.

Such new techniques offer the possibility to produce natural compounds that can be used for the preparation of a new generation of natural flavours with improved organoleptic quality.

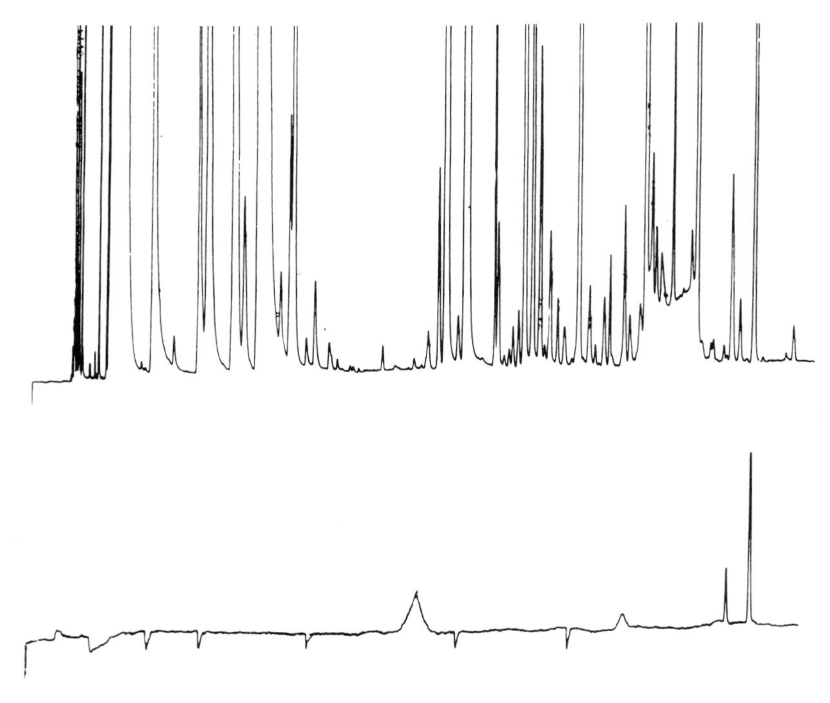

Figure 13.9 Gas chromatogram of buchu essential oil (top = FID; bottom = sulphur detector).

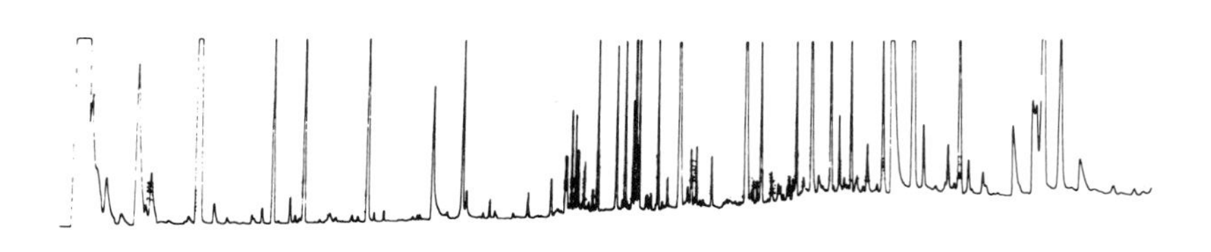

Figure 13.10 Gas chromatogram of a peach extract.

13.5 Control of the naturalness of flavours

The EC directive No. 1576/89 specifies that for many spirits only natural flavours must be used. Efforts are always made to improve the quality of these natural flavours but, as the chemical composition of plant extract is better known, it is possible to reproduce these flavours with synthetic compounds.

The so-called nature identical flavours have now a very good organoleptic quality and are cheaper than the natural ones. Therefore, it is imperative

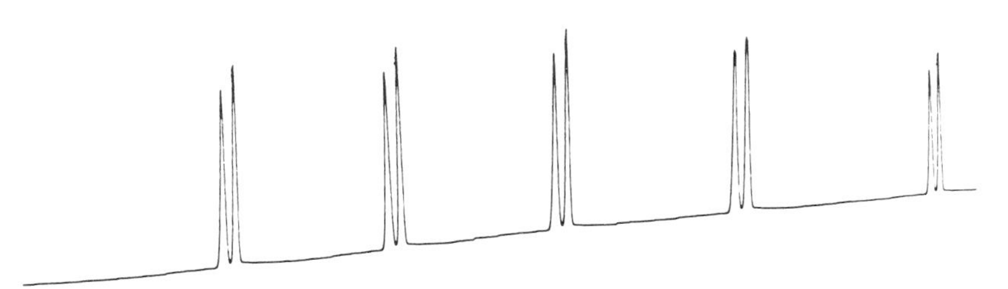

Figure 13.11 Gas chromatogram of racemic γ-lactones on a lipodex capillary column.

to develop analytical methods to detect the presence of synthetic substances in alcoholic beverages.

Volatile constituents of plants result from different enzymatic reactions, which can exhibit a high regio- or stereo-selectivity. As a consequence, some compounds exist in nature with a high enantiomer excess [15]; however, the chemical synthesis of such substances leads to racemic mixtures.

The development of chiral analysis by gas chromatography opens interesting possibilities for the detection of synthetic flavouring substances. There are two possibilities for chirality evaluation: (i) preparation of diastereoisomers using a chiral reactant and gas chromatography on the nonchiral phase; or (ii) direct gas chromatography on a chiral phase. This latter technique is better carried out using multidimensional capillary gas chromatography. The first nonchiral column allows the separation of the substance of interest, which is then transfered to a second capillary column containing cyclodextrin derivatives as the chiral stationary phase [16]. Figure 13.11 shows as an example the separation of racemic lactones from C_8 to C_{12} on a 'lipodex' column.

Studies were conducted on γ-decalactone isolated from various natural sources. It was shown that γ-decalactone occurs in nature mainly as the *R* enantiomer, except in mango where the *S* enantiomer represents 30–35% of the enantiomer [16].

Similarly natural α-ionone isolated from different plants was found to be present in approximately pure optical *R* form [17].

Studies are actually conducted in order to broaden the scope of this technique; new chiral phases are developed and the chirality of other natural flavouring components is evaluated. Unfortunately many flavouring substances have no asymmetric centre, so it is necessary to investigate the possibility of stable isotopic analysis for the differentiation between natural and nature-identical aroma compounds. Hydrogen occurs as two stable isotopes, hydrogen and deuterium. The deuterium content of any substance is expressed in per thousand of variation related to the international standard SMOW [18] according to the formula:

$$\delta\,^2\mathrm{H} = \left[\frac{[^2\mathrm{H}/^1\mathrm{H}]\ \mathrm{sample}}{[^2\mathrm{H}/^1\mathrm{H}]\ \mathrm{SMOW}} - 1\right] \times 1000$$

Table 13.5 Isotopic analysis of anethole

Sample origin	δ ^{2}H per thousand of SMOW	δ ^{13}C per thousand of PDB
Illicium verum		
China	−92	−29
North Vietnam	−87	
Foeniculum vulgare		
France	−80	−28
Australia	−84	
Synthetic (from estragole)		
USA	−50	−30

Table 13.6 Isotope content of natural and synthetic benzaldehyde

Sample origin	δ ^{2}H per thousand SMOW	δ ^{13}C per thousand PDB
Bitter almond	−118	−29.9
Plum	−146	−30.4
Benzal chloride hydrolysis	−40	−28.3
Toluene catalytic oxidation	+753	−26.7

Carbon occurs as two isotopes, ^{12}C and ^{13}C. The ^{13}C content is expressed as variation per thousand related to the international standard PDB [19] according to the formula:

$$\delta\ ^{13}C = \left[\frac{[^{13}C/^{12}C]\ \text{sample}}{[^{13}C/^{12}C]\ \text{PDB}} - 1 \right] \times 1000$$

The isotopic composition of particularly anethole was studied [20]. Results are summarized in Table 13.5.

Analysis of ^{13}C does not appear very characteristic, as in the case of vanillin, but deuterium content of anethole is a good indicator of its origin. It is particularly noticeable that the deuterium content of natural anethole is neither very dependent on the botanical species nor on the geographical area of culture.

Similar studies were conducted on benzaldehyde (Table 13.6) [21].

The chemical synthesis of benzaldehyde leads to high enrichment in deuterium, particularly during the catalytic oxidation of toluene owing to large kinetic isotopic effect. Culp and Noakes [22] have confirmed these results and have shown that similar values are obtained on cinnamaldehyde.

The usefulness of stable isotopic analysis for the control of the natural origin of flavouring substances appears very good; however, the limitation of this technique results from the sample preparation since a good isotopic measurement needs 10 mg of a pure compound.

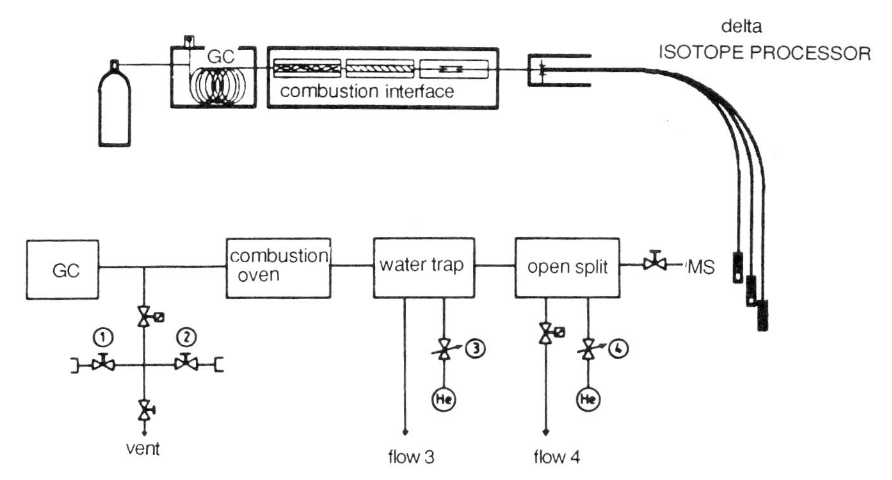

Figure 13.12 Schematic layout of the gas chromatograph isotope ratio mass spectrometer.

In order to extend the potential of stable isotopic analysis to complex flavours, the coupling of a capillary gas chromatograph with an isotopic mass spectrometer was set up. Figure 13.12 shows the layout of the system. A Finngan MAT Delta S isotopic mass spectrometer with a triple collector for masses 44, 45 and 46 is coupled on line with a capillary gas chromatograph via a combustion interface where the organic compounds are converted to CO_2. Natural extracts can be run on this system and the ^{13}C:^{12}C ratios of the different constituents can be determined with an accuracy of about 0.5‰ [23]. Figure 13.13 shows the chromatogram of a strawberry extract. The ^{13}C:^{12}C ratio trace shows a typical S shape. The reason is that the 'heavier' isotopic species elutes more rapidly from the capillary column than the 'light' species. This method has been used by the authors to measure the ^{13}C:^{12}C ratio of γ-decalactone from different origins (Table 13.7) [16]. A few off-line determinations were carried out in order to check the accuracy of the method.

It is interesting to note the difference in ^{13}C content between strawberries and the group of stone fruits where the ^{13}C content is surprisingly low. This may be due to differences occurring during the biosynthesis pathways of γ-decalactone formation in these fruits. Synthetic γ-decalactone has a higher ^{13}C content than any natural γ-decalactone including the product of microbial fermentation of natural ricinoleic acid.

13.6 Conclusion

Flavoured alcoholic beverages are 'pleasure products'. They are developed for the satisfaction of taste of the consumer who is asking for new flavour

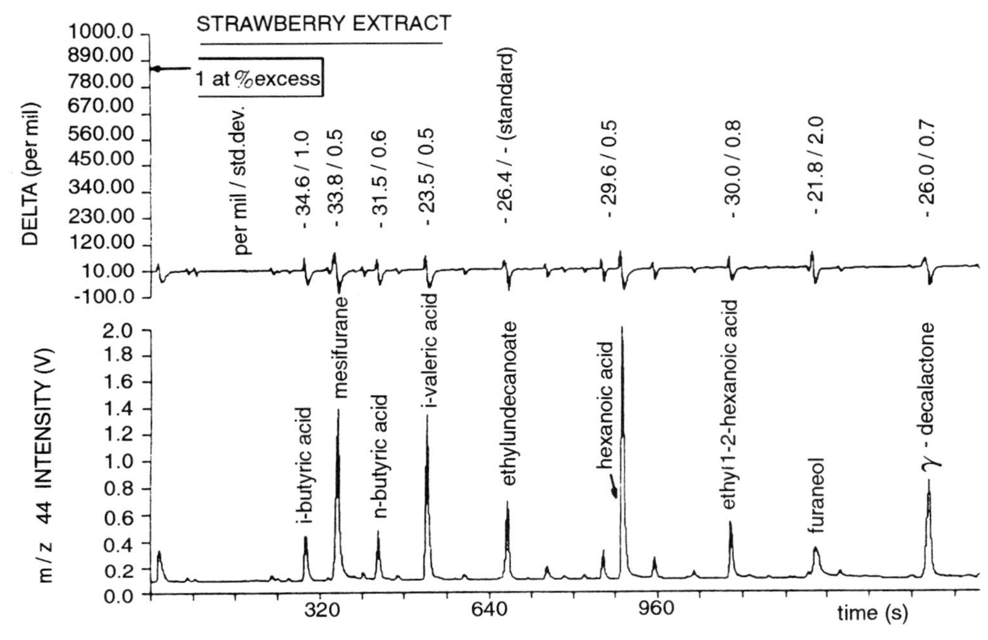

Figure 13.13 Chromatogram of a strawberry extract obtained by recording the intensity of the ionic current at mass 44 (bottom) and by recording the $^{13}C/^{12}C$ ratio (top).

Table 13.7 ^{13}C content of γ-decalactone from various sources

	δ ^{13}C percentage of PDB	
	HRCG-IRMS	Conventional
Natural		
Strawberry		
Spadeka	−28.2	
Bogota	−28.9	
Bel Ruby	−30.5	
Peach		
Italy	−40.9	
Germany	−38.5	
Plum (Mirabelle)	−39.6	
Apricot	−38.0	
Microbial (A)	−31.2	−30.8
Microbial (B)	−30.3	−30.7
Synthetic		
C	−26.9	−27.3
D	−24.4	
E	−26.0	

new taste, but whose trust remains in traditional processes where only natural flavours are used whatever the type and the strength of the flavour.

This consumer demand represents a great challenge for the industry. Methods must be developed to extract, concentrate and purify more aromatic notes. Distillation, CO_2 extraction and preparative chomatography are technologies being used; however, the resources of biotechnology are also explored in order to improve the quality of the product. The selection of new plant varieties with increased productivity and better flavour can be carried out using *in vitro* techniques. The release of flavour can be enhanced by enzymatic treatment. New processes of fermentation will be set up for the production of natural flavouring substances.

All these efforts will lead to new and better flavours; however, these efforts will be useless if no control of the 'naturalness' of the flavour components can be carried out, as the chemical industry can provide any range of flavours needed. For this reason, it is important to develop new analytical methods to allow the detection of nature identical substances even in trace quantities. Chiral analysis by gas chromatography or HPLC will continue to be used for all flavouring substances containing an asymmetric carbon. For other substances, isotopic mass spectrometry appears as a valuable tool – particularly if it can be connected on-line with multidimensional gas chromatography.

References

1. Flavouring Substances and Natural Sources of Flavourings. Council of Europe, 1981, Maison Neuve, Paris, pp. 29–30.
2. Recommended Methods of the International Organization of the Flavour Industry.
3. C. Javelot, F. Karst, V. Ladeveze, C. Chambon and B. Vladescu, in *Microbiology Applications in Food Biotechnology*, (edited by H. Nga, Y. Kilee), Elsevier, Amsterdam, 1990, pp. 101–122.
4. A. Terrier (1972) Thèse Université de Bordeaux.
5. M. Ashraf and M.K. Bhatty *J. Sci. Ind. Res.*, 1975, **18**, 236–240
6. J. Bricout *Bull. Soc. Chim. Fr.*, 1974 **9–10**, 1901–1903.
7. G.T. Carter, H.K. Schoes and E.P. Lichtenstein *Phytochem*, 1976, **16**, 615–616.
8. R. Tabacchi, J. Garnero and P. Buil *Helv. Chim. Acta*, 1974, **57**, 849–851.
9. G. Hunault, P. Desmarest and J. Du Manoir, Biotechnology, in *Agriculture and Forestry Medicinal and Aromatic Plants II* (ed. Y.P.S. Bajaj), Springer, Berlin 1989, p. 186–212.
10. D. Afchar, A. Cavé and J. Vaquette, *Plantes Medicinales et Phytotherapie*, 1980, **14** 46–50.
11. J. Rigaud, P. Etievant, R. Henry and A. Latrasse *Science des Aliments*, 1986, **6**, 213–220.
12. D. Lamparsky and P. Schudel *Tetrahedron Lett.*, 1971, **36**, 3323–3326.
13. R.J. Horvart, G.W. Chapman, J.A. Robertson, F.I. Meredith, R. Scorza, A.M. Callahan and P. Morgens *J. Agric. Food. Chem.*, 1990, **38**, 234–237.
14. US Patent No. 4.560.656.
15. A. Mosandl, C. Günther, M. Gessner, W. Deger, G. Sing and G., Heusinger, in *Bioflavour 87* (ed. P. Schreier), Walter de Gruyter, (1988) Berlin, pp. 55–74.
16. A. Bernreuther, J. Koziet, P. Brunerie, G. Krammer, N. Cristoph and P. Schreier *Z. Lebensmittel. Unter. Forsch.*, 1990, **19**, 299–301.
17. P. Werkhoff, W. Bretschneider, M. Güntert, R. Hopp and H. Surburg Chiral analysis in

flavour and essential oil chemistry. Part B, in *Flavour Science and Technology* (eds Y. Bessiere and A.F. Thomas), John Wiley & Sons, Chichester, 1990, pp. 33–37.
18. H. Craig *Geochim. Cosmochim. Acta*, (1957) **12**, 133–149.
19. H. Craig *Science*, (1961) **133**, 1833–1834.
20. J. Bricout in *Stable isotopes* (eds H.L. Schmidt, H. Forstel and K. Henziger), Elsevier Amsterdam, (1982) pp. 483–493.
21. M. Butzenlechner, A. Rossmann and H.L. Schmidt *J. Agric. Food. Chem.*, (1989) **37**, 410–412.
22. R.A. Culp and J.E. Noakes *J. Agric. Food. Chem.*, (1990), **38**, 1249–1255.
23. M. Rautenshlein, K. Habfast and W. Brand in *Proceedings of Stable Isotopes in Pediatric and Metabolic Research*, Gröningen, (1989).

14 Beer flavour

L.C. VERHAGEN

Abstract

Beer contains numerous taste and aroma compounds. Some of them are derived from the raw materials malt and hops but by far the most are formed during the brewing process, which consists of several steps. Fermentation is of particular importance. Depending on the yeast strain and the conditions more or fewer flavour compounds are formed. Hops contribute significantly to beer bitterness and beer flavour. Evidence is presented that the isometric beer bitter acids have different bitter intensities. After bottling, the beer flavour is unstable and changes considerably with time. The use of a marker compound to measure staling in research and development studies of beer flavour stability is described.

14.1 Introduction

By definition beer is a beverage that is obtained by alcoholic fermentation of a solution of water, malt and hops. Historically, beer brewing was often a spontaneous process of fermentation of starch-containing liquors but today the process is performed under strict conditions and well defined raw materials are used.

The major steps in the brewing process involve malting, brewing, wort treatment, fermentation, maturation and packaging [1]. All steps in this process play an important role in ensuring the quality of the final product as identified by the consumers in terms of colour, clarity, foam and flavour. By choosing the proper processing conditions, the brewer can monitor and control colour and ensure that the beer is bright and has a good foam head. Flavour, however, is more difficult to control because it is highly subjective and no instrumentation exists to measure the overall taste and aroma of beer. Beer flavour is affected by: (i) the choice of raw materials (e.g. barley, cereal adjuncts, hops, brewing water); (ii) processing conditions (e.g. malting, mashing, brewing, fermentation, maturation); (iii) packaging (e.g. oxygen uptake); and (iv) storage conditions (e.g. temperature, light exposure).

Depending on the choice of raw materials and the processing conditions different beers can be obtained. Pale-coloured beers are often called

Pilsner lager beers and are popular in Central Europe and the USA. In the UK ales are typical beers. They come in different strengths and colours. In recent years, low calorie beers (light or lite beers), as well as low-alcohol and nonalcohol beers have attracted much attention. The flavour of this particular class of beers is quite different from that of regular beers.

Some of the flavours of beer are present in the raw materials that are used [2]; however, most are produced during the brewing process as a result of chemical reactions, yeast metabolism [3] and sometimes, as a result of undesired accidental pick-up or staling processes.

Besides water, a main raw material for brewing is malt. Malted barley or malt contains literally hundreds of flavour compounds. Many of these are formed during kilning as a result of thermal and oxidative degradation of lipids, carbohydrates, proteins and amino acids. Depending on the kilning temperature more or less colour and flavour is formed. White malt is only slightly kilned at temperatures not exceeding 75°C. UK pale ale malt may be cured at 95°C, and dark Munich-type malts are stewed during kilning and afterwards cured at 100–110°C. Speciality malts are used in brewing to impart desired colour and flavour characteristics to beer. Prolonged roasting and kilning leads to higher levels of flavour active compounds like pyrazines [4]. Recently the use of concentrated extracts of malts has been suggested for modifying and enhancing malt flavour characteristics of beverages [5].

The laws of some European countries have precluded the use of any other raw material in beer than malted barley [6] but elsewhere unmalted cereals such as rice, maize or sugar are commonly used in addition to barley. The use of these unmalted cereals, however, creates conditions for the process that can be used in the brewhouse where malt is extracted with water in a process called mashing. Depending on the choice of raw materials and the type of beer the brewer wishes to make, basically there are two different mash procedures:

1. Decoction mashing, common in Central Europe and some parts of Germany.
2. Infusion mashing, common in the UK.

Decoction mashing is characterized by the procedure according to which portions of the mixed (mashed) raw material are slowly heated to boiling point in a separate copper and subsequently mixed with the main mash as a result of which the temperature of the mash is increased. This may be repeated several times. Decoction mashing is needed for those processes that use unmalted cereal like rice or maize. The boiling procedure breaks up the solid particles for solubilization. During the temperature rests, enzymic degradation of proteins and starch to maltose will occur.

Infusion processes, which are traditionally used in the UK, involve a temperature-programmed heating of the mash, and can only be used with

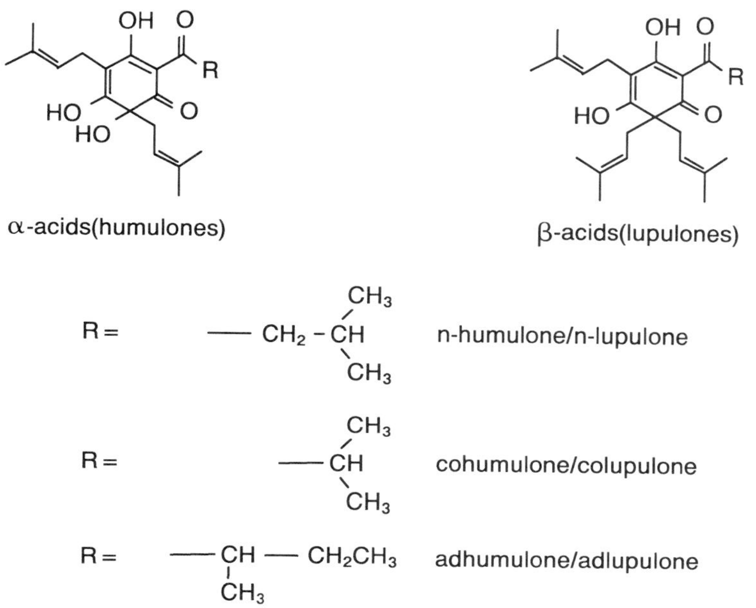

α-acids(humulones)

β-acids(lupulones)

R = $—CH_2-CH(CH_3)_2$ n-humulone/n-lupulone

R = $—CH(CH_3)_2$ cohumulone/colupulone

R = $—CH(CH_3)—CH_2CH_3$ adhumulone/adlupulone

Figure 14.1 Chemical structures of hop bitter acids.

malted cereals. The starting temperature is usually lower than in decoction mashing. The mashing procedure may affect the organoleptic character of the final product and, more particularly, the flavour stability of the beer. This will be discussed later in this chapter in more detail.

14.2 Hop flavour

14.2.1 Wort boiling

After mashing, the liquor (sweet wort) is separated from the spent grains and subsequently boiled for 1–2 h, with hops or, nowadays, hop extracts. Wort is boiled for several reasons, such as denaturing of proteins, inactivating enzymes and removing unpleasant flavour compounds. However, the main reason for boiling wort is to dissolve and convert the hop bitter acids.

The main compounds in hops are bitter resins called α-acids or humulones and β-acids or lupulones. Each group consists of three analogues, which only differ in their acyl side-chain (Figure 14.1).

During wort boiling α-bitter acids are converted into the more soluble iso-α-acids. Each α-acid gives an isomeric mixture of *cis*- and *trans*-iso-α-

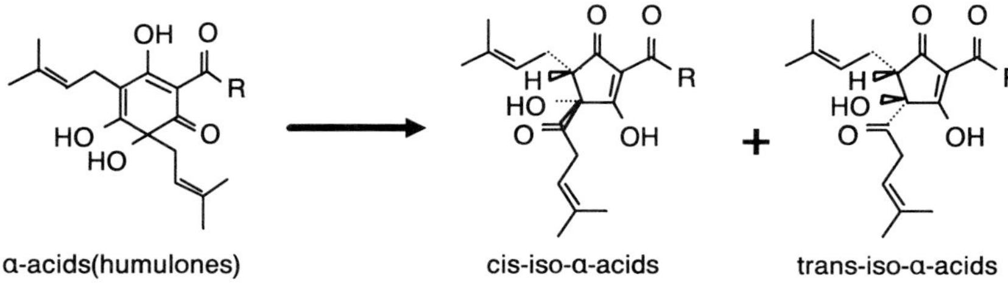

Figure 14.2 Isomerization reaction of α-acids, yielding *cis*- and *trans*-iso-α-acids.

acids (Figure 14.2). High performance liquid chromatography (HPLC) analysis reveals that isomerization in the wort copper yields a ratio of 68% *cis*- and 32% *trans*-iso-α-acid. By increasing the temperature or pressure this isomerization reaction may be speeded up. Pressure cookers for high temperature wort boiling are nowadays used in several breweries to reduce the boiling time. Reduction of boiling time, however, may affect the flavour of the final product because important flavour active compounds, such as dimethylsulphide, which is generated from its precursor *S*-methylmethionine during this step of the process, are only partly boiled out.

Different beers have a different concentration of the major bitter acids. Typical US lager beers contain levels of 12–15 mg/l, while most European lagers have about 20–25 mg/l. UK ales and stout beers may have concentrations as high as 60 mg/l. The impression of bitterness intensity depends to a great extent on the background flavour: 25 mg/l of iso-α-acids will taste much more bitter in a light lager beer than in a complex-flavoured ale or stout beer carrying substantial sweetness.

Beer bitterness is not just determined by the total concentration of iso-α-acids. From different commercial lager beers the iso-α-acid fraction was isolated and purified by means of specially developed methods [7] and redissolved in 5% aqueous ethanol. The concentrations of the iso-α-acids in the solution were subsequently determined by HPLC and, where required, adjusted to 22 ppm by dilution. These solutions were organoleptically evaluated by a trained panel in ranking tests with a solution of carefully purified *trans*-iso-n-humulone of the same concentration (two-glass test). It was concluded repeatedly that 22 ppm of *trans*-iso-n-humulone was less bitter than 22 ppm of beer iso-α-acids. As it is well known that each bitter acid in beer occurs as a mixture of *cis*- and *trans*-isomers, it was concluded, contrary to the literature [8], that there is most probably a difference in bitterness between the geometric isomers or between the iso-α-acid analogues.

By means of preparative HPLC the *cis* and *trans* isomers of iso-n- and

Table 14.1 Sensory analysis of geometric isomers of hop bitter acids

Matrix	Panel ranking[a]		Number of tasters
	trans	*cis*	
Iso-cohumulone:			
5% ethanol	22	35	19
Unhopped beer	48	66	38
Iso-n-*humulone*:			
5% ethanol	47	70	39
Unhopped beer	17	28	15

[a] Two-glass tests: 1 = least bitter; 2 = most bitter.

iso-co-humulone were separated and purified. The bitterness of solutions of these compounds in similar concentrations was subsequently evaluated organoleptically, both dissolved in 5% ethanol and in unhopped beer. It was found repeatedly that the *cis* isomers were more bitter than the corresponding *trans* isomers (Table 14.1). In a model solution, *trans*-iso-n-humulone was also more bitter than *trans*-iso-cohumulone. In unhopped beer, however, no significant difference was detected in respect of these compounds. The same applied to *cis*-iso-cohumulone and *cis*-iso-n-humulone.

These results suggest that organoleptic bitterness of beers may be controlled not only by the concentration but also by the ratio of the geometric isomers of the bitter acids. Application of specific process conditions or specific extraction methods to hops can probably be used to optimize the bitter taste of beers.

The relationship between organoleptic bitterness of beer and its analytical composition is probably much more complicated and may also involve the many hop-related minor components that can be detected in beer. These compounds may impart to the specific hop-related sensory attributes such as harshness, and bitter after-taste and so on. Interesting information about the correlation between the occurrence of specific minor hop bitter compounds in beer and particular sensory attributes may be obtained by combining new developments in analytical chemistry such as liquid chromatography-mass spectrometry (LC-MS) and capillary zone electrophoresis, techniques that have already shown promising results for the analysis of hop bitter acids [9] and modern methods for sensory analysis such as time-intensity measurement, as well as application of multivariate statistical analysis.

Hops contain not only bitter acids or resins but also essential oil as important flavour active compounds. The essential oil of hops is a complex mixture of compounds that can easily be separated into a hydrocarbon fraction (50–80% of whole oil) and an oxygenated fraction. A major part of the essential oil that is added to the wort by the hops disappears through steam distillation during boiling. Residual essential oil and related compounds, as well as volatile degradation products of bitter resins, are responsible for a significant part of the flavour of beer. Brewing trials without hops result in beer with an unpleasant, sweet flavour indicating that the hops contribute substantially to the beer flavour.

By means of gas chromatography-mass spectrometry (GC-MS) analysis, a great series of compounds that are derived from hop have been identified in beer [10]. Beers with different levels of hop flavour can be distinguished from each other by means of instrumental analysis. Irwin [11] reported that in lagers brewed with single hop varieties, a few unique flavour compounds could be detected. Significant differences were found between the concentrations of individual compounds and hop compound classes in the different beers. Correlation of sensory and instrumental data between particular hop flavour attributes in beer and certain hop oil compounds has been reported by Peppard *et al.* [12], who used the technique of partial least squares analysis, a promising statistical method to correlate sensory and analytical data, which is potentially applicable to many other areas of beer flavour research [13].

To improve the hop bouquet of beer, hops with high levels of essential oil (aroma hops) may be used and added to wort at the end of the boiling phase in order to preserve more volatiles. The UK has a long tradition of 'dry hopping'. Hops or hop extracts are added to the maturing beer either in casks or in conditioning tanks. They remain in contact with the beer for quite some time, to release original hop flavour compounds into the beer.

Alternatively, hop oil may be added to the finished beer. Although heavily flavoured beers are obtained, in practice this is difficult to reproduce and often not comparable with the flavour of kettle hopped beer. Modern extraction techniques for hops with supercritical or liquid CO_2 allow the selective isolation of bitter acids and oil, largely preserving the original compounds as a result of mild conditions and inert extraction atmosphere. These hop fractions may be used to impart a 'true' hop aroma to speciality beers.

14.2.2 Fermentation

After the wort boiling, coagulated proteins are removed and the clear wort is cooled down rapidly. This step also involves the removal of several compounds adsorbed onto the solids, such as lipids. This is particularly important to the flavour stability of the finished beer, which is a problem with beer flavour – treated in more detail later on in this chapter.

Cooled wort is aerated and pitched with yeast to start fermentation. No doubt fermentation plays a unique role in determining the beer flavour. The performance of brewing yeast (*Saccharomyces cerevisiae*) is intimately connected to beer flavour. Yeast strain and fermentation conditions are reflected in the composition of the resulting beer flavour.

Table 14.2 Yeast metabolism and beer flavour compounds[a]

Metabolism	Fermentation products
Carbohydrate	Ethanol, carbon dioxide, glycerol, organic acids
Lipids	Esters, short-chain fatty acids
Amino acids	Higher alcohols, keto acids, sulphur dioxide, hydrogen sulphide
Carbohydrate/amino acids	Diacetyl, acetoin, 2,3-butanediol

[a] Source: Quain [3].

Typically two fermentation systems are used for beer brewing: (i) top yeast fermentation; and (ii) bottom yeast fermentation, that is, strains of yeast that at the end of fermentation tend to rise to the top of the tank and form a yeast head and strains that sediment on the bottom of the vessel towards the end of fermentation.

Bottom yeast fermentations are quite common in Europe and the USA. In the UK and Germany top fermentation is used for specific beers like ales and wheat beers, although in the UK, the popularity of bottom-fermented lager beers has strongly increased.

A wide range of conditions for fermentation are in use in different countries, with differing temperature, time of fermentation, pitching rate and so on. Traditionally, infusion mashed worts are top-fermented (in the UK ales and stout beers), although this is not a prerequisite. Top fermentation is quicker because of the relative high temperatures (15–20°C), providing beers with a more intense flavour than typical lager beers that are bottom fermented at temperature of 5–10°C. For speciality beers specific strains of yeast may be used. A well known example is the production of wheat beer. A specific top-yeast strain produces relative high levels of 4-vinyl guaiacol by decarboxylation of ferulic acid. Typical concentrations of over 3 mg/l can be found in German wheat beer [14]. These concentrations are well above the taste threshold level of about 1 ppm and it is therefore a character-impact compound of wheat beer. This compound, which gives a typical phenolic flavour to beer, is recognized in bottom-fermented lager beers as an off-flavour, indicating infection with wild yeast strains or bacteria [15].

Besides alcohol and CO_2, major products of fermentation are glycerol, and new yeast mass. Yeast growth is closely linked to the formation of flavour compounds, nevertheless yeast will also adsorb several substances, especially significant amounts of hop bitter acids, thus removing them from the final beer.

Fermentation products are produced in a wide concentration range from g/l (e.g. ethanol, CO_2 and glycerol) to μg/l or less. The most important fermentation products originating from yeast metabolism that are significant to beer flavour are given in Table 14.2.

Process changes, such as temperature increases, can result in faster fermentation accompanied by increased synthesis of esters. Fermentation in pressure tanks, which is common in many modern breweries, reduces the yeast growth and thereby the ester synthesis. Top-fermented beers have a more pronounced flavour pattern owing to the higher fermentation temperature. The appearance of esters in beer accompanies the synthesis of yeast lipids. Of the multitude of esters so far identified in beer, ethyl acetate and iso-amyl acetate are quantitatively the most important ones, giving a characteristic fruity and flowery flavour.

Carbonyls are another important group of flavour substances. Flavours of several of these compounds are described as grassy, cardboard, papery and so on. During fermentation, aldehydes are reduced to their corresponding alcohols, therefore aldehydes in beer occur only in minor concentrations, except for acetaldehyde (3–10 mg/l). This is particularly important as aldehydes are flavour-active compounds with relative low taste-threshold levels in beer [16]. Important secondary yeast metabolism products of this class are also the vicinal diketones. Especially diacetyl (2,3-butanedione) has a very low taste-threshold level (0.05–0.10 mg/l) and is formed by decarboxylation of 2-acetolactic acid, a temperature- and pH-controlled reaction. Most of the generated diacetyl is reduced to 2,3-butanediol by yeast enzymes. At higher fermentation temperatures, the decomposition of 2-acetolactic acid will be faster. Sulphur dioxide (SO_2) is also an important fermentation by-product, strongly contributing to the flavour stability of beer by masking flavour-active compounds such as carbonyls. Control of native SO_2 is an important tool for brewers to improve the consistency of beer flavour and stability. Depending on fermentation conditions up to 20 ppm of SO_2 may be formed during fermentation.

14.2.3 Maturation

After the main fermentation, most of the yeast has settled and separated from the beer and a product called ‘green beer’ is obtained. It may contain a relative low CO_2 concentration and taste and aroma are still inferior compared with usual beer. Therefore maturation or lagering (German for ‘storage’) is an essential step in the process. It is carried out in closed tanks at low temperatures.

A typical storage temperature for lager-type beers is around 0°C. Higher temperatures may be used for ales. As mentioned before, traditionally ale beers in the UK are matured with hops that are added to the cask or tank (dry-hopping) to improve the natural hop flavour of beer. During maturation, a secondary fermentation by the small amount of suspended yeast, which is still in the beer after the main fermentation, generates additional flavour compounds and any CO_2 that is produced in this phase will dissolve into the beer.

Several changes in the beer flavour pattern also occur during this phase. Acetaldehyde, which is responsible for the green beer flavour, decreases strongly. Esters, the main flavour-carriers of beer, will increase in concentration, owing to a reaction of alcohols with short-chain fatty acids, and the level of vicinal diketones is reduced further. Maturation time depends on beer type and temperature. Dark beers require relatively short maturation times compared with pale-coloured lager beers. Recently the use of immobilized yeast is described in order to reduce the lagering time of beer [17].

After maturation, the beer has to be clarified. Ale beers were often stored in casks together with finings to precipitate solids. Lager beers are filtrated. The filtered bright beer can be placed either in kegs, bottles or cans. In order to ensure microbiological safety, the beer may eventually be pasteurized, although pasteurization may give some initial flavour deviations that disappear with time. For this reason, sterile filtration has replaced pasteurization in parts of Europe.

The final product is an enormously complex liquor containing numerous taste and aroma compounds. Today over 600 compounds have been identified; in the next few years this list will probably increase. Modern analytical instrumentation will enable the scientist to detect and identify many of the minor compounds that are still unknown today. In general, beers with different flavours contain only different concentrations and proportions of the same compounds, rather than novel or unique compounds. Accidental contamination of beer, however, may result in new compounds, the presence of which often leads to off-flavours [18].

Beer flavour cannot be reasonably reproduced, not even by means of a complex mixture of specific compounds. Except for certain speciality beers, no character-impact compounds have so far been discovered. Examples of correlation between analytical and sensory data of beer flavour are limited [19,20], although progress in this field is to be expected – especially in the field of hop flavour and bitter taste.

Although the identification and testing of the contribution of compounds to beer flavour are very important, it is also useful to analyse marker compounds indicating certain sensory qualities of a product, as demonstrated in the author's research into flavour stability of beer.

14.3 Flavour stability

The flavour of beer changes during storage, limiting its shelf-life. Flavour instability has been the subject of studies for a long time. Changes in flavour can already start during pasteurization owing to the (short time) thermal stress applied to the beer. Exposure to light will also result in flavour changes.

A characteristic off-flavour described as sulphury or skunky, called 'sunstruck flavour', quickly appears when beer is exposed to visible light and is attributed to the formation of 3-methyl-2-butene-1-thiol by photolysis of iso-α-acids in the presence of sulphur-containing amino acids [21]. Oxidation of beer by oxygen diffusion via the crown cork or by oxygen trapped in the container may also change the beer flavour.

A particular flavour instability of beer is recognized in the development of a papery, cardboard-like flavour, mainly detected in lager beers with rather low flavour profiles. These beers have good drinkability and a delicate flavour but are more susceptible to off-flavour detection. For a long time it has been known that, parallel to the development of stale flavour, an increase in carbonyl compounds can be found in beer. Especially higher unsaturated aldehydes with 6–12 C atoms are flavour-potent compounds with very low taste-threshold levels in beer. Although many compounds change in concentration or are formed during storage of beer, it is readily accepted that the increase of one particular compound, *E*-2-nonenal, above its threshold concentration (about 0.1 μg/l) is the major source of initial stale flavour of beer that is described by attributes such as leather-like, cucumber-like, papery and cardboard-like [22].

As a result, studies have focused on *E*-2-nonenal as a marker for staling, using this component in research and development studies on the flavour stability of beer [23], which is known to differ, not only in different beer types but also in the same beer type that is brewed in different breweries. In Figure 14.3 the formation of *E*-2-nonenal in lager beers from four different breweries is plotted as a function of storage time, showing significant differences between breweries.

It is well documented that the oxidative degradation of unsaturated fatty acids such as linoleic acid results in flavour-active carbonyl compounds with 6–12 C atoms [24]. Historically, formation of *E*-2-nonenal was attributed to the thermal decomposition of trihydroxy-linoleic acid, a compound found in beer in significant concentrations [25]. Indeed, when heating this compound in aqueous solution of $pH < 3.0$, large quantities of *E*-2-nonenal may be generated. Beer, however, has a pH of about 4.2. As a result, no decomposition of this compound in beer is found.

Several other oxidation products of unsaturated fatty acids, specifically hydroperoxides of linoleic acid (9-LOOH), have also been suggested as precursors for nonenal, although these rather labile compounds have not yet been detected in wort or beer.

In the brewing process oxidation of fatty acids is likely to occur in two different ways: via an enzymatic route as well as by autoxidation (Figure 14.4). Fatty acids, which have been liberated by the hydrolysis of lipids, are readily oxidized in the presence of the enzyme lipoxygenase, an enzyme common to many cereals and vegetables. Significant activity of this enzyme can be detected in germinating barley. During kilning of the malt, the

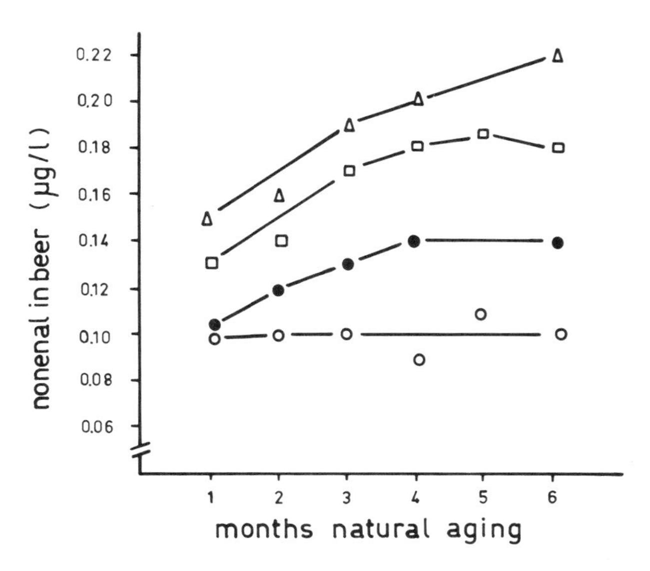

Figure 14.3 Formation of *E*-2-nonenal in different breweries as a function of storage time at room temperature [23]. △ = Brewery 1; □ = brewery 2; ● = brewery 3; ○ = pilot brewery.

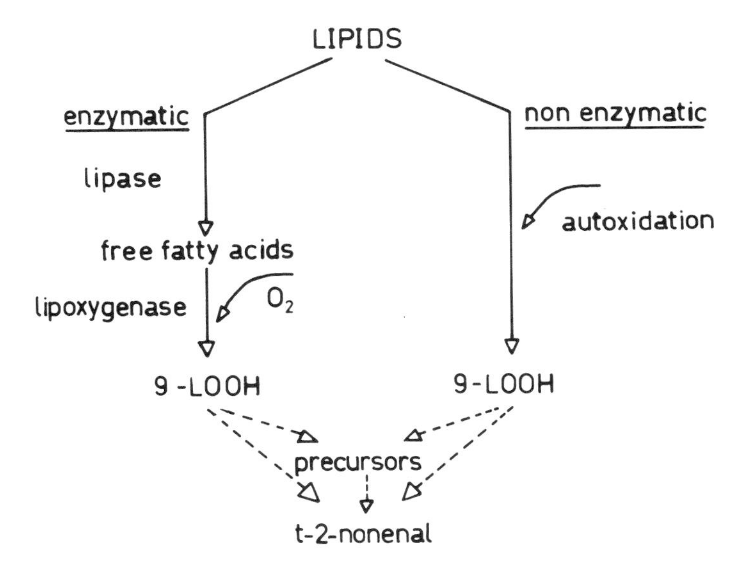

Figure 14.4 Formation of *E*-2-nonenal from precursors of lipid oxidation (9-LOOH = 9-hydroperoxide of linoleic acid).

Table 14.3 Barley variety and malt lipoxygenase activity[a]

Variety (crop 1988)	Lipoxygenase activity (ΔE_{234}/min)
Natascha (UK)	0.108
Triumph (Netherlands)	0.076
Hassan (Spain)	0.032

[a] Source: Drost *et al.* [23].

enzyme is largely destroyed; however, depending on the kilning temperature, the enzyme activity is still detectable in several malts. This particularly applies to pale-coloured malts used for pale-coloured lager beers as these malts are kilned at relatively low temperatures. In malts from different varieties of barley kilned at the same temperature in a pilot malt house, different levels of residual lipoxygenase activity were found (Table 14.3), suggesting varietal or batch dependence.

Residual lipoxygenase activity in malt may catalyse oxidation of fatty acids during mashing, resulting in a mixture of oxygenated, increased water-soluble products that may survive the brewing process. In beer (with a lower pH) these oxygenated fatty acids are expected to decompose by thermal degradation and form, among many other compounds, *E*-2-nonenal.

In order to test this hypothesis, a series of pilot-scale brewing trials were made with an experimental series of malt kilned at different temperatures (75°C and 85°C). The resulting wort was acidified to the beer pH and subsequently heated in an inert atmosphere for a certain period of time. The amount of nonenal that was generated was determined afterwards using HPLC [26]. A relatively good correlation (r = 0.914) was found between this so-called nonenal potential and the lipoxygenase activity of the malt (Figure 14.5).

In a similar way, the relationship between the nonenal potential of wort produced in production breweries and the development of free nonenal in the corresponding beers, which were stored at room temperature, was investigated. Beer from worts with a high relative nonenal potential showed a stronger increase in free nonenal on storage than beer from worts with a low nonenal potential (Figure 14.6).

The experimental beers of the brewing trials described that were stored at room temperature for 6 months were also evaluated by a trained taste panel in ranking tests at regular intervals. Although the increases of the nonenal concentration with time are organoleptically difficult to determine because of the simultaneous increase of other stale flavours, again beers from worts with high nonenal potential rated higher in the taste tests (equivalent to more staling) and *vice versa* (Figure 14.6c). This is an indication that nonenal may be used as a marker for the staling of beer and the nonenal potential of wort as a tool for early prediction of the flavour

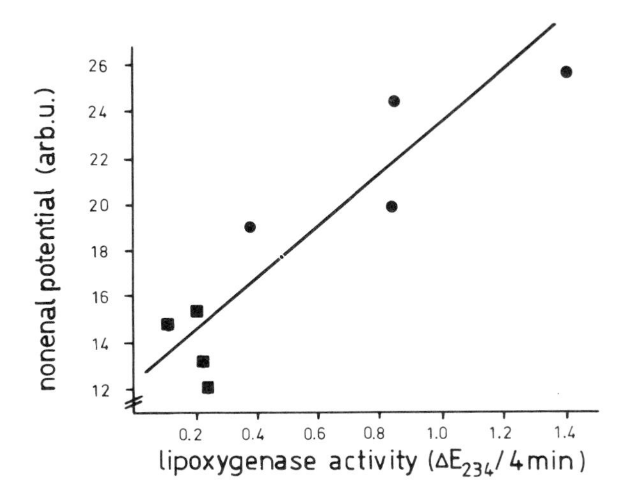

Figure 14.5 Relationship between lipoxygenase activity of malt and nonenal potential of wort [23]; (kilning temperatures: ● = 75°C; ■ = 85°C).

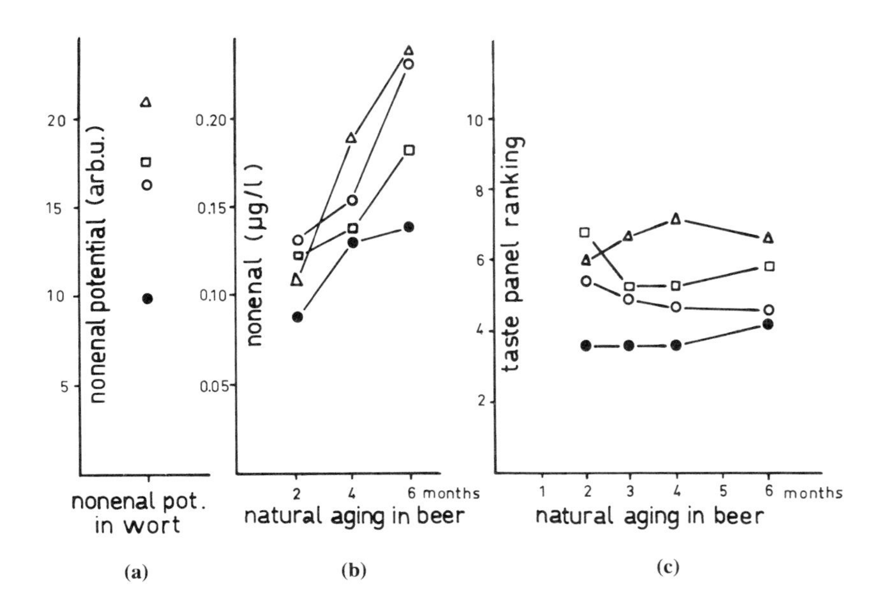

Figure 14.6 Relationships between (a) nonenal potential of wort, (b) free nonenal developed in beer in storage at room temperature and (c) panel ranking [23]. ▲ = Blank; ○ = high mashing temperature; □ = high kilning temperature; ● = high kilning plus high mashing.

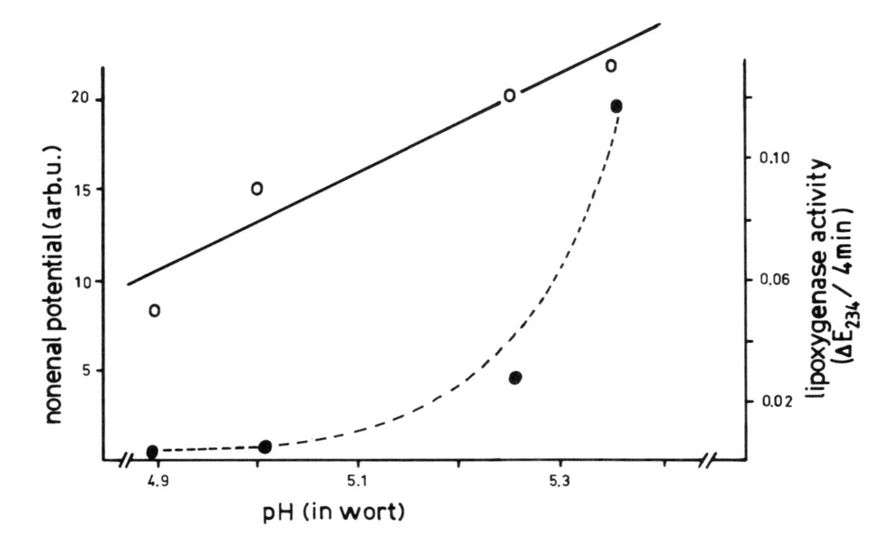

Figure 14.7 Nonenal potential in wort (○) and lipoxygenase activity (●) as a function of the mash pH [23].

stability of beer in experimental work in this field. It must, however, be realized that it is a relative method, limiting its use to a series of experiments only.

The effect of residual lipoxygenase activity in malt may be diminished by choosing mashing conditions beyond the optimum pH for lipoxygenase activity (pH = 6.0) or mash temperatures that inactivate the enzyme. Decreasing the pH in a 5 l laboratory mash indeed had a marked effect on the nonenal potential and the lipoxygenase activity in the wort (Figure 14.7).

Increasing the mash temperature also decreased the nonenal potential. However, as proper mashing is essential for brewing in respect of the yield of saccharification of starch into fermentable sugars, lowering the mash pH or strong increasing of mash temperature had unacceptable implications for the yield of extract obtained as well as for the flavour of the final beers. Nevertheless, combining slightly increased mash temperatures with high kilning temperature for malt resulted in relative flavour-stable beers, with acceptable overall flavour (Figure 14.6c).

As already explained, the lipoxygenase activity is quite high in germinating barley. As a consequence, a series of unsaturated aldehydes including nonenal is already found in malt and partly also in pitching wort, containing up to 2 μg/l of free nonenal, that is, 20 times above the taste-threshold level in beer. This *E*-2-nonenal is, however, efficiently reduced during fermentation by yeast reductases. In laboratory fermentations the

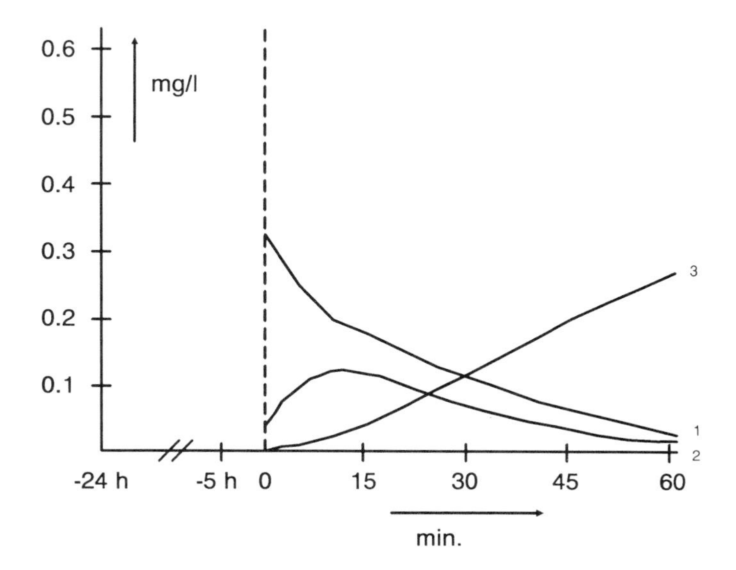

Figure 14.8 Reduction of *E*-2-nonenal added to fermenting wort as a function of time. 1 = *E*-2-nonenal; 2 = 2-nonenol; 3 = 1-nonanol.

author spiked fermenting worts 24 h after pitching with *E*-2-nonenal (0.4 mg/l). This amount of *E*-2-nonenal was reduced below 0.1 mg/l within 60 min and within 15 h below the taste threshold level of 0.11 μg/l. At first, the carbonyl function was reduced, yielding 2-nonenol; subsequently the double bond in the carbon chain was also reduced resulting in 1-nonanol, a compound much less flavour-active. Figure 14.8 shows the course of the compounds in the first 60 min, measured by gas chromatography. A further decrease of *E*-2-nonenal could be measured by applying an especially developed HPLC method [26]. From these results it was concluded that it is not to be expected that free *E*-2-nonenal from wort will pass into the beer.

It is well known that carbonyls readily form adducts with bisulphite to form flavour inactive compounds. In the literature it was suggested that *E*-2-nonenal, which is present in the wort, is partly complexed by SO_2, which is also produced during fermentation, so protecting it against reduction. Slow decomposition of this complex in beer should contribute to the increase of free nonenal in beer during storage [27]. Based on the author's work on the speed of reduction of nonenal, giving equimolar amounts of 1-nonanol, compared with the slow start of SO_2 production in fermenting wort (about 48 h after pitching) [28] and evidence about the slow rate of formation and the stability of nonenal sulphite complexes, this mechanism is not believed to be the main route for nonenal formation in beer although the author has no evidence to support this yet. Conversely, it is well known that SO_2 is definitely prolonging the shelf-life of beer in terms of flavour stability. Control of the amount of native SO_2 produced during fermentation is therefore of the utmost importance to obtain beers with high flavour stability [28]. The role of sulphite will therefore be subjected to further study in the author's research on flavour stability of beer.

14.4 Conclusions

Beer flavour is the result of a complicated process of biochemical and chemical reactions leading to a complex product with a delicate flavour balance, composed of hundreds of natural compounds. The flavour of beer, as experienced by tasters, depends strongly on conditions such as the strength of the beer, the presence of dominating (character-impact) compounds, the temperature and the sensitivities of tasters to different compounds that are known to vary dramatically between individuals.

By using the same raw materials it is possible to brew different types of beer by choosing selected process conditions and yeast strains. The rapid progress in instrumental analyses as well as sensory methods and application of multivariate statistical techniques [13] will provide us with more specific knowledge about the effect of process conditions and raw materials, as well as the significance of flavour-active compounds that play a key role in beer flavour and beer stability. This will enable the brewer to control the process, maintain quality and develop new products.

Acknowledgements

The author wishes to thank the directors of Heineken Technisch Beheer B.V. for permission to publish this paper and Dr F.J.M. Freyee for assistance in preparing the text and the diagrams.

References

1. Briggs, D.E., Hough, J.S., Stevens R. and Young, T.W., *Malting and Brewing Science.* Vols 1 and 2, Chapman and Hall, London, 1981
2. Moir, M., Effects of raw materials on flavour and aroma. *Brewers Guardian,* 1989, **118**(9), 64–67.
3. Quain, D., Fermentation and its effect on flavour and aroma. *Brewers Guardian,* 1989, **118**(10), 24–30.
4. Seaton, J.C., Malt types and beer. *Proceedings of the European Brewery Convention Congress*, Madrid, 1987, pp. 177–188.
5. Morris, T.M. and Towner, C.W., Flavour from speciality malts, in *Distilled Beverage Flavour,* (eds J.R. Piggott and A. Paterson), Ellis Horwood, Chichester, 1989, pp. 210–216.

6. Narziss, L., The German beer law, *J. Inst. Brew.,* 1984, **90**, 351–358.
7. Steenbeeke, G., Organoleptic and analytical research of hop bitter acids. PhD Thesis State University of Ghent (Belgium), 1989 (in Flemish).
8. Verzele, M., Jansen, H.E. and Ferdinandus, A., Organoleptic trials with hop bitter substances. *J. Inst. Brew.*, 1970, **76**, 25–28.
9. Vindevogel, J., Sandra, P. and Verhagen, L.C., Separation of hop bitter acids by capillary zone electrophoresis and micellar electrokinetic chromatography with UV-diode array detection. *J. High Res. Chromatogr.*, 1990, **13**, 295–298.
10. Murakami, A., Chicoye, E. and Goldstein H., Hop flavour constituents in beer headspace analysis. *J. Am. Soc. Brew. Chem.*, 1987, **45**, 19–23.
11. Irwin, A.J., Varietal dependence of hop flavour volatiles in lager. *J. Inst. Brew.*, 1989, **95**, 185–194.
12. Peppard, T.L., Ramus, S.A., Witt, C.A. and Siebert, K.J., Correlation of sensory and instrumental data in elucidating the effect of varietal differences on hop flavour in beer, *J. Am. Soc. Brew. Chem.*, 1989, **47**, 18–26.
13. Peppard, T., The use of principal component analysis in monitoring the quality of beer, in *Modern Methods of Plant Analysis, Vol. 7 Beer Analysis* (eds. H.F. Linskens and J.F. Jackson), Springer Verlag, Berlin, 1988, 264–279.
14. Narziss, L., Miedaner, H. and Nitzsche F., Ein Beitrag zur Bildung von 4-Vinyl-Guajakol bei der Herstellung von bayerischem Weizenbier. *Monatschrift f. Brauwissenschaft,* 1990, **3**, 96–100.
15. Ryder, D.S., Murray, J.P. and Stewart, M., Phenolic off-flavour problem caused by *Saccharomyces* wild yeast. *Techn. Q. Master Brew. Assoc. Am.* 1978, **15**, 79–86.
16. Meilgaard, M.C., Flavor chemistry of beer. Part 2: Flavor and threshold of 239 aroma volatiles. *Techn. Q. Master Brew. Assoc. Am.*, 1975, **12**, 151–168.
17. Gronqvist, A., Pajunen, E. and Ranta, B., Secondary fermentation with immobilized yeast-industrial scale. *Proceedings of European Brewery convention,* Zurich, 1989, 339–346.
18. Silk, N.A., Contaminants, taints and off-flavours in beer. *Anal. Proc.*, 1989, **26**, 428–429.
19. Strating, J. and Drost, B.W., Limits to beer flavour analysis, in *Frontiers of Flavour,* (ed. G. Charalambous), Elsevier Science Publishers, Amsterdam, 1987, 109–121.
20. Lukes, B.K., McDaniel, M.R. and Deinzer M.L., Isolation of aroma components from beer, wort and malt by combined sensory and analytical instrumental techniques. *Proceedings of the Institute of Brewing Convention,* Brisbane, 1988, 105–108.
21. Gunst, F. and Verzele, M., On the sunstruck flavour of beer. *J. Inst. Brew.,* 1978, **84**, 291–292.
22. Meilgard, M.C., Stale flavour carbonyls in brewing. *Brew. Dig.*, 1972, **47**, 48–57.
23. Drost B.W., van den Berg, R., Freyee, F.J.M., van der Velde, E.G. and Hollemans, M., Flavour stability. *J. Am. Soc. Brew. Chem.,* 1990, **48**, 124–131.
24. Tressl, R., Bahri, D. and Silwar, R., Bildung von Aldehyden durch Lipid oxidation und deren Dedeutung als 'Off-Flavour'-Komponenten in Bier. *Proceedings of the European Brewery Convention*, Berlin, 1979, pp. 27–41.
25. Drost, B.W., van Eerde, P., Hoekstra, S. and Strating J., Fatty Acids and Staling of Beer. *Proceedings of the European Brewery Convention*, Estoril, 1971, 451–458.
26. Verhagen, L.C., Strating, J. and Tjaden, U.R., Analysis of E-2-nonenal at the ultra trace level by high-performance liquid chromatography using precolumn derivatization and column switching techniques. *J. Chromatogr.*, 1987, **393**, 85–96.
27. Barker, R.L., Gracey. D.E.F., Irwin, A.J., Pipast, P. and Leiska, E., Liberation of staling aldehydes during storage of beer. *J. Inst. Brew.*, 1983, **89**, 411–415.
28. Gyllang, H., Winge, M. and Korch C., Regulation of SO_2 formation during fermentation. *Proceedings European Brewery Convention,* Zurich, 1989, 347–354.

15 Wine flavour

A.C. NOBLE

Abstract

The distinctive flavours of wines are affected by an enormous number of variables. However the grape variety and factors affecting wine development and berry composition exert major influences on distinctive flavours, whereas fermentation has very little overall effect. Although very little is known about specific compounds responsible for characteristic varietal flavours, in two cases compounds have been identified in grapes that are 'impact' compounds. In Muscat of Alexandria and other 'aromatic' varieties, terpenes contribute to the distinctive floral aromas; however, over 90% of the terpene content may occur as nonvolatile terpene glycosides, which can serve as flavour precursors. The amount of the bell-pepper pyrazine, 2-methoxy-3-isobutylpyrazine, has recently been correlated with the intensity of 'vegetative' Sauvignon blanc and Cabernet Sauvignon wines. In both of these instances, the 'impact' compounds have been shown to respond to climatic and/or geographical influences. The effects of specific geographical, viticultural and enological variables on wine flavour has been examined in Cabernet Sauvignon wines using partial least squares regression.

15.1 Introduction

Investigating compounds and factors that affect wine flavour is not unlike trying to piece together a jigsaw puzzle for which only few pieces are available. As for many systems, wine flavour is elicited by over 400 volatile compounds and an unquantified number of nonvolatile compounds. Quantitative and qualitative volatile data for many varieties of wines have been published, along with odour thresholds for these compounds and descriptions of their aromas in single-component systems. However, the flavour of complex systems, such as wines, cannot be predicted using additive models from information about single compounds in model systems. With a few exceptions, perceived flavour is the result of a pattern or specific ratios of many compounds, rather than being attributable to one 'impact' compound. Since so little is known about which compounds determine the varietal distinctiveness of wine aromas, at this stage, it is not possible to ask plant or yeast biotechnologists to insert a gene that will direct the grape vine or wine yeast to produce specific compounds. However, in yeast, the technology to accomplish such a genetic change already exists. For example, casein is a commonly used fining agent in wine production. The bovine β-casein gene was fused to a yeast promoter and expressed by yeast, showing the potential for production of a protein fining agent by the yeast during fermentation [1].

Although wine is the fermented product of grapes, fermentation *per se* contributes little to the distinctive aroma of varietal wines. Rather their distinct aromas are a function of other factors. Most importantly, wine flavour begins with the grape. The specific cultivar or variety of grape clearly influences wine sensory properties but the effects of geography are also important, especially climate and soil, also the vineyard management including type of trellising, irrigation, vine age and grape maturity at harvest. After this point, the winemaker can selectively influence the flavour of wines in several ways. During the primary or alcoholic fermentation, yeast (*Saccharomyces cerevisiae*) converts sugar to ethanol, carbon dioxide and several by-products, which include ethyl esters, fusel alcohols and phenethyl alcohol [2,3]. Although the production of terpenes by yeast during fermentation has been reported [4], only low levels of one monoterpene, farnesol, were produced by *S. cerevisiae*. While the esters contribute to the fruity aroma of the 'fermentation bouquet' that is ubiquitous to all young wines, the type of yeast strain influences the rate of fermentation more than differences in the distinctive flavour of wine [1]. Although the time during which the skins of the crushed grapes remain in contact with the fermenting grape must influence aroma to some extent, it also has a large effect on the extraction of phenolic compounds that elicit bitterness and astringency [5–7]. A secondary (malolactic) fermentation by *Lactobacillus* or *Leuconostoc* bacteria, which produces volatiles including the 'buttery' component – diacetyl [8], influences wine flavour but in a generic way. Similarly, holding the wines in contact with the yeast lees (*sur lie*) for an extended time after fermentation, can alter wine flavour further from that of the starting grape. Chardonnay wines aged on the lees in both stainless steel and oak were less buttery and more toasty than those with no lees contact, with inconsistent effects on fruitiness [9].

Ageing wine in oak also affects the flavour considerably, with the extraction of volatile phenolics including vanillin and eugenol [5,10] which produce vanilla and clove or spicy aromas in the finished wines. Finally, on ageing of the wine, characteristic changes in aroma occur, which, in part, are explained by hydrolysis of esters produced by yeasts during fermentation [11], although slow oxidation is a contributory factor [10,12].

Despite the enormous amount of effort spent isolating and identifying compounds that are responsible for the distinctive or varietal flavours in *Vitis vinifera* wines as reviewed by Webb [13], Schreier [14] and Rapp [15], only small advances have been made in this quest. In two interesting cases, compounds responsible for the distinctive flavour of varietal wines with 'floral' or 'herbaceous' or 'vegetative' aromas have been identified in grapes and the resulting wines. As reviewed in this chapter, in both of these examples, these components and the flavour of the finished wines reflect the influence of geographical and viticultural factors.

15.2 Terpenes

Monoterpenes have been shown to contribute to the characteristic varietal aroma of Muscat grapes [16–19] and have been examined extensively in Riesling wines as summarized previously [14,15]. At present, about 50 monoterpene compounds have been identified in *Vitis vinifera* L. grapes [18], of which the most abundant are geraniol, linalool and nerol. Lower amounts of citronellol, neroloxide, α-terpineol, diendiol-1, and various forms of linalool oxides have also been found.

In addition to the free odour-producing forms of monoterpenes, the presence of glycosidically bound monoterpenes (or monoterpene glycosides, MTG) was first suggested by Cordonnier and Bayonove [20]. Later, with the development of chromatographic techniques for the isolation of bound monoterpenes, these MTG were shown to be present at higher concentrations in grapes than the free monoterpenes [21–23]. Although MTG have been shown to be tasteless at the levels in which they are found in wine [24], they can contribute significantly to aroma upon hydrolysis [25,26], as suggested previously [21,23]. To obtain the maximum intensity of the characteristic floral aroma in Muscat and related aromatic grape varieties, at least three approaches have been considered: (i) harvesting grapes when total terpene levels are at their highest concentrations; (ii) using extended skin contact or appropriate pressing systems [27]; and (iii) hydrolysing bound terpenes enzymatically [28]. Conflicting information regarding the optimum maturity at which highest levels of terpenes are found in Muscat of Alexandria grapes has been published. Free terpenes have been reported to increase during development, then decrease in overripe grapes [29], whereas Gunata *et al.* [23] reported that the highest free linalool levels were reached at normal harvest, although nerol and geraniol increased post-harvest. In both France and Australia, bound terpenes have been reported to increase after the time at which the grapes were normally harvested [21,23].

In California for Muscat of Alexandria, concentrations of the three major terpenes, linalool, nerol and geraniol and most other monoterpenes, which were lowest at véraison, increased throughout development and continued to accumulate after harvest, as shown in Figure 15.1. Despite a plateau in sugar accumulation as the fruit reached commercial maturity,

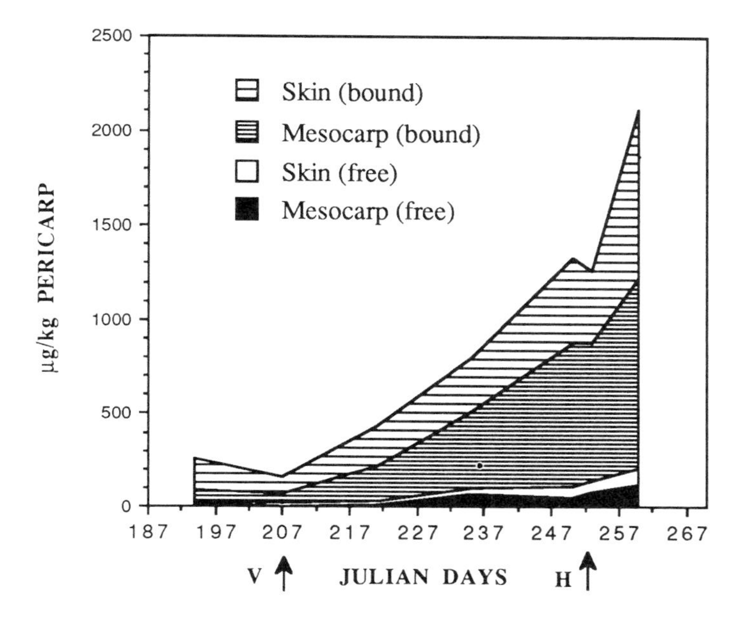

Figure 15.1 Free and bound concentrations of the three major terpenes (nerol, linalool and geraniol) in skin and mesocarp during development of Muscat of Alexandria. V and H arrows at abscissa denote véraison and harvest for wine-making respectively [31].

the continual rise in concentration of the three major bound monoterpenes is consistent with previous reports that the accumulation of monoterpenes is independent of sugar accumulation [21,29,30].

In Figure 15.2, for Muscat of Alexandria, concentrations at harvest of free and bound forms in the skin and mesocarp are provided for the three major terpenes and diendiol-1, together with the published threshold concentrations (in μg/l sugar water) [19] as indicated by an arrow for linalool, geraniol and nerol. For this vintage in California, the only free monoterpene that was found at levels near threshold levels at harvest was geraniol, whereas the bound linalool and geraniol on hydrolysis would yield concentrations four to five times threshold levels.

The relatively low terpene concentrations, and resulting low intensity of varietal aroma in the wine, may be due to the hot and dry weather conditions that occurred during development. Marais [32] suggested similarly that lack of characteristic aroma in many wines of white cultivars in South Africa is the result of high temperatures during ripening. As illustrated in Figure 15.3 for linalool, both forms of linalool and geraniol showed marked decrease in concentration at harvest, with a subsequent rise by the last sampling date. This could be a response by the vine to high temperatures that occurred between Julian days 247 to 252 when the vines

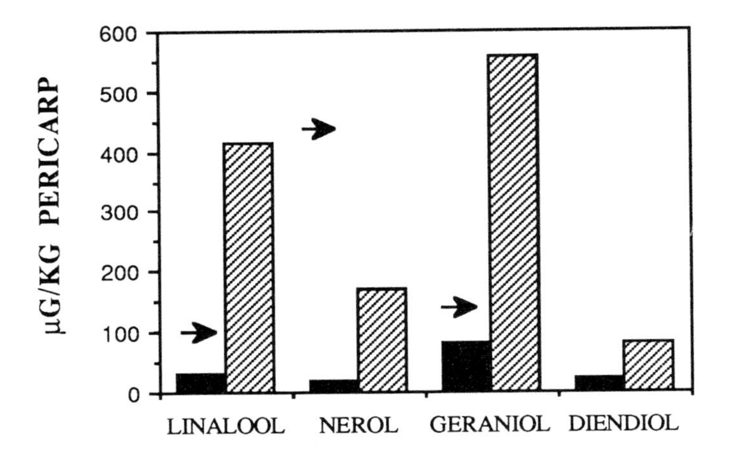

Figure 15.2 Concentrations of total free (solid shading) and bound (hatched shading) terpenes at harvest of Muscat of Alexandria grapes. Arrows denote published threshold values (mg/l) [31].

experienced several days of high temperatures over 38°C (100°F) before harvest. Despite berry dehydration, indicated by a decrease in berry weight, the total content per berry of free and bound monoterpenes had the same decreasing pattern as that shown on a concentration basis in Figure 15.3 for linalool. No corresponding increase in free linalool and geraniol was observed when bound linalool and geraniol decreased [17]. Although no correlations were made previously between accumulation of individual monoterpenes in grapes and temperature or other environmental conditions in grapes, large fluctuations between sampling dates are found in several previous reports [17,21,23,30]. It may be that transient changes in weather were also involved in the fluctuations in monoterpene concentrations noted in those studies.

Despite the low proportion of skin to mesocarp (< 10% skins by berry weight), over 42% of the major free and bound monoterpenes were found in the skin on the last sampling date in Muscat [33]. Wilson *et al.* [22] reported a similar distribution of free and bound monoterpenes between skin and juice in three Muscat varieties. In contrast, Gunata *et al.* [23] found higher levels of the free and bound monoterpenes in the skin. These contradictory results are possibly attributable to differences in grape maturity at harvest or differences in environmental conditions.

With over 90% of the terpenes present in the bound form at harvest, enhancement of wine aroma by hydrolysis of the glycosides is of considerable interest to winemakers. During fermentation and subsequent wine ageing, terpene glycosides can hydrolyse at the acid levels in wines. As shown in Figure 15.4, the concentrations of free terpenes were higher in

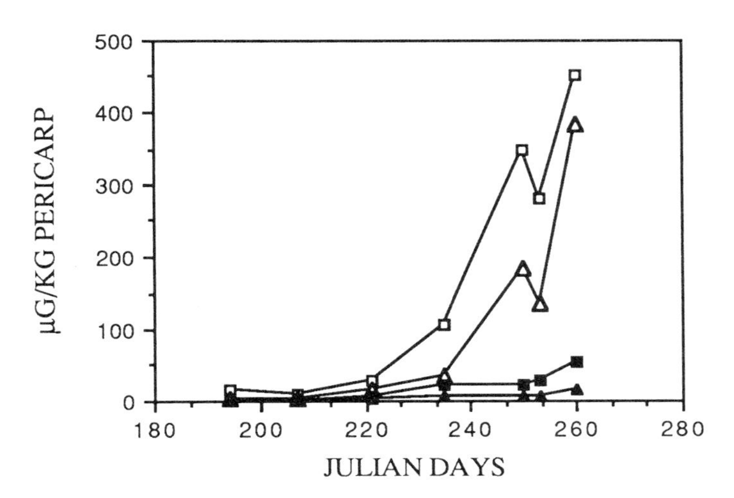

Figure 15.3 Concentrations of free (closed symbols) and bound (open symbols) linalool in skin (triangles) and mesocarp (squares) throughout development of Muscat of Alexandria grapes [31].

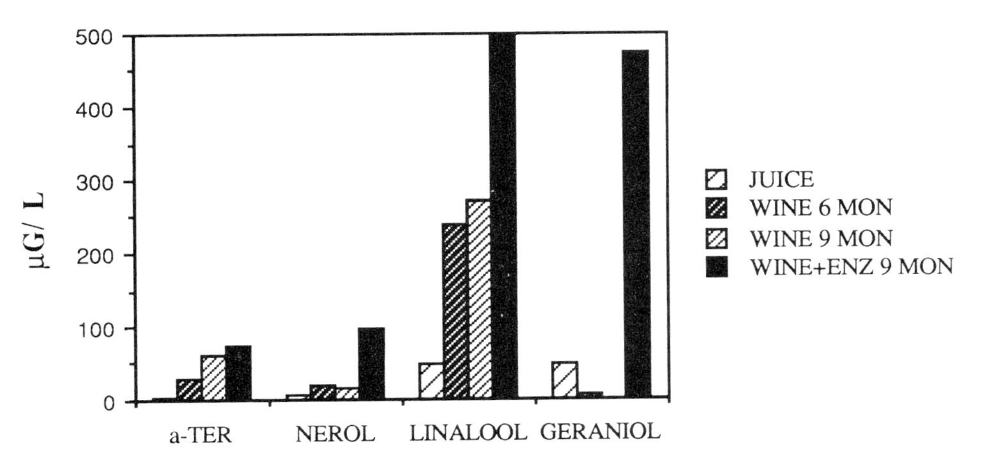

Figure 15.4 Concentration of free monoterpenes in Muscat of Alexandria juice and resulting wine after storage for 6 and 9 months, and with addition of enzyme [34].

Muscat wines (after fermentation and 6 months of storage) than in the original juice, for all compounds but geraniol. After 6 months, these increases represented hydrolysis of 16%, 27% and 34% of the total geraniol, linalool and nerol glycosides, respectively, while further increases in the free terpenes were observed after an additional 3-month storage interval [34].

To increase the rate of hydrolysis, recent studies in wines, as with other fruits, such as passion fruit and papayas, have examined the use of glycosidases to increase the concentration of the free terpenes [26,35,36].

Upon enzymatic hydrolysis by a pectinase with β-glycosidase activity (Röhm) for 1.5 months, the total concentration of the three major free monoterpenes (i.e. linalool, geraniol and nerol) was increased by about 370% (Figure 15.4), while the corresponding bound monoterpenes almost disappeared from the wines. However, during wine ageing, the free monoterpenes were interconverted as a result of hydrolysis and oxidation [37,38] resulting in an increase in compounds such as α-terpineol, as seen in Figure 15.4, and in terpene oxides [39], which are odourless or have very high thresholds.

15.3 2-Methoxy-3-isobutylpyrazine

Owing to the distinctive bell pepper aroma of 'herbaceous' or 'vegetative' Sauvignon blanc and Cabernet Sauvignon wines, it was speculated that the compound responsible was 2-methoxy-3-isobutylpyrazine (MIP) [40], which had been identified previously as the impact compound of bell peppers by Buttery *et al.* [41]. With the use of isotopically labelled MIP and selective ion monitoring mass spectrometry [42], MIP was identified conclusively and quantified in Sauvignon blanc and Cabernet Sauvignon wines [42–45]. Furthermore, the intensity of the varietal aroma or 'vegetative' character of the wines was correlated with the level of MIP [43,45,46].

Although no information is available about the site of synthesis or factors affecting the synthesis of MIP or other compounds contributing to the 'vegetative' aroma of wines, winemakers have long associated intense 'vegetative' aromas with young vines [47], cool climates [48] or less-ripe grapes [49]. Owing to the difficulty in quantifying MIP, most studies examining factors that influence this 'herbaceous' or 'vegetative' aroma rely on sensory evaluation. In descriptive analysis of California Cabernet Sauvignon wines [50], in which the intensity of specific aromas were rated quantitatively to provide a profile of the flavour of the wines, some of these commonly observed relations were verified statistically. A significant correlation between vine age and the 'berry' aroma was demonstrated, while, conversely, younger wines were shown to yield wines of higher 'vegetative' intensity. Similarly, heat summation designations were inversely correlated with the 'vegetative' aroma, warmer locations yielding less 'vegetative' wines.

Since MIP was shown to be sensitive to light, photodegrading rapidly even at low light intensity [51], it has been speculated that this may explain the tendency for wines produced from vines that have a very dense canopy to be more 'vegetative' than those from more open canopies, as demonstrated recently in a study of grapes and wines from shaded versus unshaded Cabernet Sauvignon vines [52]. In a similar experiment, berry light exposure was increased by leaf-removal treatments, resulting in a decreased intensity of the 'vegetative' aromas in Sauvignon blanc wines [53,54].

15.4 Influence of geographical factors

Clearly, environmental factors, including the climate, topography and soils of the vineyard, have a considerable influence on grape vines and on the composition of grapes, affecting the levels of compounds such as 2-methoxy-3-isobutylpyrazine or of flavour precursors such as the terpene glycosides. Soil nutrients and water-holding capacity, as well as natural precipitation or irrigation practices, for example, will influence vine vigour and the density of the canopy. Owing to the influence of factors such as these, wines from specific locations of origin are assumed to have distinctive characteristics. Few scientific studies have investigated the effects of soils and climate on wine flavour because evaluation of geographical aspects using conventionally designed experiments is tedious and difficult. Despite this, throughout the world, appellation laws delimiting grape-growing regions presume implicitly regional distinctiveness. In the USA, viticultural areas are delimited by geographical features.

Few studies have addressed this in an empirical fashion, however. In a descriptive analysis of Pinot noir wines made from grapes grown in the cool viticultural area of Carneros (at the southern end of Napa Valley), which has heavy, marine clay soils, or from grapes from warmer areas of the Napa Valley or Sonoma Valley, the wines from the Carneros area were shown to be higher in fruity characteristics, berry jam, spicy, cherry aromas, than those from either the Napa or Sonoma Valley [55]. Conversely, in a descriptive analysis of 1976 red Bordeaux wines, no clustering or distinctiveness was found on the basis of the commune of origin [56], suggesting that differences in winemaking practices eliminated any differences due to the origin of the grapes. Similarly, no differences in wine flavour due to origin of grapes were seen in 58 Chardonnay wines from five California regions [57] nor in 24 Zinfandel wines from the Sierra Foothills, coastal valleys or inland valley locations [58], although in both studies, differences were found among vintages.

More recently, partial least squares (PLS) regression has been employed to relate quantitative sensory profiles obtained by descriptive analysis to volatile composition or to geographical, viticultural and enological variables. In a study to examine factors affecting the flavour of wines, three

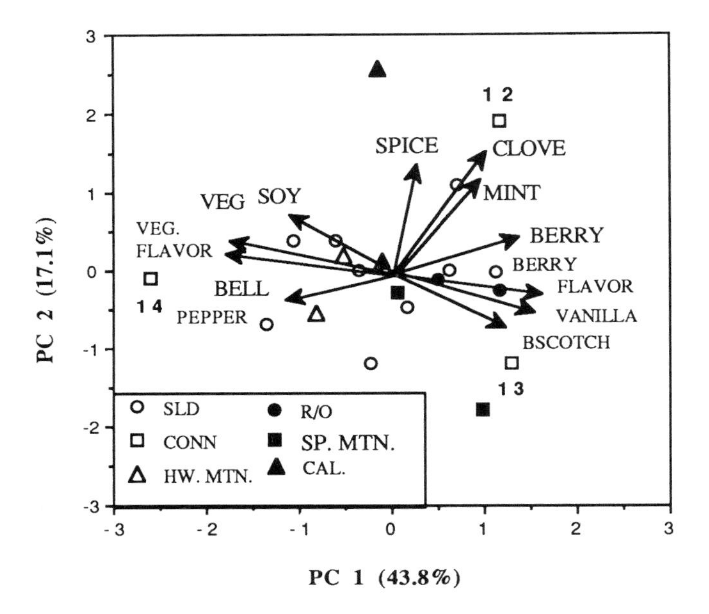

Figure 15.5 Principal component analysis of 1986 Napa Valley Cabernet Sauvignon wines from Stags Leap District (SLD), Conn Creek area (Conn), Howell Mt (Hw. Mtn), Rutherford-Oakville (R/O), Spring Mt. (Sp. Mtn) and Calistoga (Cal). Projection of sensory terms (vectors) and wine factor scores on principal components 1 and 2 [59].

vintages of commercial Cabernet Sauvignon wines from six different regions within the Napa Valley were evaluated using descriptive analysis [59], while 'soil-descriptive' data were obtained from soil pits in the test vineyards [60]. The sensory data were then related to geographical, viticultural or enological data by PLS [61]. In addition, headspace volatiles were related to the descriptive analysis data for one vintage.

As illustrated in Figure 15.5, by a principal component analysis (PCA) of the sensory data for the 1986 wines, the largest variation in wine flavour was the contrast between high in 'vegetative' aromas (and low in 'berry' notes) versus those high in 'berry' and low in 'vegetative' aromas. Similar variation in wine flavour was found for the 1987 and 1988 vintages and in previous evaluations of Cabernet Sauvignon wines [50,56]. For none of the vintages was clustering by regions found; instead, for example, there was as much variation among wines from the Conn Creek region (coded 12, 13 and 14 in Figure 15.5) as among the entire data set.

The variation in flavour among the wines in each vintage reflects differences among locations (e.g. climate, soil variables including percentage clay, percentage gravel, soil pH and soil water-holding capacity, and vine-rooting depth), as well as viticultural variables (e.g. planting density, rootstock, vine vigour, vine age, crop level, grape maturity at harvest) and

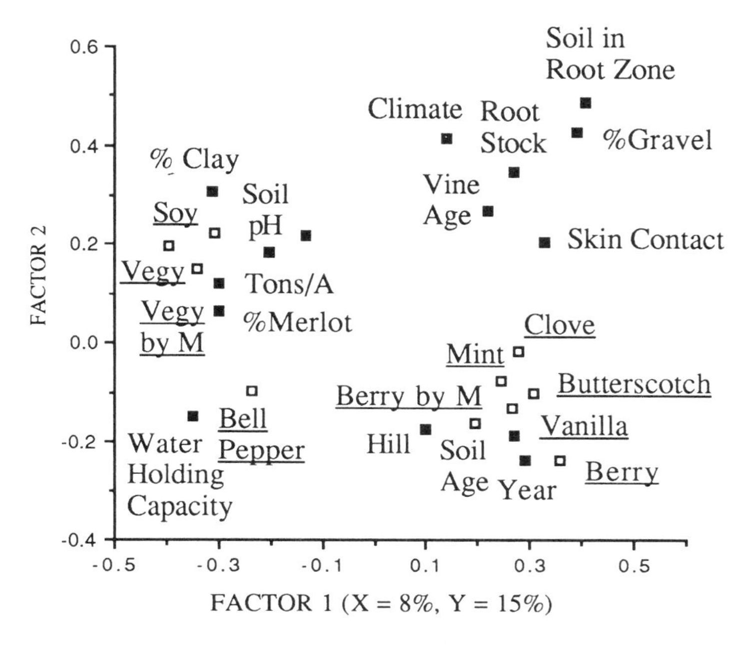

Figure 15.6 PLS of soil, viticultural and enological data (■) and sensory data (□) for three vintages. Loadings for the soil, viticulture and enological (X) and sensory terms (Y) for factors 1 and 2 [64].

enological practices (e.g. time of skin contact, age of oak barrels used and extent of oak ageing). Variation among vintages included obvious climatic differences and exposure to oak: the 1988 wines were not aged in oak barrels. To model the relationship between the sensory properties of the wines and these geographical, viticultural and enological variables, PLS was used.

In Figure 15.6, the loadings for eleven descriptive sensory terms and 18 viticultural, enological and soil-descriptive parameters are provided for the first two factors. This model, which was a one-factor solution, explained 8% of the variation in the soil variables and 15% of the variation in the sensory data. As for the PCA, the first factor contrasts 'vegetative' wines (e.g. high in bell pepper aroma, vegetative aroma and vegetative flavour by mouth) with 'fruity' wines (e.g. high in berry, vanilla and butterscotch aromas). Associated with the vegetative terms are soils with a high percentage of clay, high water-holding capacity and higher pH, higher crop yields, and higher percentages of Merlot used for blending. Associated with 'fruitier' wines are older soils with a higher percentage of gravel and a lower availability of soil fines in the rooting zone, older vines, younger wines (year) and the use of longer skin contact times.

To simplify the complex information shown in Figure 15.6, the variables

Table 15.1 Factors associated with vegetative and berry flavours of three vintages of Cabernet Sauvignon Wines

Vegetative	Berry/fruity
High percentage of clay	High percentage of gravel
High weight fines (high water-holding capacity)	High coarse : fines ratio (low soil matrix in rooting zone)
Cool climate	Old soils
A × R root stock	St. George root stock
High percentage Merlot/other	Old vines
High crop yield	Long skin-contact time

associated with 'vegetative' and 'berry or fruity' flavours are summarized in Table 15.1. Although these relationships are consistent with those demonstrated previously as already discussed, caution should be used when drawing conclusions. For example, it is speculated that addition of higher amounts of non-Cabernet Sauvignon wines (Merlot or Cabernet franc) did not increase the vegetative flavour but was performed to decrease the intensity of highly 'vegetative' wines.

Similarly, there is no evidence that the use of A × R versus St. George root stock influences the vegetative aroma through enhancement of vine vigour and resulting affect on the light penetrating canopy. However, A × R root stocks were used almost exclusively for the younger vines, which are also associated with higher 'vegetative' flavour.

A consistent relationship emerged in all analyses of the data: soils that were high in clay content and water-holding capacity were associated with 'vegetative' wines, while the 'berry' attribute was associated with soils high in gravel (and low in water-holding capacity). For example, the 'vegetative' wine 14 and 'fruity' wine 13, which were widely separated in Figure 15.5, were made by the same winery from vineyards 0.5 km apart. The alluvial clay-rich soil at site 14 was deep, high in nutrients and high in water-holding capacity, whereas site 13 was a shallow, coarse, nutrient-poor soil with a low water-holding capacity. Exclusion of these two wines did not change the model, so the overall pattern was not due solely to the influence of these extreme cases.

When headspace volatiles, which had been quantitatively analysed by gas chromatography (GC) [62], were related to the sensory data or to the soil, viticultural and enological variables by PLS, no good models were developed. By calibration, seldom was more than 1% of the variability in the sensory data predicted by the GC peaks. The poor relationship between the GC data and the sensory data is attributed speculatively to the inability of this 4 ml headspace analysis to detect the trace volatiles that are important in eliciting distinctive wine aroma. Instead, the analysis sampled the major fermentation products, which are the most volatile compounds and are present at highest concentrations.

15.5 Conclusions

Although wine flavour continues to remain an exciting mystery, many specific factors have been shown to have a large effect on the composition of 'impact' compounds and on wine flavour. As the influence of soil, climate, grape cultivation and winemaking practices on wine flavour is better understood, models can be developed to optimize the production of distinctive flavour notes.

Acknowledgements

The contributions of my collaborators, Prof. D.L. Elliott-Fisk (Dept. Geography), and Theresa Spears and Seung Park, are gratefully recognized.

References

1. Bisson, L.F., Noble, A.C. and Kunkee, R.E., *Sensory Effects and Fermentation Differences of 11 Commercial Wine Yeasts in Chenin Blanc Wines*. Paper presented at Annual Meeting of American Society of Enologists and Viticulturists, Los Angeles, June 1990.
2. Suomalainen, H., Yeast and its effect on the flavour of alcoholic beverages. *J. Inst. Brew.*, 1971, **77**, 164–170.
3. Suomalainen, H. and Lehtonen, M., Production of aroma compounds by yeast. *J. Inst. Brew.*, 1979, **85**, 149–156.
4. Hock, R., Benda, I. and Schreier, P., Formation of terpenes by yeast during alcoholic fermentation. *Z. Lebensm. Unters. Forsch.*, 1984, **170**, 450–452.
5. Singleton, V.L. and Noble, A.C., Wine flavor and phenolic substances, in *Phenolic, Sulfur, and Nitrogen Compounds in Food Flavors*, (ed. G. Charalambous), American Chemical Society, Washington DC, 1976, pp. 47–70.
6. Schmidt, J.O. and Noble, A.C., Investigation of the effect of skin contact time on wine flavor. *Am. J. Enol. Vitic.*, 1983, **34**, 135–138.
7. Noble, A.C., *Effect of Skin Contact on the Composition of Wine: Nonvolatiles and Volatiles*. Grape and Wine Centennial Symposium, University of California, Davis, 1982, pp. 330–335.
8. Bertrand, A., Smirou-Bonnamour, C. and Lonvaud-Funel, A., Aroma compounds formed in malolactic fermentation, in *Flavor Research of Alcoholic Beverages*, (ed. L. Nykänen and P. Lehtonen), Foundation for Biotechnical and Industrial Fermentation Research, Helsinki, 1984, pp. 39–49.
9. La Follette, G., Influence of extended yeast lees contact on Chardonnay wines, MS Thesis, University of California, Davis, 1990.
10. Puech, J., Extraction of phenolic compounds from oak wood in model solutions and evolution of aromatic aldehydes in wines aged in oak barrels. *Am. J. Enol. Vitic.*, 1987, **38**, 236–238.
11. Ramey, D. and Ough, C.S., Volatile ester hydrolysis or formation during storage of model solutions and wines. *J. Agric. Food Chem.*, 1980, **28**, 928–993.
12. Simpson, R., Aroma and compositional changes in wine with oxidation, storage and aging. *Vitis,* 1978, **17**, 274–287.
13. Webb, A.D. and Muller, C.J., Volatile compounds of wines and other fermented beverages, in *Advances in Applied Microbiology*, (ed. D. Perlman), Academic Press, New York, 1972, pp. 75–146.

14. Schreier, P., Flavor composition of wines: A review. *CRC Food Sci. Nut.*, 1979, **12**, 59–111.
15. Rapp, A. and Mandery, H., Wine aroma. *Experientia,* 1986, **42**, 873–884.
16. Webb, A.D., Kepner, R.E. and Maggiro, L., Gas chromatographic comparison of volatile aroma materials extracted from eight different Muscat flavoured varieties. *Am. J. Enol. Vitic.*, 1966, **17**, 247–254.
17. Terrier, A., Boidron, J.N. and Ribéreau-Gayon, P., Teneurs en composes terpeniques des raisins de *V. vinifera. C.R. Acad. Sci. Ser. D.*, 1971, **275**, 941–994.
18. Strauss, C.R., Wilson, B., Gooley, P.R. and Williams, P.J., Role of monoterpenes in grape and wine flavor, in *Biogeneration of Aromas*, (ed. T. Parliment and R. Croteau), American Chemical Society, Washington DC, 1986, pp. 222–242.
19. Ribéreau-Gayon, P., Boidron, J.N. and Terrier, A., Aroma of Muscat grape variety. *J. Agric. Food Chem,* 1975, **23**, 1042–1047.
20. Cordonnier, R. and Bayonove, C., Mise en evidence dans la baie la raisin var Muscat d'Alexandria de monoterpenes lies revelables pour une ou plusiers enzymes du fruit. *C.R. Acad. Sci. Ser. D*, 1974, **278,** 3387–3390.
21. Wilson, B., Strauss, C.R. and Williams, P.J., Changes in free and glycosidically-bound monoterpenes in developing Muscat grapes. *J. Agric. Food Chem.,* 1984, **32**, 919–924.
22. Wilson, G., Strauss, C.R. and Williams, P.J., The distribution of free and glycosidically-bound monoterpenes among skin, juice, and pulp fractions of some white grape varieties. *Am. J. Enol. Vitic.,* 1986, **37**, 107–111.
23. Gunata, Y.Z., Bayonove, C., Baumes, R. and Cordonnier, R.E., The aroma of grapes. The localization and evolution of free and bound fractions of some grape aroma components c.v. Muscat during development and maturation. *J. Sci. Food Agric.,* 1985, **36,** 857–862.
24. Noble, A.C., Strauss, C.R., Williams, P.J. and Wilson, B., Contribution of terpene glycosides to bitterness in Muscat wine. *Am. J. Enol. Vitic.*, 1988, **39**, 129–131.
25. Noble, A.C., Strauss, C.R., Williams, P.J. and Wilson, B., Sensory evaluation of nonvolatile flavour precursors in wine, in *Flavour Science and Technology* (Proceedings of the Fifth Weurman Flavour Research Symposium, Oslo), (eds M. Martens, G. Galen and R. Russwurm Jr.), John Wiley & Sons, Chichester, 1987, pp. 383–391.
26. Shoseyov, O., Bravdo, B.A., Siegel, D., Goldman, A., Cohen, S., Shoseyov, L. and Ikan, R., Immobilized endo-β-glucosidase enriches flavor of wine and passion fruit juice. *J. Agric. Food Chem.*, 1990, **38**, 1387–1390.
27. Kinser, G. and Schreier, P., The influence of different pressing systems on the composition of volatile constituents in unfermented grape musts and wines. *Am. J. Enol. Vitic.*, 1980, **31**, 7–13.
28. Aryan, A.P., Wilson, B., Strauss, C.R. and Williams, P.J., The properties of glycosidase of *Vitis vinifera* and a comparison of their β-glucosidase activity with that of exogenous enzymes: An assessment of possible applications in enology. *Am. J. Enol. Vitic.*, 1987, **38**, 182–188.
29. Hardy, P.J., Changes in volatiles of Muscat grapes during ripening. *Phytochem.*, 1970, **9**, 709–715.
30. Williams, P.J., Strauss, C.R., Wilson, B. and Dimitriadis, E., Recent studies in grape terpene glycosides, in *Progress in Flavor Research*, (ed. J. Adda), Elsevier Science Publishers, Amsterdam, 1985, pp. 349–357.
31. Park, S.K., Distribution of free and bound forms of monoterpenes during maturation of Muscat of Alexandria and Symphony grapes. MS Thesis. University of California, Davis, 1989.
32. Marais, J., Terpene concentrations of *V. vinifera* L. cv. Gewurztraminer as affected by grape maturity and cellar practices. *Vitis*, 1987, **26**, 231–245.
33. Park, S.K., Morrison, J.C., Adams, D.O. and Noble, A.C., Distribution of free and glycosidically bound monoterpenes in skin and mesocarp of Muscat of Alexandria grapes during development. *J. Agric. Food Chem.*, 1991, **39**, 514–518.
34. Park, S.K., Morrison, J.C., Adams, D.O. and Noble, A.C. Changes in free and glycosidically bound terpenes as a function of fermentation, wine aging and enzyme treatment. 41st Annual Meeting American Society of Enology and Viticulture, Technical Abstract, 1990, #4.

35. Schwab, W., Mahr, C. and Schreier, P., Studies on the enzymic hydrolysis of bound aroma components from *Carica papaya* fruit. *J. Agric. Food Chem.*, 1989, **37**, 1009–1012.
36. Gunata, Z., Brillouet, J.-M, Voirin, S., Baumes, R. and Cordonnier, R., Purification and some properties of an α-L-arabinofuranosidase from *Aspergillus niger*. Action on grape monoterpenyl arabinofuranosylglucosides. *J. Agric. Food Chem.*, 1990, **38**, 772–776.
37. Williams, P.J., Strauss, C.R., Wilson, B. and Massy-Westropp, R.A., Studies on the hydrolysis of *Vitis vinifera* monoterpene precursor compounds and model monoterpene β-D-glycosides rationalizing the monoterpene composition of grapes. *J. Agric. Food Chem.*, 1982, **30**, 1219–1223.
38. Rapp, A., Wine aroma substances from gas chromatographic analysis, in *Wine Analysis. Modern Methods of Plant Analysis*, (ed. H. Linskens and J. Jackson), Springer Verlag, Berlin, 1988, pp. 29–65.
39. Simpson, R., Aroma composition of bottled aged white wine. *Vitis*, 1979, **18**, 148–154.
40. Bayonove, C., Cordonnier, R. and Dubois, P., Etude d'une fraction caracteristique de l'arome du raisin de la variete Cabernet Sauvignon: mise en evidence de la 2-methoxy-3-isobutylpyrazine. *C.R. Acad. Sci. Ser. D*, 1975, **281**, 75–78.
41. Buttery, R.G., Seifert, R.M., Guadagni, D.G. and Ling, L.C., Characterization of some volatile constituents of bell peppers. *J. Agric. Food Chem.*, 1969, **17**, 1322–1327.
42. Harris, R.L.N., Lacey, M.J., Brown, W.V. and Allen, M.S., Determination of 2-methoxy-3-alkoxypyrazines in wine by GC/MS. *Vitis*, 1987, **26**, 201–207.
43. Allen, M.S., Lacey, M.J., Harris, R.L.N. and Brown, W.V., Contribution of methoxypyrazines to Sauvignon blanc wine aroma. *Am. J. Enol. Vitic.*, 1991, **42**, 109–112.
44. Lacey, M.J., Allen, M.S., Harris, R.L.N. and Brown, W.V., Methoxypyrazines in Sauvignon blanc grapes and wines. *Am. J. Enol. Vitic.*, 1991, **42**, 103–108.
45. Allen, M.S., Lacey, M.J., Harris, R.L.N. and Brown, W.V., Sauvignon blanc varietal aroma. *Aust. Grapegrower Winemaker*, 1988, 52–56.
46. Allen, M.S., Lacey, M.J., Brown, W.V. and Harris, R.L.N., *Contribution of Methoxypyraxines to the Flavour of Cabernet Sauvignon and Sauvignon Blanc Grapes and Wines*. Seventh Australian Wine Industry Technical Conference, 1989, pp. 114–117.
47. Ough, C.S. *Intense Flavors of Monterey Grapes and Wines*. Wine Institute Progress Report, 1979.
48. Slingsby, R.W., Kepner, R.E., Muller, C.J. and Webb, A.D., Some volatile components of *Vitis vinifera* variety Cabernet Sauvignon wine. *Am. J. Enol. Vitic.*, 1980, **31**, 360–363.
49. Augustyn, O., Rapp, A., and Wyk, C.J. v., Some volatile aroma components of *Vitis vinifera* L. cv. Sauvignon blanc. *S. Afr. J. Enol. Vitic.*, 1982, **3**, 53–60.
50. Heymann, H. and Noble, A.C., Descriptive analysis of commercial Cabernet Sauvignon wines from California. *Am. J. Enol. Vitic.*, 1987, **38**, 41–44.
51. Heymann, H., Analysis of methoxypyrazines in wines. 1. Development of a quantitative procedure. *J. Agric. Food Chem.*, 1986, **34**, 268–271.
52. Morrison, J.C. and Noble, A.C., The effects of leaf cluster shading on composition of Cabernet Sauvignon grapes and on fruit and wine sensory properties. *Am. J. Enol. Vitic.*, 1990, **41**, 193–200.
53. Smith, S., Codrington, I.C., Robertson, M. and Smart, R.E., *Viticultural and Oenological Implications of Leaf Removal for New Zealand Vineyards*. Proceedings of the Second International Cool Climate Viticultura and Oenology Symposium. 1988, pp. 127–133.
54. Arnold, R.A. and Bledsoe, A.M., The effect of various leaf removal treatments on the aroma and flavor of Sauvignon blanc wine. *Am. J. Enol. Vitic.*, 1990, **41**, 74–76.
55. Guinard, J.-X and Cliff, M., Descriptive analysis of Pinot noir wines from Carneros, Napa and Sonoma. *Am. J. Enol. Vitic.*, 1987, **38**, 211–215.
56. Noble, A.C., Williams, A.A. and Langron, S.P., Descriptive analysis and quality ratings of 1976 wines from Bordeaux communes. *J. Sci. Food Agric.*, 1984, **35**, 88–98.
57. Ohkubo, T., Noble, A.C., and Ough, C.S., Evaluation of California Chardonnay wines by sensory and chemical analyses. *Sci. des Aliments*, 1987, **7**, 573–587.
58. Noble, A.C. and Shannon, M., Profiling Zinfandel wines by sensory and chemical analyses. *Am. J. Enol. Vitic.*, 1987, **38**, 1–5.

59. Spears, T.A., Evaluation of the effects of soil and other geographic parameters on the composition and flavor of Cabernet Sauvignon wines from Napa Valley. MS Thesis, University of California, Davis, 1990.
60. Noble, A.C. and Elliott-Fisk, D.L., *Evaluation of the Effects of Soil and Other Geographical Parameters on Wine Composition and Flavor: Napa Valley, California.* Proceedings of the Fourth Symposium International d'Oenologie, (eds P. Ribereau-Gayon and A. Lonvaud), University of Bordeaux II, France, 1990, pp. 37–45.
61. Elliott-Fisk, D.L. and Noble, A.C., The diversity of soils and environments in Napa Valley, California and their influence on Cabernet Sauvignon wine flavor, in *Viticulture in Geographic Perspective* (Proceedings of the 1991 Miami Meeting of the Association of American Geography (ed. H.J. de Blij), 1992, pp. 45–71.
62. Shimoda, M., Shibamoto, T. and Noble, A.C., Headspace volatiles of Napa Valley Cabernet Sauvignon wines. *J. Agric. Food Chem.*, 1993, **41**, 1664–1668.

16 Flavour of distilled beverages

H. MAARSE and F. VAN DEN BERG

Abstract

The quality of distilled alcoholic beverages is affected largely by the presence of volatile compounds formed during the production of the wine or its distillate and during maturation. The quality of grape brandies is influenced by the following factors: climatological conditions, grapes, vinification, distillation and maturation. The first factor cannot be manipulated. To improve or at least to maintain the quality of a product, insight into the effect of each of the other factors is needed. The quality of malt whiskies is similarly influenced by the following factors: climatological conditions, wort production step, fermentation, distillation, maturation and blending. The mechanisms of formation of flavour compounds from wood constituents during maturation has been the subject of many studies but has not yet been fully resolved. Most investigations have been concerned with the maturation of grape brandies, some others with that of whiskies, but no reports on the maturation of rum have been published recently. Governmental institutions and other groups are interested in the development of methods by which the authenticity and origin of a product can be ascertained. Some applications to whisky and grape brandy samples will be described.

16.1 Introduction

The quality of distilled alcoholic beverages is affected largely by the presence of volatile compounds formed during the production of wine or its distillate and during maturation. Nonvolatile compounds originating from the wood of the barrels in which the beverages have been matured also play some role.

The volatile compounds are listed in the TNO publication *Volatile Compounds in Food – Qualitative and Quantitative Data* [1]. In Table 16.1 an overview of the number of volatile compounds in distilled alcoholic beverages is presented.

Undoubtedly, more compounds will be detected in the future but it is more important to know which compounds contribute to the quality of a

Table 16.1 Number of volatile compounds in distilled alcoholic beverages[a]

Product	Number	Product	Number
Whisky		Grape brandies	
Malt	227	Cognac	486
Scotch blended	184	Armagnac	77
Bourbon	127	Weinbrand	181
Irish	88		
Canadian	86	Rum	550
Japanese	109		

[a] Source: Maarse and Visscher [1].

product in a positive or negative way. The mechanisms of their formation should be known or studied.

In this chapter the following subjects will be discussed: (i) factors influencing the quality of grape brandies; (ii) factors influencing the quality of whiskies; (iii) maturation of distilled beverages; and (iv) characterization of distilled beverages.

16.2 Factors influencing the quality of grape brandies

16.2.1 Introduction

The flavour and quality of grape brandies are influenced by several factors. The most relevant factors in the route from grape to end-product are compiled in Table 16.2. Except for the weather conditions, which strongly affect the quality of the grape musts [2], all other factors can be

Table 16.2 Factors influencing the quality of grape brandies

Stage	Input and variation
Grapes	Climatological conditions
	Grape variety
	Ripeness of grapes
	Sanitary state of grapes
	Method of harvesting
Vinification	Fermentation
	Yeast variety
	Storage conditions
Distillation	Method of distillation
	Presence of yeast
	Processing
Maturation	Barrel
	Wood extracts

manipulated. In order to be able to improve or at least maintain the desired quality level and flavour characteristics, insight into the effect of each of these factors is needed. The most relevant factors will be discussed.

16.2.2 Grape variety

The grape variety is the basis of the flavour of a wine distillate. Large differences in flavour and composition between wines of different cultivars have been noticed [3–6].

The bulk of cognac wine distillates (*c.* 90%) is produced from wines obtained by fermentation of Ugni blanc grapes. The remaining part is produced from Folle blanche and Colombard grapes. The latter two cultivars have been replaced gradually by the Ugni blanc variety during the past 20 years. The motives for this shift were mainly of an economic nature. Ugni blanc grapes are less sensitive to rot and give a significantly higher yield.

To evaluate the effect of this agricultural preference distillates of wines derived from each of these cultivars will be compared.

16.2.2.1 Comparison of wine flavour from Colombard, Folle blanche and Ugni blanc cultivars. During the harvest of 1987 small batches of the cultivars Ugni blanc, Colombard and Folle blanche originating from the Charante area were collected. Vinification took place under identical conditions. From each of the wines first a Broullis and subsequently a Bonne Chauffe distillate was prepared on a laboratory scale. Distillation was performed on a small copper pot still. The distillates were fractionated on a glass plate column. The fractions were subjected to sensory analysis and then selected. The selected fractions were analysed by means of gas chromatography.

The results of the gas chromatographic analysis of the medium-volatile compounds of the Bonne Chauffe distillate are presented in Figure 16.1.

Sensory evaluation of the fractions of the distillates indicated that the Ugni blanc distillate possessed the most neutral flavour, while the Colombard distillate was considered to have the most fruity and ester-like flavour.

The results of this investigation confirmed that the difference in grape variety affects quality and flavour of wine distillates.

16.2.3 Ripeness of the grapes

The flavour components of a wine distillate are largely fermentation products. However, flavour components originating from the grapes may not be neglected. Besides the grape variety, the ripeness of the grape at the time of harvesting also has its effect on the flavour of the wine distillate.

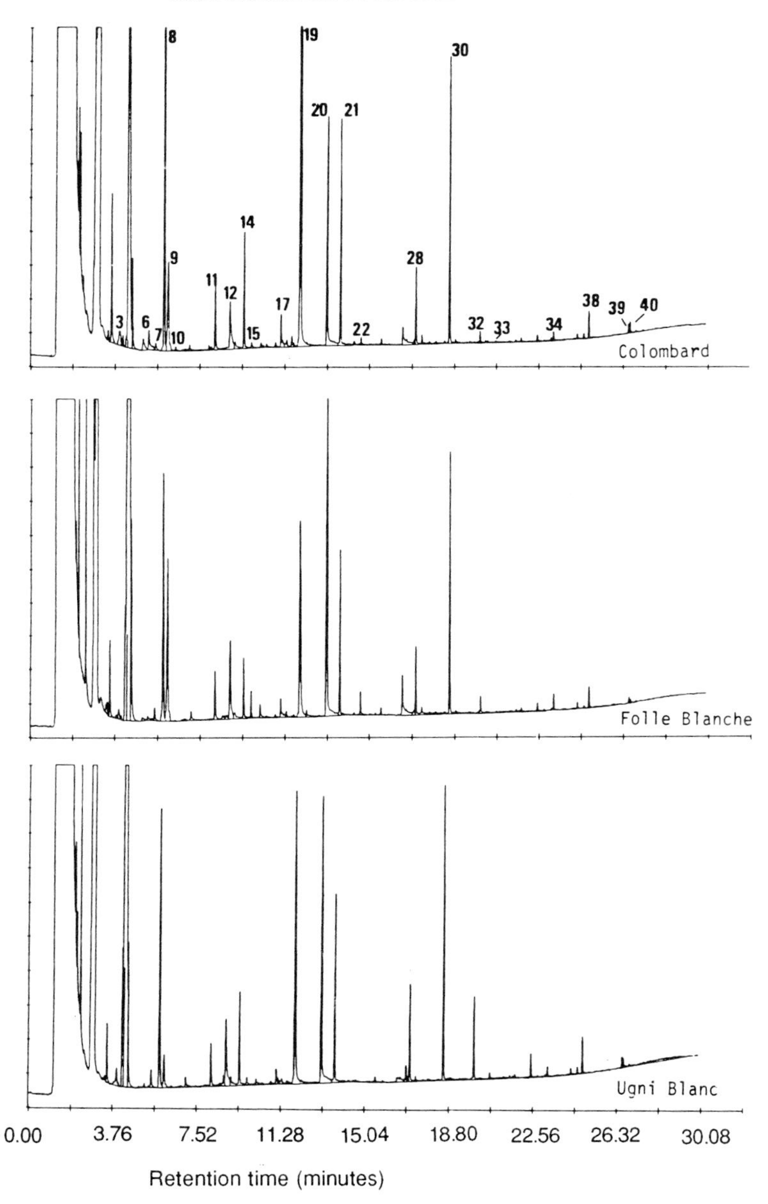

Figure 16.1 Gas chromatograms of medium volatiles of wine distillates of different grape varieties: 3 = ethyl butanoate; 6 = furfuryl alcohol; 7 = ethyl 3-methylbutanoate; 8 = hexanol + 1,1-diethoxyisobutane; 9 = isoamyl acetate; 10 = 2-heptanone; 11 = benzaldehyde; 14 = ethyl hexanoate; 15 = hexyl acetate; 17 = octanol; 19 = 2-phenylethanol; 20 = diethyl succinate; 21 = ethyl octanoate; 22 = phenethyl acetate; 28 = ethyl decanoate; 30 = internal standard; 32 = ethyl dodecanoate; 33 = isoamyl decanoate; 34 = ethyl tetradecanoate; 38 = ethyl hexadecanoate; 39 = ethyl 9,12-octadecadienoate; 40 = ethyl 9-octadecanoate.

Van Rooyen *et al.* [7], for example, demonstrated that the maturity of Colombard grapes is related to the concentration of several volatile constituents such as 2-phenethyl acetate, 2-phenylethanol and hexanoic acid.

16.2.4 Sanitary state of the grapes

Infection with botrytis is one of the most common microbial contaminations of grapes. Botrytis produces such substances as sotolon (a lactone with a honey-like and maderized odour) [8] and 1-octen-3-ol (mushroom odour) [9]. These substances have a negative effect on the aroma of wine distillates.

Increased contents of aldehydes such as 2-propenal (acrolein), which is formed by bacteria from glycerol, and acetaldehyde can lend a pungent and smoky odour to distillates [9]. Increased concentrations of acetaldehyde in combination with ethanol are responsible for the formation of 1,1-diethoxyethane. This acetal can also affect negatively the flavour of a distillate.

These observations indicate that infected grapes are unsuitable for the preparation of wine distillates.

16.2.5 Method of harvesting

Most grapes in the Charante area are harvested mechanically, which implies a potentially higher exposure to oxygen. In the production of normal wine, sulphur dioxide can be added to the grapes to avoid oxidation. In the case of wine production for distillation purposes this should be avoided. In Figure 16.2, gas chromatographic fingerprints of wines produced from grapes harvested by different methods clearly illustrate the effect of oxidation on the composition of volatiles [10]. It may be concluded from these results that strong oxidation should be avoided. Good logistics resulting in a short period between harvesting and processing reduces the effects of oxidation.

16.2.6 Fermentation

The flavour of wines and distillates is strongly influenced by the type of yeast applied [4,5]. Cavazza *et al.* [4] characterized six strains of *Saccharomyces cerevisiae* on the basis of their production of volatile compounds. In this investigation each of the musts from four different

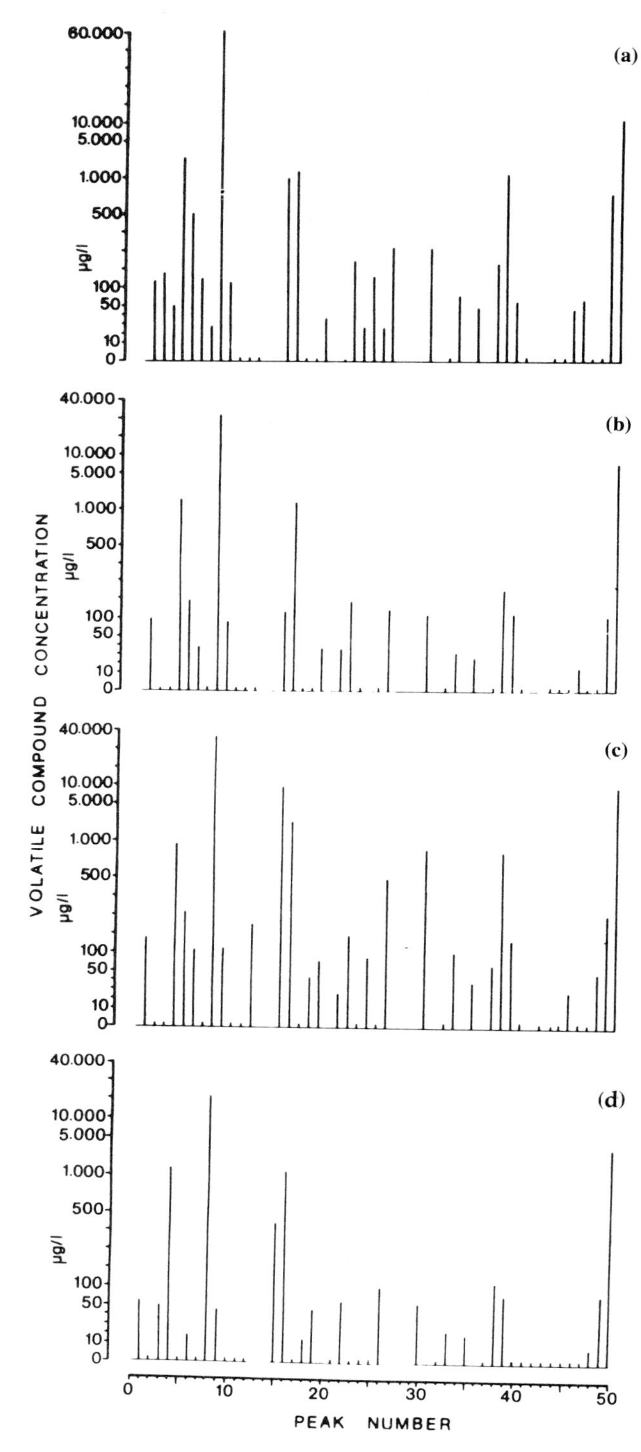

Figure 16.2 Profiles of volatile components of wines harvested under different conditions [10]. (a) Hand picking; (b) mechanical harvesting (in air); (c) mechanical harvesting (plus SO_2); (d) mechanical harvesting (plus O_2).

cultivars (Riesling, Müller Thurgau, Chardonney and Traminer) were fermented with the *S. cerevisiae* strains CH 101, Sau 201, GTR 304, MUT 501, SY 612 and RM 1515. In each fermentation product, 38 volatiles were quantified. Statistical data analysis showed that the concentrations of important flavour compounds such as 2-phenethyl lactate, ethyl butanoate, butanoic acid, hexanoic acid (fruity aroma), ethyl hydroxybutanoate (winy flavour), hexanol (green leaf-like flavour) and 2-phenylethanol (rosy smell) differed significantly from batch to batch depending on the yeast strain applied.

In practice, the application of dry yeasts is very limited in the production of wine in the Charante area. Dry yeasts are applied to start the first batch of the season and to remedy fermentation problems. This implies that the natural yeast strongly influences the fermentation process of Charante wines. Park [11] and Park and Bertrand [12] have shown that many yeast varieties are present on Charante grapes (Table 16.3).

The *S. cerevisiae* variety is found, in most cases, to be the most dominant species during fermentation. These studies indicate clearly that the quality and the sensory properties of wines produced through traditional fermentations vary widely. In a recent study Galy *et al.* [13] have compared the formation of volatiles in wines and wine distillates produced with three types of dry yeasts (M 1107, SM 102 and Fermivin) by the traditional method of fermentation. Characteristic sensory and analytical differences were noticed. This investigation indicates that more insight into fermentation mechanisms is needed to enable selection of specific types of yeasts yielding a product possessing the desired sensory properties.

Table 16.3 Identified yeast varieties found on Charante grapes[a]

Saccharomyces cerevisiae	*Pichia membranaefaciens*
S. rosei	*P. kluyveri*
S. uvarum	*P. polymorpha*
S. kluyveri	*Debaryomyces phaffii*
S. chevalieri	*Metschnikowia pulcherrima*
S. capensis	*Hanseniaspora uvarum*
S. bayanus	*H. osmophila*
S. pretoriensis	*H. valbyensis*
S. cidri	*Saccharomycodes ludwigii*
S. diastaticus	*Nadsonia elongata*
S. globosus	*Hansenula anomala*
S. prostoserdovii	
Torulopsis stellata	*Candida valida*
Torulopsis etchellsii	*Candida sake*
Torulopsis lactis condensi	*Candida intermedia*
Torulopsis candida	*Rhodotorula glutinis*

[a] Source: Park [11].

It may be concluded that an inventory of commercially available yeast strains can be of help to producers of wines and wine distillates to improve the quality of their products and to maintain a constant quality.

16.2.7 Influence of yeast contact and storage time on volatile compounds in wine

During storage of wine with its yeasts changes in composition of the volatile constituents occur. Postel and Adam [14], for example, have shown that during 10 months' storage of wine on its yeasts the concentration of acetaldehyde increased considerably, the concentration of 1-hexanol increased and the concentration of 2-phenylethanol decreased in the second half of the storage period. The contents of all other higher alcohols and of methanol did not change. The contents of acetic acid esters (e.g. ethyl, isoamyl, hexyl, 2-phenethyl) diminished during storage. Among the higher esters only ethyl hexanoate, ethyl decanoate and ethyl dodecanoate showed a slight increase.

16.2.8 Influence of quantity of yeast during distillation on the composition of volatiles in wine distillates

The presence of yeast in wine during distillation strongly affects the quantitative composition of several volatiles such as higher esters and acids [15]. In Figure 16.3, the gas chromatograms of Brouillis distillates of a wine without yeast deposit and of the same wine distilled with an 8% yeast deposit are presented. Distillates of wine distilled with yeast were found to possess a strong perfume-like flavour. The intensity of the fresh, fruity, flower-like flavour is unchanged by the presence of yeast.

16.2.9 Influence of the method of distillation

In general, two systems of distillation are applied to wine distillates: (i) continuous (plate column); and (ii) batch (pot still) distillation. The applied system of distillation of wine plays a part in the composition of volatile constituents [16,17]. The composition of volatile constituents of wine distillates produced on a continuous distillation unit and on a pot-still apparatus are presented in Figure 16.4. Large differences in concentrations between both distillates can be noticed. Some of those differences are caused by the distillation system but the actual processing is just as important for the quality and composition of the product.

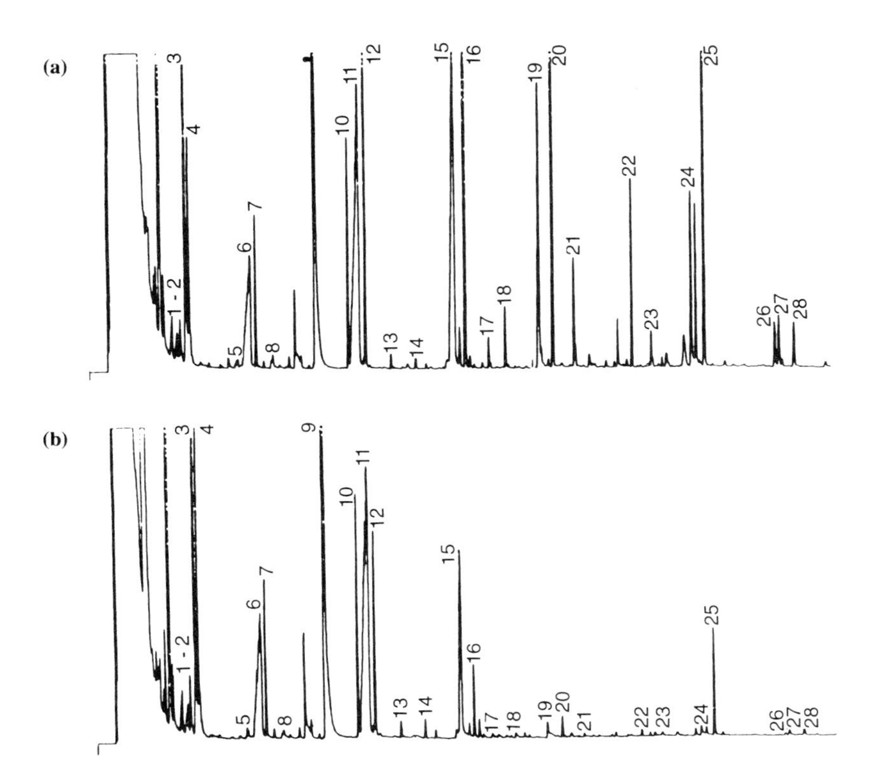

Figure 16.3 Gas chromatograms of the medium-volatile constituents of Broullis distillates produced from wine with (a) 8% and (b) no yeast: 1 = ethyl 2-methylbutanoate; 2 = ethyl 3-methylbutanoate; 3 = 1,1-diethoxyisobutane; 4 = isoamyl acetate; 5 = 1,1-diethoxypentane; 6 = hexanoic acid; 7 = ethyl hexanoate; 8 = benzyl alcohol; 9 = phenethyl alcohol; 10 = diethyl succinate; 11 = octanoic acid; 12 = ethyl octanoate; 13 = phenethyl acetate; 14 = vitispirane; 15 = decanoic acid; 16 = ethyl decanoate; 17 = isoamyl octanoate; 18 = oxo ester (MW 200); 19 = dodecanoic acid; 20 = ethyl dodecanoate; 21 = isoamyl decanoate; 22 = ethyl tetradecanoate; 23 = isoamyl dodecanoate; 24 = ethyl hexadecenoate; 25 = ethyl hexadecanoate; 26 = ethyl octadecadienoate; 28 = ethyl octadecenoate; 27 = ethyl octadecanoate.

16.2.10 Conclusions

These are as follows:

1. The grape varieties and their ripeness affect the flavour of wine distillates.
2. The selection of yeast strains is important in connection with the quality of wine distillates.
3. The presence of yeast deposits during distillation imparts a strong perfume-like odour to the distillate.

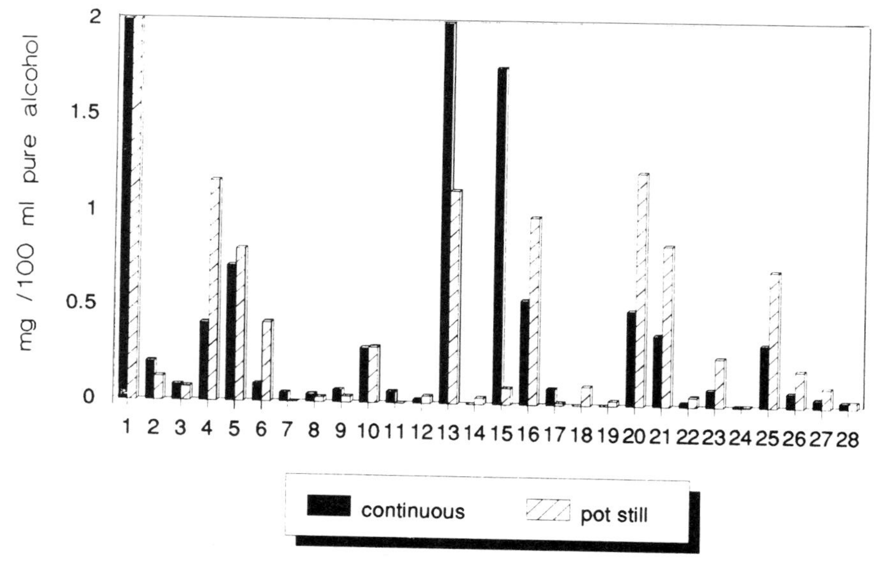

Figure 16.4 Histogram representing the composition of wine distillates produced by means of continuous distillation and by means of pot-still distillation: 1 = hexanol; 2 = ethyl 3-methylbutanoate; 3 = *cis*-3-hexen-1-ol; 4 = 1,1-diethoxy isobutane; 5 = isoamyl acetate; 6 = amyl acetate; 7 = benzaldehyde; 8 = heptanol; 9 = 1-octen-3-ol; 10 = ethyl hexanoate; 11 = isoamyl 2-methylpropanoate; 12 = octanol; 13 = 2-phenylethanol; 14 = isobutyl hexanoate; 15 = diethyl succinate; 16 = ethyl octanoate; 17 = 2-phenethyl acetate; 18 = isoamyl hexanoate; 19 = 2-phenethyl isobutanoate; 20 = ethyl decanoate; 21 = ethyl dodecanoate; 22 = isoamyl decanaoate; 23 = ethyl tetradecanoate; 24 = isoamyl dodecanoate; 25 = ethyl hexadecanoate; 26 = ethyl 2,9-octadecedienoate; 27 = ethyl 9-octadecenoate; 28 = ethyl octadecanoate.

4. The distillation system and the actual processing are important for the quality of the distillate.

16.3 Factors influencing the quality of Scotch whisky

16.3.1 Introduction

Whiskies are produced in many countries (e.g. Scotland, Ireland, USA, Japan, Canada) but the most popular type is Scotch whisky [18]. Most of these Scotch whiskies are blends of malt and grain whisky.

The flavour of newly distilled whisky is influenced by many factors during the route from selection of the raw materials to the final distilled product. The different production steps are listed in Table 16.4. To obtain a product of constant quality, a thorough insight into the reactions taking place during production is of utmost importance.

Table 16.4 Factors influencing the quality of Scotch whisky

Stage	Input and variation
Grain	Climatological conditions Barley variety Water
Malting	Malting process Kilning procedure (peat smoke) Yeast strains
Fermentation	Fermentation conditions Yeast varieties *Lactobacillus* Microbial contaminations
Distillation	Operator's skill Company policy
Maturation	Type of casks Period of maturation
Blending	Company policy Marketing aspects

Quality is not only related to sensory and visual characteristics but also to health aspects. In this context, the formation and occurrence of ethyl carbamate is considered relevant.

For a detailed review of the influence of the different production steps on the flavour of whisky the reader is referred to a publication by Paterson and Piggott [19]. In this chapter, attention will be paid to just some of these factors.

The contribution of compounds formed during the maturation of the distillate will be discussed in section 16.4.

16.3.2 Peated malt

The kilning process has a major influence on the formation of compounds and precursors of compounds, contributing to a large extent to the typical malt flavour. During this process the germinated barley is exposed to heat, peat smoke and gaseous sulphur dioxide. Maillard reactions and pyrolysis products of the peat result in the formation of phenols, pyrazines and thiazols. The ratios between these compounds depend on the type of peat, the sulphur dioxide dosage and burning conditions.

The content of total phenols is used to determine to what extent the malt has been peated [19]:

- lightly peated 1.0–5.0 mg/l total phenols;
- intermediately peated 5.0–15.0 mg/l total phenols;
- heavily peated 15.0–50.0 mg/l total phenols.

The occurrence and formation of phenols have been extensively studied [19, 20]. Nevertheless, phenols are not regarded the most important flavour contributors for the peated malt aroma in whisky [19]. Pyridines and thiazoles have lower threshold values and are considered to contribute more to the final flavour.

Viro [21] has studied the occurrence of pyridine derivatives in two samples of whisky. The analytical results are presented in Table 16.5. These constituents are described in sensory analysis as astringent, green, hazelnut-like, earthy, buttery, caramel-like, rubbery, roasted, fatty and bitter [22]. In sensory assessments, whisky sample B was preferred to whisky A (Table 16.5). Viro suggests therefore that the formation of these compounds should be discouraged by steering the production process.

Leppänen *et al.* [23] have quantified several polysulphides and thiophenes in several commercial samples of Scotch whisky. Organic sulphurous compounds are mostly pyrolysis reaction products formed during distillation. Leppänen *et al.* established correlations between different flavour characteristics and the quantity of dimethyl disulphide (DMDS) and 2-methylthiophene (Figure 16.5). The results of the sensory evaluation of these groups of samples are presented in Table 16.6. The results indicate that the concentrations of these compounds have an effect on different flavour characteristics of whisky.

16.3.3 Mashing, fermentation and distillation

The procedure of mashing and fermentation can be strongly influenced by environmental conditions (e.g. temperature, pH, hygienic status and oxygen concentration). These are, together with the yeast influence during the fermentation stage, important for the desired or undesired formation of higher alcohols, fatty acids, esters, carbonyls and phenols.

Table 16.5 Contents of some pyridines in whisky[a]

	Whisky sample (μg/l)	
Compound	A	B
Pyridine	4.8	2.8
2-Methylpyridine	1.7	0.06
3-Methylpyridine	0.57	0.13
3-Ethylpyridine	0.36	0.04
2,6-Dimethylpyridine	0.49	0.04
3,4-Dimethylpyridine	0.19	0.04
2-Acetylpyridine	0.17	0.05

[a] Source: Viro [21].

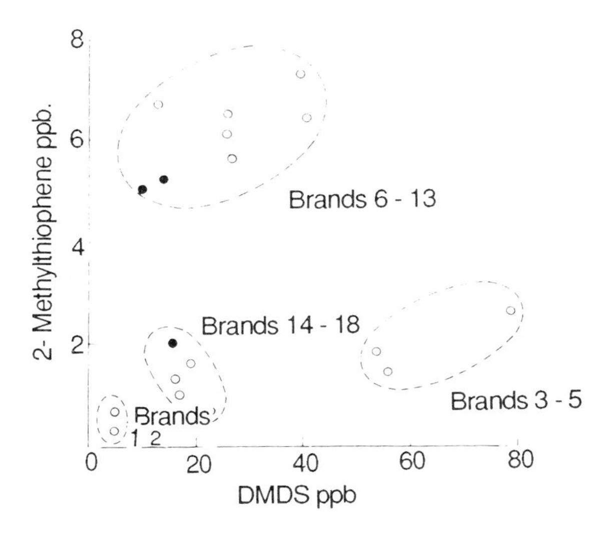

Figure 16.5 Whisky samples grouped according to their contents of DMDS and 2-methylthiophene (○ = ordinary, ● = well-matured blends) [25].

Table 16.6 Sensory evaluation of groups of whisky with varying concentrations of sulphurous compounds[a] (see Figure 16.5)

Brands	Average levels (μg/l)		Flavour description
	DMDS	2-MT	
1–2	5	0.9	Light and neutral
3–5	63	3.4	Slightly bitter, roasted tinge
6–13	29	2.8	Heavy, prolonged distillation
14–18	19	1.5	Intermediate flavour

[a] Source: Aylott *et al.* [25].

In the production of Scotch whisky two categories of yeasts are used, namely primary and secondary yeasts. The primary yeasts (*Saccharomyces cerevisiae*), known as distiller's yeast, are selected according to criteria such as: (i) ethanol production and tolerance; (ii) product flavour; and (iii) resistance to other micro-organisms. The secondary type of yeasts is brewer's or baker's yeast. They are applied because they are capable of producing more flavour compounds such as fatty acids and fatty acid esters [18,19].

Apart from the effect on the alcoholic fermentation, which is dominated by the added yeasts, other types of yeast and bacteria enter the fermentation system via raw materials especially lactic acid bacteria, which are well suited to the distillery environment.

Geddes and Riffkin [24] reported that lactobacilli in controlled fermentation led to reduced levels of acetaldehyde, higher levels of fatty acids and corresponding levels of ethyl esters. Sensory assessment led to the conclusion that the presence of lactobacilli is essential to the development of the perceived quality parameters of Scotch malt whisky. The quality of the actual distillate is, apart from the factors mentioned before, strongly influenced by the distiller's skill and the company's policy.

With regard to these aspects in relation to the flavour quality of whisky a comprehensive overview is given by Paterson and Piggott [19].

16.3.4 Ethyl carbamate

Ethyl carbamate is a naturally occurring compound in most fermented foods and beverages. However, its carcinogenic properties have led to some concern with regard to its presence, even at trace levels, in alcoholic beverages. Since Canada introduced regulatory limits in 1985, research on this topic focused on the development of methods of determination and the pathways of its formation has been given priority in many laboratories all over the world. A group of scientists of the Research Centre of United Distillers has recently reported, in a series of four publications, the results of a profound study on ethyl carbamate (25–28). Their results enabled the definition of selection criteria for raw materials and process adjustments to tackle problems related to the presence of ethyl carbamate in whisky. The most relevant results and conclusions from these studies are referred to in this section.

Aylott *et al.* [25] reported that the potential concentration of ethyl carbamate is related to the amount of so-called 'measurable cyanide' (MC) (e.g. cyanide, cyanate, lactonitrile, cyanohydrin, copper cyanide complex anions) in freshly distilled grain spirits. Storage experiments showed that after 3 months of maturation in oak casks, the ethyl carbamate content is stabilized to a constant level. The correlation between the initial MC concentrations in 221 samples of new-make grain spirit with the ethyl carbamate concentration after 3 months of maturation is presented in Figure 16.6. These results demonstrate that monitoring of initial MC levels is a useful guide for predicting final ethyl carbamate levels in mature spirits. On the basis of this information Aylott *et al.* have studied many factors that were found to affect the concentration of MC in newly distilled grain spirits and the potential ethyl carbamate concentration. The factors reported are compiled in Table 16.7. Furthermore, it was noticed that the MC precursor, which was identified as epihetrodendrin, is located predominantly in the acrospires of the malt.

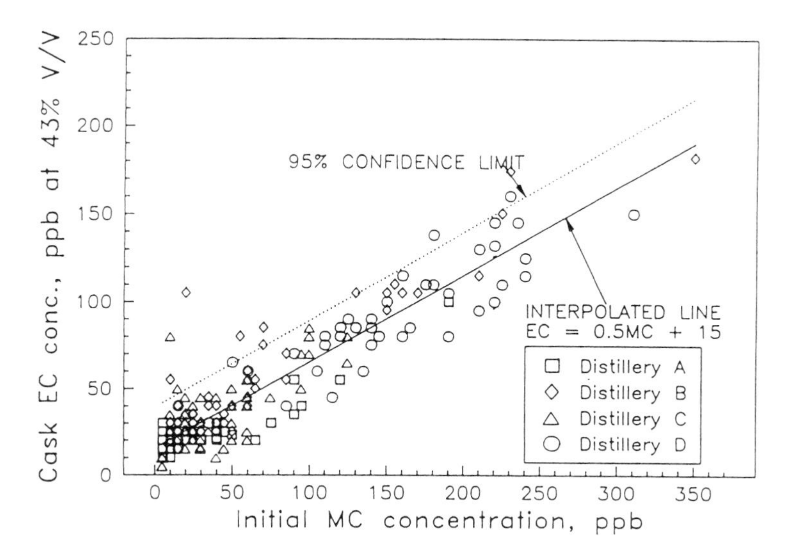

Figure 16.6 Correlation between initial MC concentrations in 221 samples of 'new-make' spirit produced in four different distilleries with ethyl carbamate concentrations in the same spirit sampled after 3 months of maturation. Ethyl carbamate concentrations are normalized to 43% v/v alcoholic strength and MC concentrations are quoted as received at 94% v/v distillation strength [25].

The results of these studies enable producers of whisky to reduce the concentration of ethyl carbamate in their products by:

- selecting low MC producing barley varieties;
- adjusting process parameters; and
- analysing the MC components.

The results of preliminary studies suggest that in Scotch malt whisky and in Bourbon whiskey, conversion of MC into ethyl carbamate was essentially complete within the first 24 h in casks.

16.3.5 Conclusions

These are as follows:

1. Phenols are not considered to be major contributors to the flavour of whisky.
2. Too high levels of pyridines have a negative effect on consumer acceptance.
3. The flavour character of whisky is dependent on the level of sulphurous compounds.

Table 16.7 Factors that affect the concentration of MC in newly distilled grain spirits and the EC concentration after light exposure[a]

Factor	Summary of observations
Barley	
Variety	73 barley varieties could be classified into 4 categories: High MC concentration (800–1000 μg/kg) in 22% of the varieties Intermediate MC concentration (500–800 μg/kg) in 48% of the varieties Medium/low MC concentration (100–500 μg/kg) in 18% of the varieties Low concentration (0–100 μg/kg) in 12% of the varieties
Crop year	MC concentration is more related to the variety than to the crop year
Germination	
Time	MC and EC concentration increases with germination time
Moisture	Within the range 42–50% moisture: an increase of 4%; germination moisture content may more than double the MC potential
Air supply	Ventilation during germination increases the MC level
Wet casting	Hydrostatic pressure has a significant retarding effect on MC formation
Hormonal additives	Application of gibberellic acid leads to significantly higher MC concentrations
Kilning	No effect
Kernel size	Larger kernels produce more MC
Mashing	Temperature strongly affects MC formation

[a] Source: Aylott *et al.* [25]; Mackenzie *et al.* [26]; Cook *et al.* [27]; McGill and Morley [28].

4. Extensive combined sensory and analytical research is needed to improve insight into the relative importance of groups of constituents to the flavour of whisky.
5. The formation of ethyl carbamate is affected by many factors, which can be controlled.

16.4 Maturation

16.4.1 Introduction

The contribution of wood constituents to the flavour of alcoholic beverages has recently been reviewed by Maga [29], covering aspects such as wood composition, various types of wood used, oak wood composition and oak flavour properties, the effect of wood on spirit flavour and the practice of accelerated wood ageing. The last two subjects only will be discussed in this chapter, giving special attention to the oak lactones and the aromatic aldehydes as well as to the use of wood extracts.

16.4.2 Oak lactones

In the group of constituents that originate from the barrels used for the ageing of alcoholic beverages, the oak lactones are prominent. They can be found in the literature with the following names:

- 3-methyl-4-octanolide;
- 4-hydroxy-3-methyloctanoic acid lactone;
- 5-butyl-4-methyl-dihydro-2(3*H*)-furanone;
- whisky lactone; and
- *Quercus* lactone.

One of the isomers was identified in whisky by Suomalainen and Nykänen [30] in 1969, and the *cis* and *trans* isomers were independently found by Masuda and Nishimura [31] in oak wood and in whisky in 1971. Günther and Mosandl [32,33] reported, in one of their interesting publications on enantiomers in fruit and beverages, on the *cis/trans* diastereomers of oak lactone. Their results were not in agreement with those of Masuda and Nishimura [31] and Kepner *et al.* [34]. Günther and Mosandl assigned the *trans* and *cis* configuration on the basis of proton-NMR measurements and found a *cis/trans* ratio of 3:1 for the lactones in a 62% v/v ethanol extract of oak wood. This ratio is opposite to that found by others, which can be explained by their reverse *cis/trans* assignment. Having evaluated the applied techniques and reported results the authors supports the conclusions of Günther and Mosandl [32].

Assuming that the identification of Günther and Mosandl is indeed correct, all concentrations determined by others [35–38] are incorrect and should be interchanged. In this context it should be mentioned that Kepner *et al.* [34] concluded that the 'relative amounts of the two isomers obtained from the synthesis were in reverse order from that expected from thermodynamic considerations of the stabilities of the products'.

Recently Günther and Mosandl [33] determined the configurations of the isomers by preparing diastereomeric di-esters of the 3-methyloctan-1,4-diols obtained by reductive cleavage of the lactones by $LiAlH_4$. They found that oak wood contained two of the possible four isomers (Figure 16.7): 77% (3*S*,4*S*) configurated *cis* and 23% (3*S*, *4R*) configurated *trans*. A flavour description of the four isomers is given in Table 16.8. Knowing that the flavour of many compounds is dependent on their concentration it is to be regretted that taste as well as odour were described of only one, rather high, concentration.

Sharp [39] mentions in his paper on whisky maturation some sensory work with oak lactone carried out at Seagram. Most probably they used the *cis/trans* lactone mixture. Assessing samples of oak lactone in 40% alcohol in five concentrations ranging from 5.3 mg/l to 0.1 mg/l they found that the flavour character depended on the concentration of the lactones. At the

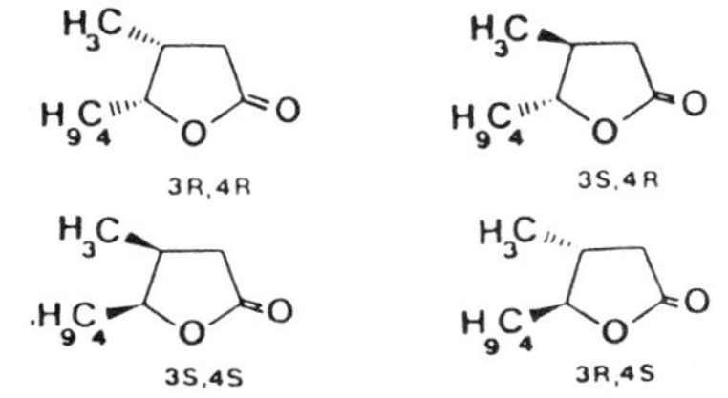

Figure 16.7 Stereoisomers of 3-methyl-4-octanolide [33].

Table 16.8 Flavour descriptions of the isomers of 3-methyl-4-octanolide[a]

		Flavour description	
Stereo-isomers	Occurrence in wood	Taste of 10 mg/l isomer in aqueous sucrose	Odour of 10% isomer in ethanol
cis-3*r*,4*R*	–	Creamy soft, coconut-like	Sweet-woody, fresh, coconut-like
cis-3*S*,4*S*	+	Herbal, coconut-like	Weakly coconut-like, weakly musty earthy after-odour: hay note
trans-3*S*,4*R*	+	Sweet-creamy, fatty, coconut-like	Herbal celery note, weakly coconut-like, clearly green walnut note
trans-3*R*,4*S*	–	Weakly herbal spicy	Strongly coconut-like, after-odour: reminiscent of celery, soup flavouring

[a] Source: Günther and Mosandl [32].

highest concentration (5.3 mg/l) the coconut odour prevailed over a weak oak wood-like flavour. At the lowest of three concentrations (2.1, 1.1 and 0.5 mg/l) the oak wood-like flavour was more pronounced.

The odour threshold value of the *cis* and *trans* oak lactone and their concentrations in matured cognacs and whiskies have been determined by several workers. A compilation of their results is presented in Table 16.9. Assuming that the assignment of Günther and Mosandl [32] of *cis* and *trans* was correct the authors have corrected the data from the literature, that is reversed the concentrations for the *cis* and *trans* isomers reported in the literature.

16.4.3 Aromatic aldehydes

Lignin degradation products such as vanillin and vanillin-like aldehydes and acids are generally regarded as important contributors to the maturation flavour of distilled spirits. This contribution cannot be explained by their

Table 16.9 Odour threshold values and concentrations in wood, whisky, rum and cognac of oak lactones[a]

	Oak lactone		
	cis	*trans*	Reference
Odour threshold	0.067	0.79	[35]
Value (mg/l)	0.05		[36]
Concentration in wood (mg/kg)			
Heartwood	570	50	[37]
Heartwood charred	1750	140	[37]
Sapwood	350	30	[37]
Sapwood charred	920	100	[37]
Concentration in spirits (mg/l)			
Scotch whisky	0.70–1.42	0.26–0.75	[35]
Bourbon whisky	3.84	0.39	[35]
Canadian whisky	0.95	0.07	[35]
Irish whiskey	0.58	0.21	[35]
Jamaica rum	1.21	0.05	[35]
Cognac	0.17–0.43	0.14–0.22	[35]
	0.74	0.25	[38]

[a] The assignments *cis* and *trans* have been interchanged on the basis of the results of Günther and Mosandl [32].

taste threshold values since these are generally higher than the concentrations of the aldehydes found in brandies (Table 16.10). On the basis of these data, their contribution to the flavour of matured beverages seems to be doubtful. In model solutions, evaluating only equal amounts of all compounds, Maga [40] found strong synergistic effects between these compounds, resulting in much lower threshold values of the mixtures compared with the values for the individual compounds (Table 16.10).

Table 16.10 Synergistic effects of aromatic aldehydes in 40% ethanol[a]

Compounds	Taste threshold values (mg/l)
Vanillin (V)	0.1
Syringaldehyde (Y)	15
Sinapaldehyde (I)	50
Ferulic acid (FA)	30
Vanillic acid (VA)	25
Syringic acid (YA)	10
Synapic acid (IA)	100
V/Y	2
FA/VA/YA/IA/V	4
FA/VA/YA/V/Y/I	2

[a] Source: Maga [40].

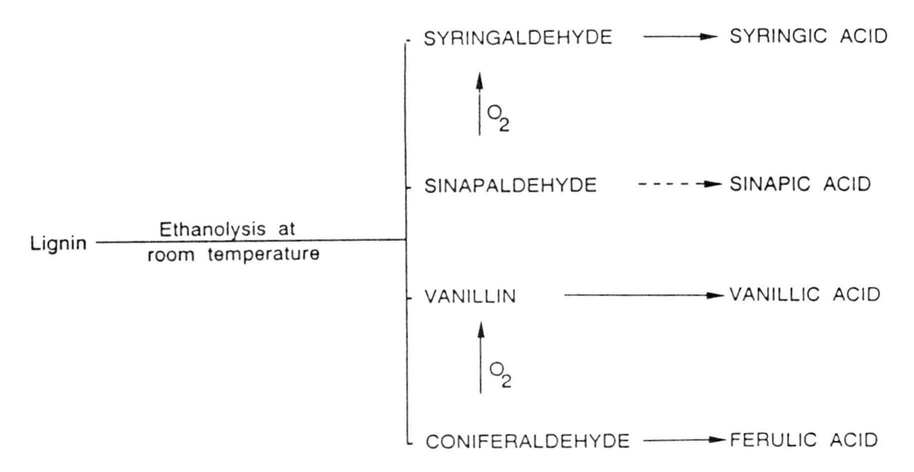

Figure 16.8 Model of degradation of lignin according to Puech [41].

Taking these effects into account the aromatic aldehydes most likely contribute to the woody flavour.

The degradation reaction of lignin has been studied by many workers and has resulted in the degradation model depicted in Figure 16.8. In this model some degradation pathways are presented indicating the influence of oxygen. The influence of oxygen on the degradation of lignin during the extraction of oak wood with alcoholic solutions varying in oxygen concentration has been studied by the authors. Some results will be described in the next section.

16.4.4 Wood extracts

Alcohol losses of 2–4% on an annual basis have been reported for grape brandies and whiskies matured in wooden barrels [42]. This means that much money could be saved by shortening the ageing period, particularly because also fewer barrels and less storing capacity would be required. Maga [29] has reviewed the abundance of approaches to reach this goal. Here attention will be paid to the use of wood extracts only.

In preparing wood extracts the wood is pre-treated to remove bitter-tasting compounds such as tannins. This allows of larger additions of wood extracts, without giving the beverage an intensively bitter, astringent taste. One such pre-treatment has been reported by Litchev [43] who applied heat treatment to improve the quality of wood extracts. Firstly the wood was heated with water at 70°C, then for 10 h at 120°C, and finally for 80 h at 125°C. The pre-treated wood was extracted with 40% alcohol. One of the advantages of this pre-treatment is that the concentration of tannins in the extracts is lower than in the untreated wood.

The influence of oxygen on the degradation of lignin during the extraction of oak wood with alcohol–water mixtures varying in oxygen concentration has been studied by the authors [44]. It was noticed that the presence of oxygen had a strong effect on the degradation of lignin. A higher oxygen concentration resulted in higher concentrations of vanillin, syringaldehyde, coniferaldehyde, vanillic acid and syringic acid. More sinapaldehyde was found in the extracts prepared in an atmosphere free of oxygen. The latter extracts were found to have a more harmonic, cognac-like and less astringent flavour.

Although data are lacking on the production and use of wood extracts, one assumes that they are being applied to some extent by many producers.

Puech [45] concluded his study on oak wood extracts applied to the ageing of grape brandies with the warning that they should be used with caution because they do not possess the chemical characteristics of spirits aged traditionally in barrels. That is why many producers do not consider a more extensive use of wood extracts. They fear that changing traditional production procedures might be detrimental to the quality of their products.

16.4.5 Conclusions

These are as follows:

1. Oak lactones are prominent compounds in the maturation flavour.
2. The original assignment of *cis*- and *trans*-oak lactone was not correct; the concentration of *cis*-lactone is always higher than that of the *trans* isomer.
3. Although aromatic aldehydes occur in concentrations below the threshold values they contribute to the maturation flavour because of synergistic effects.
4. Extracts of pre-treated wood could largely improve the economies of grape brandy production.

16.5 Characterization of distilled beverages using instrumental analysis

Organizations that are protecting the interests of groups of producers, consumer organizations and governmental institutions are interested in the development of methods by which authenticity and origin of a product can be ascertained.

For this purpose statistical analysis and, in particular, multivariate analysis are playing an ever-increasing role in the evaluation of instrumen-

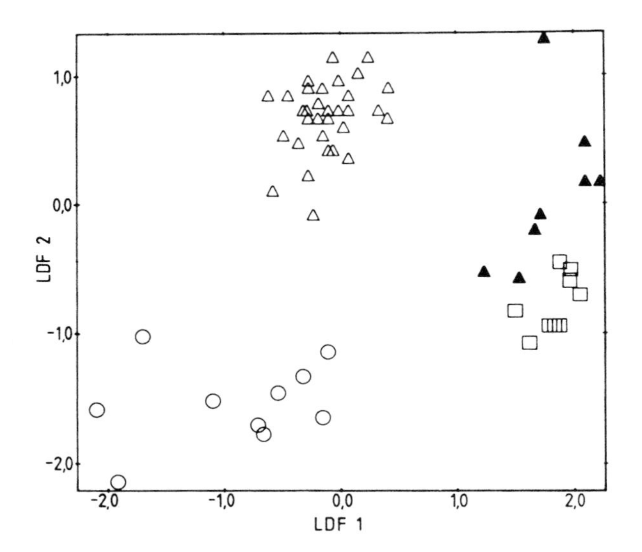

Figure 16.9 Scatter diagram of the two linear discriminant functions for the separation of cognacs (△), armagnac (○) and brandy (□). Suspect cognacs (▲) are included [47].

tal data on distilled beverages. In this section, an overview of different applications is presented.

Cantagrel [46] has described a method for the characterization of cognac by means of principal component analysis based on gas chromatographic data and several standard parameters. The pH, content of ethyl esters and isoamyl esters of aliphatic acids contribute strongly to the differentiation of brandy and cognac. Similar results were obtained by van der Schee *et al.* [47], who were able to characterize Cognac, Armagnac and brandies by means of discriminant analysis based on the content of volatile constituents (Figure 16.9). Maarse and van den Berg [44] described a method for characterization of wine distillates as to their origin. This method is applied to selection of raw materials in the production of brandy.

Principal component analysis has successfully been applied by Headley and Hardy [48] to differentiate Canadian, blended, Scotch and Tennessee whiskies based on gas chromatographic data.

Alvarez and Cabezudo [49] applied several supervised stepwise discriminant analysis (SDA), statistical isolinear multi-category analysis (SIMCA), nearest neighbour analysis (KNN), unsupervised principal component analysis (PCA) and cluster analysis (CA) classification techniques to analytical data for different whisky samples in order to distinguish between genuine whisky of certain brands and less expensive (Spanish) brands.

The use of characterization methods is best illustrated in a paper by Herranz *et al.* [50] who investigated whisky that is sold in Spanish establishments and concluded on the basis of the results of multivariate analysis that 28% of the samples investigated were Spanish whisky sold as Scotch whisky.

In summary it is possible to characterize distilled beverages on the basis of their origin by multivariate statistical analysis of instrumental data. These methods could be used to detect frauds.

References

1. Maarse, H. and Visscher, C.A., *Volatile Compounds in Food-Qualitative and Quantitative Data*, 6th edn., TNO- CIVO Food Analysis Institute, Zeist, Netherlands, 1989.
2. Cantagrel, R., Mazerolles, G., Vidal, J.P., Lablanque, O. and Boulesteix, J.M., L'Assemblage: Une étape importance dans le procès d'élaboration des cognacs. *Symposium Les eaux-de-vie traditionelles d'origine viticole*, 1990, Bordeaux 26–30 June.
3. Baumes, R., Cordonnier, R., Nitz, S. and Drawert, F., Identification and determination of volatile constituents in wines from different vine cultivars. *J. Sci. Food Agric.*, 1986, **37**, 927–943.
4. Cavazza, A., Versini, G., Dalla Serra, A., and Romano, F., Characterization of Six *Saccharomyces cerevisiae* Strains on the Basis of Their Volatile Compounds Production, as Found in Wines of Different Aroma Profiles. *Seventh International Symposium on Yeasts*, Wiley, Chichester 1989, pp. 163–167.
5. Houtman, A.C. and de Plessis, C.S., Influence du cépage et de la souche de levure. *Bull. O.I.V.*, 1985, **58**, 235–236.
6. Polo, M.C., Martin-Cordoro, P. and Cabezudo, M.D., Influence des charactéristiques vartiétales de mout de cépages different sur la fermentation alcoolique par une seule souche de levure sélectionnée. *Bulletin O.I.V.*, 1984, **57**, 312–321.
7. Rooyen, P.C. van, Marais, J. and Ellis, L.P., Multivariate analysis of fermentation flavour profiles of selected South African white wines, in *Progress in Flavour Research*, (ed. J. Adda), Elsevier 1984, pp. 359–385.
8. Nishimura, K. and Masuda, M., Identification of some flavour and characteristic compounds in alcoholic beverages, in *Flavour Research of Alcoholic Beverages, Instrumental and Sensory Analysis*, (eds L. Nykänen and P. Lethonen), Foundation of Biotechnical and Industrial Fermentation, 1984, pp. 111–120.
9. Yunome, H., Zenibuyashi, Y., and Date, A., Characteristic components of botrytised wine, sugars, alcohols, organic acids and other factors. *Hakkokoguka*, 1981, **59**, 169–175.
10. Carnacini, A., Amati, A., Capella, P., Casalini, A., Galassi, S. and Riponi, C., Influence of harvesting techniques, grape crushing and wine treatment on the volatile components of white wines. *Vitis*, 1985, **24**, 257–267.
11. Park, Y.H., Contribution à l'étude des levures de Cognac. I. Etude et classification des levures de Cognac. *Connaissance Vigne Vin*, 1974, **3**, 253–278.
12. Park, Y.H. and Bertrand, A., Contribution à l'étude des levures de Cognac. II. Etude des produit volatiles formes au cours de la fermentation par le levures de Cognac. *Connaissance Vigne Vin*, 1974, **4**, 343–372.
13. Galy, B., Lurton, L., and Cantagrel, R., Aspects oenologiques de la vinification Charantaise, Phases Prefermentaire et Fermentaire, *Symposium Les eaux-de-vie traditionelles d'origine viticole*, 26–30 June, 1990, Bordeaux.
14. Postel, W. and Adam, L., Einfluss des Hefekontakts und der Lagerdauer auf die flüchtige Stoffe des Weines. *Mitt. Klosterneuburg*, 1987, **37**, 54–56.
15. Postel, W. and Adam, L., The influence of the quantity of yeast in wine on the volatiles of grape wine brandies, in *Distilled Beverage Flavour, Recent Developments*, (ed. Piggott, J.R. and Paterson, A.), Ellis Horwood, Chichester, 1989, pp. 149–150.
16. Bertrand, A., Role of the continuous distillation process on the quality of Armagnac, in *Distilled Beverage Flavour, Recent Developments*, (ed. Piggott, J.R. and Paterson, A.), Ellis Horwood, Chichester, 1989, pp. 97–115.

17. Bertrand, A., Comparison analytique des eau-de-vie d'Armagnac obtenues par destillation continue et double chauffe. *Connaisance Vigne Vin*, 1988, **22**, 89–92.
18. Sharp, R., Pert, D.M.K. and Canaway, P.R., Some aspects of shelf-life stability in whisky, in *The Shelf life of Food and Beverages*, (ed. Charalambous, G.), Elsevier, Amsterdam, 1986, pp. 205–224.
19. Paterson, A. and Piggott, J.R., The contribution of the process to flavour in Scotch malt whisky, in *Distilled Beverage Flavour, Recent Developments*, (ed. Piggott, J.R. and Paterson, A.), Ellis Horwood, Chichester, 1989, pp. 151–169.
20. Howie, D. and Swan, J.S., Compounds influencing peatiness in Scotch malt whisky flavour. *Biotechn. Industr. Ferment. Res.*, 1984, **3**, 279–290.
21. Viro, M., N-Hetrocyclic aroma compounds in whisky, *Biotechn. Industr. Ferment. Res.*, 1984, **3**, 227–233.
22. Maga, J.A., Pyridines in food. *J. Agric. Food Chem.*, 1981, **29**, 895–898.
23. Leppänen, A., Ronkainen, P., Denslow, J., Laasko, R., Lindeman, A. and Nykänen, L., Polysulphides and thiophenes in whisky, in *Flavour of Distilled Beverages, Origin and Development*, (ed. J.R. Piggott), Ellis Horwood, Chichester, 1983, pp. 206–214.
24. Geddes, P.A. and Riffkin, H.L., Influence of lactic acid bacteria on aldehyde, ester and higher alcohol formation during Scotch whisky fermentations, in *Distilled Beverage Flavour, Recent Developments*, (eds J.R. Piggott and A. Paterson), Ellis Horwood, Chichester, 1989, pp. 193–199.
25. Aylott, R.I., Cochrane, J.C., Leonard, M.J., MacDonald, J.S., Mackenzie, W.M., McNeish, A.S. and Walker, D.A., Ethyl carbamate formation in grain spirits. Part I: Post-distillation ethyl carbamate formation in maturing grain whisky. *J. Inst. Brew*, 1990, **96**, 213–221.
26. Mackenzie, W.M., Clyne, A.H. and Macdonald, L.S., Ethyl carbamate in grain based spirits, Part II. The identification of cyanide related species involved in ethyl carbamate formation in Scotch grain whisky. *J. Inst. Brew*, 1990, **96**, 223–232.
27. Cook, R., McCaig, N., McMillan, J.M.B. and Lumsden, W.B., Ethyl carbamate in grain based spirits. Part III. The primary source. *J. Inst. Brew*, 1990, **96**, 233–244.
28. McGill, D.J. and Morley, S., Ethyl carbamate in grain spirits. Part IV Radiochemical studies. *J. Inst. Brew*, 1990, **96**, 245–246.
29. Maga, J.A., The contribution of wood to the flavour of alcoholic beverages. *Food Rev. Intern.*, 1989, **5**, 39–99.
30. Suomalainen, H. and Nykänen, L. Investigation on the aroma of alcoholic beverages. *Nährungsmiddelindustrien*, 1970, **23**, 1–15.
31. Masuda, M. and Nishimura, K., Fagaseae. Branched nonalactones from some *Quercus* species, *Phytochemistry*, 1971, **10**, 1401–1402.
32. Günther, C. and Mosandl, A., Stereoisomere Aromastoffe XII. 3-Methyl-4-octanolid-'Quercuslacton, Whiskylacton'. Stuktur und Eigenschaften der Stereoisomer, *Liebigs Ann. Chem.*, 1986, 2112–2122.
33. Günther, C. and Mosandl, A., Stereoisomere Aromastoffe XV. Chirospezifische Analyse natürlicher Aromastoffe: 3-Methyl-4-octanolid-'Quercus, Whiskylacton'. *Z. Lebensm. Unters. Forsch.*, 1987, **185**, 1–4.
34. Kepner, R.E., Webb, D.A. and Muller, C.J., Identification of 4-hydroxy-3-methyloctanoic acid gamma-lactone [5-butyl-4-methyldihydro-2-(3H)-furanone] as a volatile component of oak-wood aged wines of *Vitis vinifera* var. 'Cabernet Sauvignon', *Am. J. Enol. Viticult.*, 1972, **23**, 103–105.
35. Otsuka, K., Zenibayashi, Y., Itoh, M. and Totsuka, A., Presence and significance of two diastereomers of beta-methyl-gamma-octalactone in aged distilled liquors, *Agric. Biol. Chem*, 1974, **38**, 485–490.
36. Salo, P., Lehtonen, M. and Suomalainen, H., The development of flavour during ageing of alcoholic beverages. *Proceedings of the Nordic Symposium on Sensory Properties of Foods*, 18–20 March, 1976, pp. 87–108.
37. Maga, J.A., Formation and extraction of *cis*- and *trans*-beta-methyl-gamma-octalactone from *Quercus alba*, in *Distilled Beverage Flavour, Recent Developments*, (eds J.R. Piggott and A. Paterson), Ellis Horwood Ltd., Chichester, 1989, pp. 171–176.
38. Schreier, P., Drawert, F. and Winkler, F., Composition of neutral volatile constituents in grape brandies, *J. Agric. Food Chem.*, 1979, **27**, 365–372.

39. Sharp, R., Analytical techniques in the study of whisky maturation, in *Current Developments in Malting, Brewing and Distillery*, (eds F.G. Priest and I. Campbell), Institute of Brewing, London, 1983, pp. 143–156.
40. Maga, J.A., Flavor contribution of wood in alcoholic beverages, in *Progress in Flavour Research*, (ed. J. Adda), Elsevier, Amsterdam, 1984, pp. 409–416.
41. Puech, J.L., Extraction and evolution of lignin products in armagnac matured in oak, *Am. J. Enol. Viticult.*, 1981, **32**, 111–114.
42. Swan, J.S., Maturation of potable spirits, in *Handbook of Food and Beverage Stability*, (ed. J.G. Charalambous), Academic Press Inc., Orlando, 1986, pp. 801–833.
43. Litchev, V.A., Bases scientifiques de technologie d'elaboration du Cognac de Bulgarie, Thesis, University of Moscow, 1978, pp. 1978, (French translation).
44. Maarse, H. and Van den Berg, F., Current issues in flavour research, in *Distilled Beverage Flavour, Recent Developments*, (ed. J.R. Piggott and A. Paterson), Ellis Horwood, Chichester, 1989, pp. 1–16.
45. Puech, J.L., Phenolic compounds in oak wood extracts used in the ageing of brandies, *J. Sci. Food Agric.*, 1988, **42**, 165–172.
46. Cantagrel, R., Application of multidimensional analysis to the characterization of cognac in relation to other grape brandies and wine spirits, *Nineteenth International Congress de la Vigne et du Vin*, Santiago, 1986.
47. Schee, H.A. van den, Kennedy, W.H.B., Bouwknegt, J.P. and Hittenhausen-Gelderblom, R.C., A case of Cognac adulteration. *Z. Lebensm. Unters. Forsch.*, 1989, **188**, 11–15.
48. Headley, L.M. and Hardy, J.K., Classification of whiskies by principal component analysis, *J. Food Sci.* 1989, **54**, 1351–1358.
49. Alvarez, P.J. and Cabezudo, M.D., Application of several statistical classification techniques to the differentiation of whisky brands, *J. Sci. Food Agric.*, 1988, **45**, 347–358.
50. Herranz, A., de la Serna, P., Barro, C., Martin, P.J. and Cabezudo, M.D., Application of the statistical multivariate analysis to the differentiation of whiskies of different brands, *Food Chem.*, 1989, **31**, 73–81.

17 Cocoa flavour

B.D. BAIGRIE

Abstract

Cocoa makes a unique contribution to the flavour of chocolate. Cocoa-specific flavour is thought to reside largely in the volatile aroma fraction: the complex composition of this fraction is the consequence of several important post-harvesting operations. Pod ripeness at time of harvest and post-harvesting pod storage can both significantly affect the course of the fermentation stage and, ultimately, the flavour of the cocoa. Cocoa fermentation is crucial not only for the primary production of important flavour volatiles but also for the formation of many nonvolatile chemicals, the so-called cocoa flavour precursor. Bean drying is performed to minimize growth of moulds, to reduce levels of volatile and nonvolatile acids within the bean and to reduce astringency by promoting tanning reactions. Finally, cocoa roasting converts the flavour precursors formed during fermentation into two main classes of flavour-active compounds, aldehydes and pyrazines, via a series of complex chemical reactions, thus completing the spectrum of chemical compounds that comprise cocoa flavour.

17.1 Introduction

Chocolate flavour is universally liked and accepted across a wide span of religious, cultural and ethnographic extremes. Approximately 2.4 million tonnes of cocoa are produced annually [1] and consumed in a variety of products, for example in beverages, baked goods, ice cream and, most commonly, as chocolate confectionery. Although the terms cocoa and chocolate are often used interchangeably they are not synonomous. Commercial cocoa is the fermented, dried beans of the cocoa plant *Theobroma cacao*. The beans are roasted to develop cocoa flavour, shelled and then ground to a fine homogenous mass called cocoa liquor, which is the basic starting material of all chocolate processes [2]. This material may be treated with alkali and then pressed to give cocoa butter and cocoa press cake, which is subsequently milled to give cocoa powder. The cocoa liquor may be mixed with cocoa butter and sugar to give dark chocolate or with milk, sugar and other ingredients (including added flavours) to give milk chocolate. Chocolate is usually subjected to further processing operations, for example, conching, refining and ageing, which may significantly alter its flavour. Although cocoa flavour and chocolate flavour are different, there is no doubt that one of the key attributes of chocolate confectionery that accounts for its widespread appeal is the unique contribution of cocoa to its flavour complex. What then is cocoa flavour and what are the factors that influence the development of cocoa flavour? A critique of these aspects will form the content of the rest of this chapter.

17.2 Components of cocoa flavour

The full flavour of cocoa is necessarily a combination of taste and aroma. The basic taste sensations of cocoa are acidity, bitterness and astringency. Although the major organic acid of unfermented cocoa is citric acid [3], other important contributions to acidity in fermented and roasted cocoa come from lactic acid, oxalic acid and succinic acid [4]. In addition, when cocoa has been fermented under highly aerobic conditions or because cocoa has been dried too rapidly, as is often the case with Asian cocoas, then acetic acid predominates and has a major impact on the perceived acidity of the cocoa [4–6].

Major contributions to the bitterness of cocoa come from the xanthines, caffeine and theobromine. However, unlike tea and coffee where caffeine predominates and is present in the range of 1–4%, in commercial cocoa caffeine is present as the minor alkaloid (typically 0.1–0.7%) whereas theobromine predominates and is present at concentrations in the range 0.8–1.6% [7]. Recent research has shown that theobromine is not the sole contributor to the bitterness of cocoa. Diketopiperazines formed via the thermal decomposition of proteins have been found in roasted cocoa and although the diketopiperazines were shown to be bitter in their own right, the bitterness specific to cocoa was obtained only when theobromine and diketopiperazine were combined in the molar ratio of *c.* 2:1 [8,9]. It was noted that complexation between the diketopiperazine and theobromine occurred in aqueous solution and that this modified the sensory properties of both components and resulted in the specific cocoa bitterness.

Cocoa is a well-known source of polyphenols and it is these compounds that are responsible for the astringency of cocoa. Astringency may often be confused with bitterness, particularly since many astringent compounds are also bitter. However, these two sensations can be distinguished from each other by experienced taste panels and are readily quantified [10]. The major classes of polyphenol found in cocoa are phenolic acids, flavonol glycosides, anthocyanins, catechins and procyanidins. The phenolic acids and flavonol glycosides are present only in trace amounts. This is in contrast to the situation in coffee where the chlorogenic acids, which comprise a group of quinic acid esters, form a quantitatively important fraction of green and roasted coffee beans. Chlorogenic acids are thought to be the astringent components of coffee [11].

The anthocyanins of cocoa, which are responsible for the characteristic purple pigmentation of unfermented cocoa (Forastero varieties), constitute only 4.5% of the total polyphenols in cocoa but are used as the basis of the conventional monitor of raw cocoa quality, the cut test.

The major constituents of the phenolic fraction of cocoa are the catechins and procyanidins [12]. The catechins comprise mainly of the pentahydroxyflavan, (−)-epicatechin with only trace amounts of other catechins such as the epimeric (+)-catechin, (+)-gallocatechin and (−)-epigallocatechin. Research has shown that the procyanidins are oligomers of catechin units; three dimers (B1, B2, B5) and a trimer (C1) have been well characterized, and further oligomers up to the octameric are also known to be present [13,14]. It is certain that the procyanidins are responsible for the astringency of cocoa and may possibly contribute also to its bitterness.

The necessary amelioration of cocoa bean astringency to hedonically acceptable levels during the course of the fermentation, drying and roasting stages is usually rationalized in terms of reactions of catechins and procyanidins, either with themselves or other bean constituents – particularly the proteins/peptides or amino acids. The exact mechanism of these reactions, which are generally referred to as tanning, is poorly understood.

The compounds contributing to the basic tastes of cocoa are not unique to that foodstuff and the main taste sensations of acidity, bitterness and astringency are common to many other foods and thus cannot be regarded as being primarily responsible for the uniqueness of cocoa flavour.

Cocoa specific flavour is thought to reside mainly in the volatile or aroma fraction, that is, in those substances that are primarily detected by nasal receptors rather than by oral taste buds. Since the pioneering analytical investigation of the cocoa volatiles fraction by Mohr [15], many other groups have contributed to the identification of the compounds present in this fraction [16–26] and, to date, 462 compounds have been identified in cocoa aroma. This compares favourably with tea aroma (467 compounds) but is somewhat less complex than coffee aroma in which 655 compounds have been identified. It should be recognized, however, that this number represents the cumulative total of compounds found in cocoas from different countries and in cocoas processed in many different ways. Most commercial cocoa does not have this degree of complexity.

Classification of the constituents of cocoa aroma by their functional group/ring structure into 18 groups (Table 17.1) has been carried out by Flament [27] who showed that pyrazines, amines, esters and acids were the

Table 17.1 Classification of cocoa, coffee and tea flavour constituents into chemical classes

	Number of identified compounds		
Functionality	Cocoa	Coffee	Tea
A Hydrocarbons	39	50	37
B Alcohols	25	20	46
C Aldehydes	22	28	55
D Ketones	24	70	57
E Acids	51	20	21
F Esters	58	29	55
G Lactones	7	8	16
H Phenols	6	42	19
I Furans	19	99	9
J Thiophenes	–	26	1
K Pyrroles	18	67	10
L Oxazoles	15	27	2
M Thiazoles	9	28	7
N Pyridines	12	13	23
O Pyrazines	94	79	22
P Amines and N-containing products	45	24	18
Q Sulphides and S-containing products	10	16	5
R Other compounds	8	9	14
Total	462	455	467

[a] Source: Flament [27].

classes containing the most components unique to cocoa. Half of the 94 pyrazines identified in cocoa were found to be unique to that food when compared with coffee and tea. Caution should be exercised in interpreting the data from Table 17.1 since published quantitative data concerning flavour volatiles in cocoa are somewhat sparse and neither the sensory properties nor the wide variation in the taste thresholds of individual compounds has been taken into account. Hence the most abundant groups may not necessarily be those that contain the components that are most characteristic of cocoa or which have the greatest flavour impact.

For example, in quantitative terms, acetic acid may dominate the volatile composition of cocoa to a remarkable extent with levels in the range 1000–2000 ppm being observed [4,28] yet it is not an essential component of good cocoa flavour. In terms of perception thresholds, the family of substituted pyrazines found in cocoa has published odour thresholds in the range from 2×10^{-6} ppm to 1×10^{2} ppm [29] but many of the cocoa pyrazines are present at concentrations well below their thresholds. In her review on the sensory properties of Maillard reactions Fors [30] has compiled sensory data on 118 pyrazines, including the 94 found in cocoa, yet none has been ascribed cocoa or chocolate-like descriptors for both odour and taste and only five have had the term 'chocolate' applied to them although only as an ancillary descriptor. The most common descriptor used for the main pyrazines of cocoa is nutty [31].

Table 17.2 Compositions of eight model roasting mixtures

Mixture	Composition (w/w)
1.	Fructose/L-valine/L-leucine (2:1:1).
2.	Fructose/L-valine/L-leucine/(-)epicatechin (2:1:1:0.01).
3.	Glucose/glycyl-L-leucine/phenylalanine/(-)epicatechin (1:1:0.1:0.02).
4.	Glucose/glycyl-L-leucine/(-)epicatechin (1:1:0.02).
5.	Glucose/glycyl-L-leucine/phenylalanine (1:1:0.1).
6.	Glucose/glycyl-L-leucine/L-leucine/(-)epicatechin (1:0.8:0.1:0.02).
7.	Glucose/glycyl-L-leucine/phenylalanine/(-)epicatechin (1:0.8:0.1:0.02).
8.	Glucose/glycyl-L-leucine/L-leucine (1:1:0.1).

Nevertheless there appears to be a general consensus that pyrazines, acids and aldehydes are among the most important constituents of cocoa aroma. Indeed, analysis of the ratios of specific pyrazines and aldehydes has been suggested as a fermentation index and as an aid to estimating the optimum degree of roasting [32,33]. It has also been demonstrated by several groups [34–36] that levels of pyrazines increase with degree of roast and concomitant chocolate flavour development.

The compound 5-methyl-2-phenylhex-2-enal, identified in cocoa by Van Praag [20], was claimed to possess 'a deep bitter persistent cocoa note': other authors have suggested that certain saturated aldehydes, for example, isobutyraldehyde in conjunction with dialkyl sulphides gave cocoa or chocolate-like aroma [37,38]. These claims, however, have not been validated substantially. Thus, despite the obvious importance to cocoa flavour of those classes of compounds formed primarily during the roasting of cocoa, no character-impact compound(s) have been unequivocally identified.

There are many claims in the literature of authentic cocoa aroma being generated by roasting simple model mixtures comprising amino acids or peptides and reducing sugars with and without epicatechin [37,39–42]. A re-examination of eight of the most promising model mixtures (Table 17.2) in terms of cocoa aroma development on roasting was undertaken in our laboratories. In this study, the mixtures were roasted and stored for different periods then their aroma was assessed by a trained panel of expert assessors. The major aroma volatiles generated on roasting were collected using headspace sampling techniques [43] and identified by gas-chromatography-linked mass spectrometry.

Although many of the model systems generated complex mixtures of flavour volatiles with characteristic roasted aromas, none of the mixtures produced a realistic roasted cocoa bean aroma when assessed sensorially by an expert panel. The major aroma volatiles produced were carbonyl containing compounds, aldehydes and ketones derived from Strecker

Table 17.3 Major headspace volatiles identified in roasted model mixtures

Compound	Model mixture							
	1	2	3	4	5	6	7	8
3-Methylbutanal	*	*						
Diacetyl	*							
Undecane	*							
Methyl pyrazine	*							
2,5-Dimethylpyrazine	*	*						
2,6-Dimethylpyrazine	*	*						
Ethylmethylpyrazine	*	*						
Furfural	*		*	*	*	*	*	*
Benzaldehyde			*		*		*	
Heptanal					*			
Terpinyl ether		*		*		*	*	
5-Methylfurfural			*	*	*	*	*	*
2-Acetylfuran	*	*			*	*	*	*
C_6-Pyrazine	*	*						
Dimethylisobutyl pyrazine	*							
2-Methylpropanoic acid	*							
Isoamylmethylpyrazine	*							
Isoamyldiethylpyrazine	*							
Isoamyldimethylpyrazine	*	*						
C_3-Hydroxypyrazine	*							
4-Methylpentanoic acid	*							
Phenylacetaldehyde					*			
Acetophenone					*			
3-Phenylfuran					*			
2-Acetylpyrrole	*	*	*				*	
C_{11}-Pyrazine		*						
5-Methyl-2-phenyl-2-hexenal			*		*	*	*	*
Nonanoic acid	*							
N-Methyl-2-formyl-pyrrole				*	*		*	*
4-Methoxypropiophenone	*							
Methyl-di-*tert*-butylphenol	*							

degradation of the amino acid/peptide component or from Maillard reaction of the carbohydrate moiety, and pyrazines (Table 17.3). It was also noted that higher numbers and intensities of pyrazines were produced with mixtures containing fructose rather than glucose, a result in agreement with previous studies [44].

In contrast with authentic cocoa aroma, alcohols, esters and fatty acids were almost totally absent from the model systems indicating that their origin in cocoa is probably via microbiological rather than chemical pathways. Thus although volatiles generated by the Maillard reaction (or Strecker degradation) undoubtedly make a significant contribution to cocoa aroma, the contribution to the cocoa aroma complex of the fermentation volatiles appears to be equally important and further highlights the importance to flavour development of the early stages of cocoa processing.

The flavour of cocoa cannot therefore be attributed to only volatile or nonvolatile components or to one or two chemical classes within either of these broad divisions but rather arises as a consequence of balanced contributions from a wide range of volatile and nonvolatile constituents. If the necessary balance between the key chemical components is not obtained this can manifest itself in cocoa as a major flavour defect or in extreme cases as a severe depression of perceived cocoa flavour [26].

The key to maintaining the full complexity of the cocoa flavour profile and the relative balance of flavour components lies in how the cocoa is grown and processed.

17.3 Cocoa cultivation and processing

The species *Theobroma cacao* is a tropical tree of Central and South American origin that is grown throughout the world within 20° latitude of the equator. The tree, which grows in the understory of a tall forest (i.e. requires shade), flowers and fruits throughout the year – both flowers and fruits being borne, somewhat unusually, directly on the trunk and main branches. Cocoa is harvested in two annual crops: mature fruit pods weigh about 0.5 kg and contain up to 50 seeds or beans enclosed in a mucilaginous pulp. Factors such as genotype can affect significantly the flavour of cocoa; for example fine cocoa varieties such as Criollo, Criollo Forastero hybrids (Trinitarios) or Nacional cocoa differ radically in flavour from Amelonado or Amazonian Forasteros. However, the latter now constitute the vast bulk of world cocoa production and it is other variables, primarily within harvesting, fermentation and drying practices that influence flavour in bulk cocoa.

17.3.1 Harvesting

Pods are usually harvested when mature, in the case of Amelonado cocoa when the pod colour changes from green to golden yellow. It has been established that pod ripeness at harvesting can affect the flavour of the final cocoa [45]. After pods are cut from the trees they are piled in heaps and stored. Harvesting practices vary between different cocoa producing countries: in West Africa pods are usually harvested when fully ripe and stored for several days before the beans are removed; in Malaysia pods are harvested less ripe and are split immediately after harvesting.

Immature pods contain less pulp sugar than mature pods but fruits which are stored dehydrate and lose moisture from the pulp. Both conditions affect the subsequent fermentation process, the former adversely; but the latter is more beneficial for cocoa flavour development and overrides the former. Pod storage for 12 days prior to processing was shown recently to effect a significant improvement in the flavour of Malaysian cocoa [46].

17.3.2 Fermentation

After storage the pods are split and the beans and pulp are scooped out and placed in a large heap, pit or box depending on the producing country. Fermentation then ensues. Application of the correct fermentation regime to cocoa is crucial if good flavour is to be developed. In a recent field trial [46] the optimum fermentation conditions for best cocoa flavour were found to be a full 5 days fermentation with one turn only after 48 h.

Characterization of the microbial changes taking place during fermentation has demonstrated the involvement of three key microbial groups, namely yeasts, lactic acid bacteria and acetic acid bacteria [47–49]. Following the initial colonization of pulp, yeasts proliferate rapidly and dominate the microbial population during the first 2 days of the fermentation. During this time the pulp sugars are metabolized to ethanol, and pectin in the pulp is broken down. The native citric acid of the pulp is simultaneously metabolized by lactic acid organisms but the initial growth of these species is suppressed as oxygen levels become depleted. Conditions of low pH and low oxygen initially favour yeast growth over lactic organisms.

As the pulp is broken down and drains away, the fermenting mass is rendered more permeable to oxygen, so establishing suitable conditions for the growth of acetic acid bacteria. The oxygen content of the mass of beans is increased by turning the heap. After 48 h the microflora become dominated by acetobacter and the ethanol is oxidized to acetic acid. This reaction is exothermic and produces heat sufficient to elevate the temperature in the middle of a heap to 50°C. The by-products of pulp fermentation, particularly the acetic acid, permeate the testa and cotyledon and it is a combination of the acetic acid and heat that kills the bean. Towards the end of the fermentation other organisms such as bacilli, moulds and actinomycetes begin to flourish and dominate the fermentation microflora. A summary of the changes in microflora during fermentation is shown in Figure 17.1.

The ingress of acetic acid into the cotyledon, as well as causing bean death, causes a major disruption of the cellular organization, resulting in the release of previously segregated enzymes and their mixing with other components of the bean. Various hydrolytic and oxidative reactions ensue: (i) glycosidases cleave reducing sugars from the anthocyanins; (ii) invertase converts sucrose to glucose and fructose; (iii) proteinases hydrolyse storage proteins to peptides and free amino acids; and (iv) polyphenoloxidase converts polyphenols to quinones, which are removed via complexation with amino acids and proteins [12,50–52].

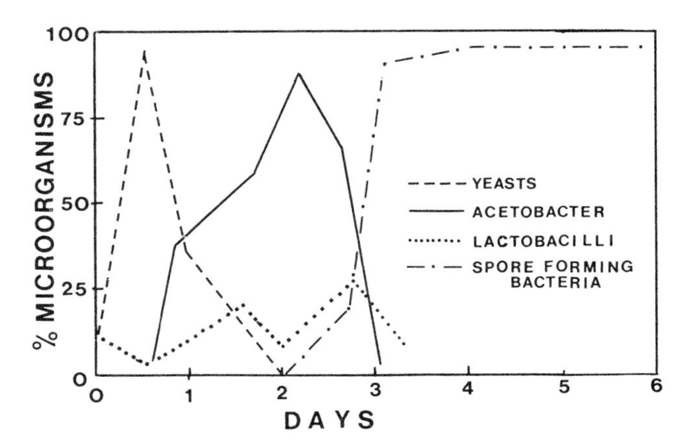

Figure 17.1 Changes in microflora during the course of cocoa fermentation. From Rombouts [47].

These reactions release the chemicals that, during roasting, degrade and/or react to give many of the chemicals that contribute to the cocoa flavour complex. However, the diffusion of fermentation metabolites into the bean is also a primary prerequisite for the development of true cocoa flavour since many of its components originate via this route and not through roasting. During fermentation, soluble components also diffuse out of the bean resulting in a net loss of total solids.

17.3.3 Drying

After the fermentation is complete the wet beans, now completely free from pulp, are dried to a moisture content of 6–7%. Drying stabilizes the microbiological activity of the beans and prevents any further mould growth. In addition, enzymic reactions occurring within the bean are gradually terminated. In practice, fermented cocoa is naturally dried in the sun (typically in West Africa) or artificially dried with hot air from wood or gas fired dryers (typically in Brazil or Malaysia). Despite the fact that the latter method has been implicated as an extraneous source of smoky or hammy flavour defects in cocoa [53], it is also thought to be a factor contributing to other flavour defects endemic to Malaysian cocoa [4].

In a comparative study of sun drying with hot air drying (60°C) it was noted that, irrespective of fermentation treatment, sun drying resulted in a more hedonically acceptable cocoa [46]. The rationale proposed for this observation suggested that sun drying allowed efficient diffusion of free liquid (containing volatile and nonvolatile acids) from within the bean to the surface where it evaporated. The process of acid removal from the interior of the bean continued until the free liquid was exhausted and drying of the bean commenced. With hot air drying the shell lost moisture faster than it could be replenished by diffusion from the cotyledon. As a result case hardening occurred, resulting in evaporation of residual free liquid from inside the bean with concomitant deposition of volatile and nonvolatile acids inside the bean. Thus artificially dried cocoa was often found to be excessively acidic and also suffered from the associated defect of excessive fruitiness caused by higher than normal levels of certain acetate esters.

Another advantage of sun drying over artificial drying is that the more regulated loss of water achieved in the former allows polyphenol oxidase activity to continue well into the drying stage. Continual removal of the polyphenols via oxidation and complexing with protein during drying results in a cocoa that is significantly less astringent and bitter. In countries where, for climatic reasons, sun drying is not a viable proposition, an improved artificial drying regime has been designed. In this regime beans are dried by blowing ambient air for *c.* 80 h until 20% moisture is reached and then heated with air at 60°C for *c.* 15 h to a moisture level of 7.5% [46].

Cocoa that has been fermented and dried now possesses many of the flavour volatiles, particularly acids, esters and alcohols, necessary for good cocoa flavour (Table 17.4).

17.3.4 Roasting

After drying, cocoa is bulked ready to be exported to the consuming countries. At this stage good quality cocoa will possess an acidic, cheesy, floral or nutty aroma; however, development of the typical intense aroma of cocoa requires the application of a controlled high-temperature roasting step. The roasting stage can be applied to whole beans, cocoa nib (shelled beans broken into small pieces) or cocoa mass and there are several commercial units, each purporting to possess unique advantages, available to perform this operation [2]. Cocoa roasting is usually carried out at temperatures between 110 and 200°C and for periods as long as 45 min.

During roasting a series of complex reactions occur between some components, the so-called flavour precursors, formed during fermentation. Studies of cocoa roasting conducted by two groups [54–56] have shown that amino acid levels are reduced by 50% and that almost total loss (92%) of reducing sugars occurs over the same time period. The generic name by which such reactions between amino acids and reducing sugars is known is the Maillard reaction and it is cardinal to almost all processed food products that require production of aromas and the development of colour. Many reviews of Maillard chemistry can be found in the literature [57–59], although the scheme proposed by Hodge [60] is still regarded as one of the

Table 17.4 Flavour volatiles identified in naturally fermented, dried unroasted cocoa

Acids	*Alcohols*	*Esters*	*Hydrocarbons*
Acetic	Pentan-1-ol	Isobutyl acetate	Toluene
Propanoic	Pentan-2-ol	3—Methyl-2-butyl acetate	Myrcene
2-Methylpropanoic	2-Methylbutanol	Isoamyl acetate	Limonene
Butanoic	3-Methylbutanol	Ethyl octanoate	Methylnaphthalene
2-Methylbutanoic	Heptan-2-ol	2,3-Butanediol monoacetate	
3-Methylbutanoic	Linalool	Ethyldecanoate	
Pentanoic	Linalool oxide	Ethyl benzoate	
2-Methylbut-zenoic	1-Phenylethanol	Ethyl phenyl acetate	
4-Methylpentanoic	2-Phenylethanol	2-Phenylethyl acetate	
Hexanoic	Nonan-2-ol	Ethyl dodecanoate	
Heptanoic	2,3-Butanediol		
Octanoic	Benzyl alcohol		
Nonanoic			
Decanoic			
Aldehydes	*Ketones*	*Pyrazines*	*Others*
3-Methylbutanal	Diacetyl	Methylpyrazine	Dimethyl disulphide
Pentanal	Heptan-2-one	2,3-Dimethylpyrazine	n-Pentylfuran
Hexanal	Acetoin	2,5-Dimethylpyrazine	Dimethyltrisulphide
Octanal	Nonan-2-one	2,6-Dimethylpyrazine	γ-Butyrolactone
Nonanal	2-Acetylfuran	Ethylpyrazine	
Decanal	Acetophenone	Trimethylpyrazine	
Benzaldehyde	2-Acetylpyrrole	Tetramethylpyrazine	
2-Phenyl-2-butenal		Ethyldimethylpyrazine	

best general summations of the complex series of chemical reactions known to occur under this title.

That the Maillard reaction is the source of many of the volatiles key to aroma of cocoa and that these volatiles are generated during roasting of cocoa, has been recognized by many workers [61–64]. However, alternative routes for the generation of some of these key volatiles also exist. Aldehydes are known to be formed in the cold via Strecker degradation of amino acids, with orthoquinones providing the dicarbonyl moiety [65]. Recent work by Rizzi [66] has demonstrated the general formation of pyrazines via the acid-catalysed reaction of acyloins (e.g. acetoin) with ammonia at temperatures (22°C) significantly lower than those necessary for cocoa roasting. It is also known that tetramethylpyrazine may be produced microbiologically: it has been found in other fermented foods [67] and, more specifically, has been shown to be produced by *Bacillus subtilis*, a species that is present in the later stages of cocoa fermentation [68].

Moreover, tetramethylpyrazine has been shown to be present at similar levels in cocoa butter from unroasted and roasted cocoa, respectively; other pyrazines including dimethylpyrazine and trimethylpyrazine have been identified in cocoa butter expelled from unroasted cocoa [25,69]. Hence non-thermal routes to the formation of some key classes of cocoa volatiles are strongly implicated.

17.3.5 Cocoa lipids

No discussion about the flavour of cocoa would be complete without considering the role played by the cocoa lipids. Cocoa lipid, or cocoa butter as it is known colloquially, is the major single constituent of cocoa, comprising some 54% of the dry weight of the bean. It consists of a mixture of trisaturated, mono-unsaturated, di-unsaturated and tri-unsaturated triglycerides, with the mono-unsaturated triglycerides comprising about 80% of the total. The fatty acids involved are almost entirely stearic, palmitic and oleic with a small amount of linoleic. A study of the role of lipid in the formation of aroma volatiles concluded that they played a minor role, except when flavour defects were involved [70]. However, although pure cocoa butter is tasteless and odourless, the distribution of aroma compounds between lipid and nonlipid phases is such that volatile flavour resides primarily in the cocoa butter. Hence it is an important factor affecting the perception of chocolate flavour. The unique melting characteristics of the fat contribute in a major way to the release of flavour in the mouth and also to the specific mouthfeel attributes of chocolate.

17.4 Conclusion

The uniqueness of cocoa flavour arises from the fact that it is one of the few foods that is both fermented and roasted. To achieve this unique taste and aroma complex requires a combination of microbiological, enzymic and thermal pathways. The specific post-harvesting operations described in this paper are vital to the generation of the chemical moieties participating in these pathways and are responsible for the variation and complexity of cocoa flavour.

References

1. Gill and Duffus Group, *Cocoa Statistics*. Gill & Duffus Group, St Dunstan's House, London, 1989.
2. Beckett, S.T., *Industrial Chocolate Manufacture and Use*, Blackie, Glasgow, 1988.
3. Biehl, B. and Adomako, D., Cocoa fermentation – control acidification and proteolysis. *Lebensmittelchemie und Gerichtlicht Chemie*, 1983, **37**, 57–63.
4. Baigrie, B.D. and Rumbelow, S.J., Investigation of flavour defects in Asian cocoa liquors. *J. Sci. Food Agric.*, 1987, **39**, 357–368.
5. Chong, G.F., Shepherd, R. and Poon, Y.C., Mitigation of cocoa bean acidity – fermentary investigations, in *International Conference on Cocoa and Coconuts, Kuala Lumpur. Malaysian Agricultural Research and Development Institute*, 1978, pp. 387–414.
6. Chick, W.H., Mainstone, B.J., and Wai, S.T., *Mitigation of cocoa acidity in peninsular Malaysia. Proceedings of the Eighth International Cocoa Research Conference, Cartegena, Colombia*, Cocoa Producers Alliance, Lagos, 1982, pp. 759–4.
7. Egan, H., Kirk, R.S. and Sawyer, R., *Pearson's Chemical Analysis of Foods*, Churchill Livingstone, Edinburgh, 1981, p. 301.

8. Pickenhagen, W., Dietrich, P., Keil, B., Polonsky, J. Nouaille, F. and Lederer, E., The identification of the bitter principle of cocoa. *Helv. Chim. Acta*, 1975, **58**, 1078–1086.
9. Van der Greef, J., Tas, A.C., Nijssen, L.M., Jetten, J. and Höhn, M., Identification and quantitation of diketopiperazines by liquid chromatography-mass spectrometry, using a moving belt interface. *J. Chromatogr.*, 1987, **394**, 77–88.
10. Lea, A.G.H. and Arnold, G.M., The phenolics of cider: Bitterness and astringency. *J. Sci. Food Agric.*, 1978, **29**, 478–483.
11. Clifford, M.N. and Ohiokpehai, O., Coffee astringency, *Anal. Proc.*, 1983, **20**, 83–86.
12. Forsyth, W. and Quesnel, V.C., The mechanism of cacao curing. *Advances in Enzymology*, 1963, **25**, 457–492.
13. Thompson, R.S., Jacques, D., Tanner, R.J.N. and Haslam, E., Plant proanthocyanidins. Part 1: Isolation, structure and distribution in nature. *J. Chem. Soc. Perkin 1*, 1972, 1387–399.
14. Cros, E., Villeneuve, F. and Vincent, J.C., Evaluation of polyphenols of cocoa during fermentation in relation to quality, *Proceedings of the Ninth International Cocoa Research Conference*, Lome, Togo, pp. 651–655.
15. Mohr, W., Untersuchung über das kakao-aroma mit hilfe der gaschromatographie unter besonderer berücksichtigung des conchierens von schokoladen massen. *Fette, Seife, Anstrichm.*, 1958, **60**, 661.
16. Bailey, S.D., Mitchell, D.G., Bazinet, M.L. and Weurman, C., Studies on the volatile components of different varieties of cocoa beans. *J. Food Sci.*, 1962, **27**, 165–170.
17. Dietrich, P., Lederer, E., Winter, M. and Stoll, M., Sur l'ârome de cacao. I. *Helv. Chim. Acta*, 1964, **47**, 1581–1590.
18. Flament, I., Willhalm, B. and Stoll, M., Sur l'ârome de cacao. III. *Helv. Chim. Acta*, 1967, **50**, 2233–2243.
19. Marion, J.P., Muggler-Chavan, F., Viani, R., Bricout, J., Reymond, D. and Egli, R.H., Sur la composition de l'ârome de cacao. *Helv. Chim. Acta*, 1967, **50**, 1509–1516.
20. Van Praag, M., Stein, H. and Tibbetts, M., Steam volatile aroma constituents of roasted cocoa beans. *J. Agric. Food Chem.*, 1968, **16**, 1005–1008.
21. Van Der Waal, B., Kettenes, D., Stoffelsma, J., Simpa, G. and Semper, A.T.J., New volatile components of roasted cocoa. *J. Agric. Food Chem.*, 1971, **19**, 276–280.
22. Vitzthum, O.G., Werkhoff, P. and Hubert, P., Volatile components of roasted cocoa: the basic fraction. *J. Food Sci.*, 1975, **40**, 911–916.
23. Carlin, J.T., Lee, K.N., Hsieh, O.A., Hwang, L.S., Ho, C.T. and Chang, S.S., Cocoa butter flavours. *Manuf. Conf.*, 1982, **6**, 64–73.
24. Gill, M.S., Macleod, A.J. and Moreau, M., Volatile components of cocoa with particular reference to glucosinolate products. *Phytochemistry*, 1984, **23**, 1937–1942.
25. Carlin, J.T., Lee, K.N., Hsieh, O.A., Hwang, L.S., Ho, C.T. and Chang, S.S., Comparison of acidic and basic volatile compounds of cocoa butters from roasted and unroasted cocoa beans. *J. Am. Oil Chem. Soc.*, 1986, **63**, 1031–1036.
26. Baigrie, B.D., Rumbelow, S.J. and McHale, D., Flavour defects of cocoa, in *Flavour Science and Technology*, (eds M. Martens, G.A. Dalen and H. Russwurm Jr), John Wiley & Sons, Chichester, 1987, pp. 133–141.
27. Flament, I., Coffee, cocoa and tea. *Food Reviews Internat.*, 1989, **5**(3), 317–414.
28. Dimick, P.S. and Hoskin, J.M., Chemicophysical aspects of chocolate processing – a review. *Can. Inst. Food Sci. Techn. J.*, 1981, **14**, 269–282.
29. Ziegleder, G., New knowledge of cocoa formation and its modification through technical processes. *Lebensm. Gerichtl. Chemie.*, 1983, **37**, 63–69.
30. Fors, S., Sensory properties of volatile Maillard reaction products and related compounds – a literature review, in *The Maillard Reaction in Foods and Nutrition*, (eds G.R. Waller and M.S. Feather), ACS Symposium Series No. 215, 1983, pp. 185–286.
31. Tressl, R., Bildung von aromastoffen durch Maillardreaktion. IX. *Proceedings of Colloque Scientifique International sur le Café*, London, 16–20 June 1980, ASIC (Paris), 1981, pp. 55–76.
32. Ziegleder, G., Highly volatile cocoa flavour constituents as indicators during cocoa processing. *CCB Review for Chocolate, Confectionary*, 1982, 7, 7–18, 20–22.
33. Ziegleder, G., Possibilities for the analytical estimation of cocoa aroma. *Susswaren*, 1984, pp. 422–426.

34. Reineccius, G.A., Keeney, P.G. and Weissberger, W., Factors affecting the concentration of pyrazines in cocoa beans. *J. Agric. Food Chem.*, 1972, **20**, 202–206.
35. Ziegleder, G., Gaschromatographische rostgradbestimmung von kakao über methylierte pyrazine. *Deutsch Lebensm. Rund.*, 1982, **78**, 77–81.
36. Baigrie, B.D., unpublished data, 1988.
37. Gramshaw, J., Studies on cocoa and chocolate aroma. *Cadbury Ltd, Research Report*, 1979.
38. Lopez, A.S. and Quesnel, V.C., The contribution of sulphur compounds to chocolate aroma, in *Proc. Int. Kongr. Kakao-Schokoladeforsch.*, 1974, pp. 92–104.
39. Rohan, T.A., The flavour of chocolate, its precursors and a study of their reaction. *Gordian*, 1969, pp. 443–447, 500–501, 542–544, 587–590.
40. Jee, M.H., Volatile constituents of model systems related to cocoa and chocolate. PhD Thesis, University of Leeds, 1977.
41. Lane, M.J. and Nursten, H., The variety of odours produced in Maillard systems and how they are influenced by reaction conditions, in *The Maillard Reaction in Foods and Nutrition, ACS Symposium Series 215*, 1983, pp. 141–57.
42. Ney, K.H., Cocoa aroma – model roasting experiments. Reaction products of sugars and amino acids. *Gordian*, 1985, 88–92.
43. Baigrie, B.D., Laurie, W.A. and McHale, D., Artifact formation during headspace analysis, in *Progress in Flavour Research 1984*, (ed. J. Adda), Elsevier, Amsterdam, 1985, pp. 577–582.
44. Koehler, P.E. and Odell, G.V., Factors affecting the formation of pyrazine compounds in sugar-amino reactions. *J. Agric. Food Chem.*, 1970, **18**, 895–898.
45. Lewis, J.F. and Lee, M.T., The influence of harvesting, fermentation and drying practices and their implication for malaysian practices. Seminar on the latest development in raw cocoa bean quality improvement through processing and the Malaysian cocoa grading system, Tawau, Sabah, 1985.
46. Duncan, R.J.E., Godfrey, G., Yap, T.N., Pettipher, G.L. and Tharumarajah, T., Improvement of Malaysian cocoa bean flavour by modification of harvesting, fermentation and drying methods – The Sime-Cadbury Process. *Cocoa Growers' Bulletin*, 1989, **42**, 42–57.
47. Rombouts, J.E., Observations on the microflora of fermenting cacao beans in Trinidad. *Proc. Soc. Appl. Bact.*, 1952, **15**, 103–111.
48. Ostovar, K. and Keeney, P.G., Isolation and characterisation of microorganisms involved in the fermentation of Trinidad's cocoa beans. *J. Sci. Food Agric.*, 1973, **38**, 611–617.
49. Carr, J.G., Davies, P.A. and Dougan, J., Cocoa fermentation in Ghana and Malaysia, Part 1. Research Report, University of Bristol Research Station, Long Ashton, Bristol, 1979.
50. Purr, A., Springer, R. and Morcinek, H., The enzymatic transformations within cacao beans during the fermentation process, in *Proceedings of the Eighth Inter-American Cacao Conference, Trinidad and Tobago*, 1960, pp. 65–70.
51. Purr, A., Enzymic reaction in cocoa beans during fermentation under vacuum infiltration. *Susswaren*, 1972, **16**, 543–550.
52. Biehl, B., Enzymologische und cytologische probleme der kakaoaufbereitung. *Annales de Technologie Agricole*, 1972, **21**, 435–455.
53. Maravalhas, N., Defectos do cacao brasileiro, suas causes e meios de correcaos. *Cacua Actualidades*, 1965, **2**, 19.
54. Rohan, T.A. and Stewart, T., The precursors of chocolate aroma: changes in free amino acids during the roasting of cocoa beans. *J. Food Sci.*, 1966, **31**, 202–205.
55. Rohan, T.A. and Stewart, T., The precursors of chocolate aroma: changes in sugars during the roasting of cocoa beans. *J. Food Sci.*, 1966, **31**, 206–209.
56. Mohr, W., Rohrle, M. and Severin, Th., On the formation of cocoa aroma from its precursors. *Fette, Seife, Anstrichmittel*, 1971, **73**, 515–521.
57. Mauron, J., The Maillard reaction in food, in *Maillard Reactions in Food (Progress in Food and Nutrition Science Vol. 5),* (ed. C. Eriksson), Pergamon Press, Oxford, 1981, pp. 5–35.

58. Baltes, W., Chemical changes in food induced by the Maillard reaction. *Food Chemistry*, 1982, **9**, 59–73.
59. Hoskin, J.C. and Dimick, P.S., Role of non-enzymatic browning during chocolate processing – a review. *Process Biochemistry*, 1984, **19**, 92–104.
60. Hodge, J.J., Chemistry of browning reactions in model systems. *J. Agric. Food Chem.*, 1953, **1**, 928–934.
61. Rohan, T.A., The flavour of chocolate. *Food Proc. and Marketing*, January 1969, January 12.
62. Mohr, W., Uber das roesten von kakaobohnen I. *Fette, Seife, Anstrichm.*, 1970, **72**, 695–699.
63. Mohr, W. and Roehrle, M., Uber das roesten von kakaobohnen II. *Fette, Seife, Anstrichm.*, 1970, **17**, 703–707.
64. Keeney, P.G., Various interactions in chocolate flavour. *J. Am. Oil Chem. Soc.*, 1972, **49**, 567–572.
65. Motoda, S., Formation of aldehydes from amino acids by polyphenol oxidase. *J. Ferment. Tech*, 1979, **57**, 395–399.
66. Rizzi, G.P., New aspects on the mechanism of pyrazine formation in the Strecker degradation of amino acids, in *Flavour Science and Technology*, (eds M. Martens, G.A. Dalen and H. Reisswurm Jr), John Wiley & Sons, Chichester, 1987, pp. 23–28.
67. Kosuge, T., Zenda, H., Tsuji, K., Yamamoto, T. and Narita, H., Distribution of tetramethylpyrazine in fermented foodstuffs. *Agric. Biol. Chem.*, 1971, **35**, 693.
68. Zak, D.L., Ostovar, K. and Keeney, P.G., Implication of *Bacillus subtilis* in the synthesis of tetramethylpyrazine during fermentation of cocoa beans. *J. Food Sci.*, 1972, **37**, 967–968.
69. Baigrie, B.D., unpublished data, 1984.
70. Hansen, A.P. and Keeney, P.G., Comparison of carbonyl compounds in mouldy and non-mouldy cocoa beans. *J. Food Sci.*, **35**, 37–40.

18 Cheese flavour

J. BAKKER and B.A. LAW

Abstract

The development of the texture and flavour of cheese takes place over a long period of time, starting during cheese-making and continuing during the maturation period. The main processes are the fermentation by the starter culture and the enzymic modifications of the constituents in the milk. Enzymes can be indigenous to the milk, produced by adventitious organisms, starter bacteria, secondary organisms growing during maturation or added with the rennet. Their activity towards lactose, lipids and proteins will be described in relation to the development of texture and flavour. Proteolysis influences the texture but also leads to the formation of flavour peptides and free amino acids, which form the precursors for the development of aroma compounds. Lipolysis is important for the development of the typical flavour of mould-ripened cheeses, while the starter bacteria have a direct impact on the flavour of fresh cheeses such as cottage cheese.

18.1 Introduction

Cheese is one of the earliest fermented foods, and references to cows and milk are found in Sanskrit writings of the Sumerians *c.* 4000 BC and in Babylonian records dating around 2000 BC. The great advantage of cheese was that much of the nutritional value of the milk could be preserved for a long period of time [1].

Many different ways of cheese-making have been developed in the world but, although there are over 400 varieties of cheese – generally named after their place of origin, there are only about 18 distinctly different types [2]. The common factor is that all natural cheeses are made from milk. There are three main classes: soft, blue-veined and hard-pressed cheeses. These classes vary widely in moisture content and, as a consequence, in keeping quality, method of ripening and their sensory characteristics. In this chapter the various stages in cheese-making are described briefly with emphasis on the processes important for flavour formation. Although some of these processes influencing flavour formation are poorly understood, some new methods developed for the acceleration of cheese-ripening will also be discussed.

18.2 Cheese-making

Cheese-making starts with lactic fermentation of the milk. Some cheeses, such as cottage cheese, may be consumed fresh, and their flavour depends mostly on this fermentation process. In other varieties, considerable changes in flavour and texture are obtained after a ripening period, allowing the formation of many compounds influencing the sensory properties of the product.

The milk used for cheese-making is a complex biological substrate with a fairly bland aroma. Milk varies in its physical and chemical properties and contains enzymes that also contribute to the cheese-making processes. Additionally, the natural floras present in the milk may play a role. Treatment of the milk will influence the microflora and the resulting enzyme activity, with obvious effects on the development of flavour. In the UK, milk for cheese-making in factories is usually pasteurised, to destroy potential pathogenic organisms [2].

The lactic fermentation is induced by the inoculation of the milk with a starter culture, usually consisting of streptococci and/or lactobacilli. The lactose is converted to lactic acid, lowering the pH value of the curd, and small amounts of carbon dioxide, acetic acid, acetaldehyde, acetoin and diacetyl are also produced, depending on the type of starter.

An addition of rennet, which contains the enzyme chymosin, aids gel formation, curd syneresis and protein breakdown. The fresh curd is rather bland in flavour and has a texture that is considerably different from the ripened product.

During the maturation of cheese, proteolysis of the proteins in the curd, lipolysis of the fats and other enzymic processes bring about major changes in the texture and aid the development of the flavour of cheese [3]. Cheeses with a surface flora and those inoculated with spores of blue mould will develop quite different sensory characteristics. Although micro-organisms play an important role in maturation through their enzymic system, it is believed that some of the flavour products are probably formed in a purely chemical way [4].

In Figure 18.1, a schematic presentation of the main components in the milk is shown, indicating the various pathways by which these components are broken down during the various stages of cheese-making. It can be seen that cheese is a very complex product, and despite many publications, the chemical basis of cheese flavour and aroma has not yet been elucidated. Despite this lack in understanding, successful attempts have been made to influence the period of time required for the maturation of cheese, thus it is

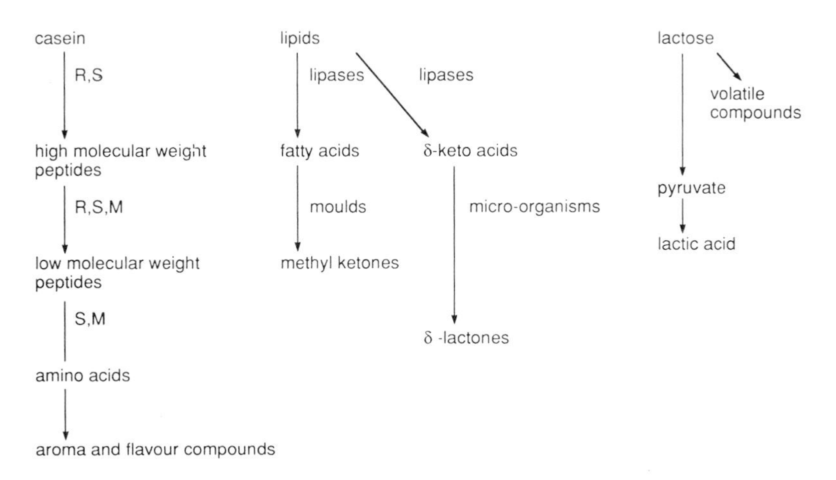

Figure 18.1 Diagram showing the breakdown of the main components in cheese. R = enzymes from rennet; S = starter enzymes; M = milk enzymes.

now possible to accelerate the processes during maturation. This will be discussed in more detail later in this chapter.

18.3 Lactic acid fermentation

18.3.1 Starter cultures

The fermentation of lactose into lactic acid by the starter bacteria influences the sensory quality of the cheese at the end of the maturation stage. The drop in pH due to the production of lactic acid changes the physical and chemical properties of the substrate, thus influencing the growth of the microflora [5]. The various pathways of carbohydrate metabolism in lactic acid bacteria has been reviewed by Kandler [6].

The lactic acid bacteria used as starters in the production of cheese belong to the genera *Streptococcus, Leuconostoc* and *Lactobacillus* [1]. The homofermentative organisms (the streptococci and some lactobacilli) produce mainly lactic acid from lactose or glucose as a substrate, and produce close-texture cheeses [1]. The heterofermentative bacteria, to which group the leuconostocs belong, produce lactic acid, acetic acid and carbon dioxide in molar ratios 1:1:1 [6]. The citrate-fermenting species, such as *Streptoccus diacetylactus* also produce carbon dioxide. The texture of these cheeses would be more open, allowing the required aeration to grow moulds in the blue-veined cheeses, or forming the 'eyes' typical for the appearance of Gouda and Edam.

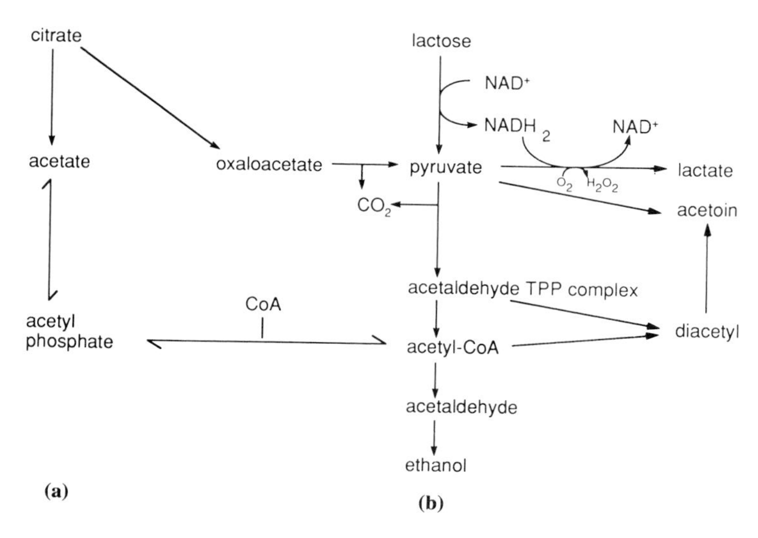

Figure 18.2 Alternative pathways of carbohydrate metabolism in lactic streptococci. (a) *Streptococcus diacetylactis*; (b) group N streptococci.

Streptoccus diacetylactus produces diacetyl, a by-product of the pyruvate metabolism, forming the basis of cottage cheese flavour [1]. Figure 18.2 shows the alternative pathways of the carbohydrate metabolism in lactic streptococci. The formation of acetoin and diacetyl is low when hexoses are the only carbon sources, but the production is significant if additional pyruvate originating from the breakdown of organic acids such as citrate is available [6].

The oxidation-reduction balance is lowered as a consequence of the development of the lactic acid flora [7] to a potential of about −130 mV or less. This balance determines to what extent flavour compounds such as diacetyl and acetaldehyde are formed, rather than the more reduced flavourless compounds such as acetoin, butyleneglycol or ethanol [6]. The existence of reducing conditions during maturation has been shown to continue to have a significant effect on the formation of key aroma compounds [8].

18.3.2 Effect on maturation

A common defect in cottage cheese is caused by the reduction of diacetyl to the flavourless acetoin by diacetyl reductase, an enzyme formed by some starters [1]. Excessive amounts of acetaldehyde can give a harsh flavour; for a balanced flavour the ratio of diacetyl to acetaldehyde should be between 5:1 and 3:1 [9].

The production of large concentrations of ethanol can lead to the formation of off-flavours during prolonged periods of maturation. Fruity flavoured esters are formed by reaction with butyric or hexanoic acids, formed by lipolysis [10]. These fruity flavours are often associated with lipolytic rancidity [1].

Cheddar cheese can be made under controlled bacteriological conditions, and the starter bacteria alone can mediate in the development of a balanced and typical flavour [11]. It has been suggested that their contribution may be largely indirect by creating the correct conditions for the development of flavour compounds, rather than by contributions of their cellular metabolism or their residual enzymes [12]. No relationship could be found between the concentrations of released intracellular starter enzymes and the flavour intensity of Cheddar cheese. However, this was later shown to be caused by a rate-limiting effect of proteolytic activity in cheese [13] and the peptidases of lactic starters are now acknowledged to be significant in determining cheese flavour intensity.

It has been suggested that the nonstarter flora, which could be heat-resistant bacteria from the raw milk or post-pasteurization contaminants from the creamery, also contribute to the maturation [11]. However, attempts to define the most important components of these floras have not been successful [14], although the inclusion of different reference floras affected the formation of several potential flavour compounds.

The drop in pH value to about 5.2 as a result of the production of lactic acid affects the texture as well as the flavour of the cheese. The acidity of the curd aids in dissolving the phosphates and calcium. The calcium level plays an important role in the cohesion of the body of the cheese. For example Emmental contains 0.9–1.0% calcium, while a soft cheese contains 0.2–0.3% calcium. Softening of the texture in soft cheeses such as Brie and Camembert are, in part, attributed to an increase in pH during maturation, due to lactic acid being used as a substrate for the surface flora [15].

18.4 Lipids

18.4.1 Role of lipids

The lipid content of a cheese has an important effect in the development of a good texture and it is well known that a higher fat content leads to a less firm and more elastic body [16]. The composition of the fat can also influence the texture, for example in Emmental it has been observed that a more unsaturated fat results in a softer, less-firm structure [17].

Lipids also play an essential part in the solubilization and retention of the more hydrophobic flavour compounds and flavour precursors, and consequently affect the perception of cheese flavour [5].

An important aspect of lipids is the formation of flavour compounds by two separate processes: lipolysis and oxidation. Cheese made from skimmed milk does not develop a full flavour, indicating the importance of these processes [18]. The formation of free fatty acids from substrates other than lipids and their contribution to the sensory characteristics will also be discussed in this section.

18.4.2 Formation of aroma and flavour compounds

Lipolysis leads to the formation of free fatty acids, which make important contributions to the flavour of Parmesan and Romano cheeses [19] but the contribution of free fatty acids to Cheddar cheese flavour is less clear. Many papers are published, reporting conflicting results. However, there seems to be general agreement that excessive levels of free fatty acids produced by heat-resistent psychotroph lipases are organoleptically described as giving rancid, soapy off-flavours [20].

The lipolytic activity of the lipase depends on the source of the enzyme. Mesophilic lactic streptococci produce lipases with a weak activity against triglycerides but they hydrolyse mono- and diglycerides [21]. The substrate available to the starter lipases depends on the extent of partial triglyceride breakdown by milk lipase and heat-resistent lipase excreted by heat-resistant psychotroph bacteria.

Volatile fatty acids, considered more important to the general background of the taste rather than the aroma of cheese, since they did not influence the aroma at low pH values [22], can be produced by starter streptococci in most cheese varieties. Volatile fatty acids can be produced from amino acids via oxidative deamination by *Streptococcus diacylactus* [23] but, since milk fat is required for the production of volatile acids other than acetic acid [24], it is not certain whether volatile fatty acids can be produced via such a pathway in cheese.

In Swiss cheeses, fermentation of lactic acid and residual sugars by propionic acid bacteria forms a vital part of the formation of the typical aroma and flavour. Figure 18.3 shows the pathways used by *Propionibacterium shermani*. The fermentation proceeds via pyruvate, which can lead to the formation of acetic acid and carbon dioxide, which are important for the formation of the eyes in the cheese. In the presence of catalytic amounts of methylmalonyl-CoA, propionic acid is also formed [25]. In Swiss cheese flavour, production of volatile fatty acids from lipases is not significant in a desirable flavour, but higher fatty acids formed by the starter and propionic acid bacteria may contribute to the typical flavour.

In cheeses matured with actively growing moulds, such as blue-veined cheeses and surface-mould cheeses such as Camembert, the production of

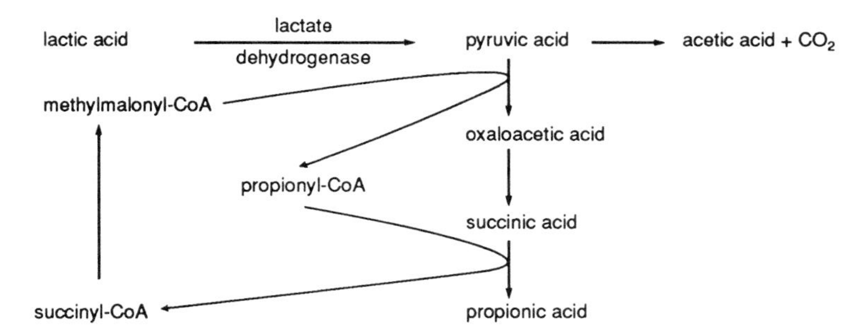

Figure 18.3 Pathways of *Propionibacterium shermani* leading to the formation of propionic acid, acetic acid and carbon dioxide.

methyl ketones and fatty acids is very important to the typical aroma and flavour of these products. These classes of compounds are formed by the mycelium and the spores of *Penicillium roqueforti*; the metabolic pathway of their formation is shown in Figure 18.4. The rate-determining step in the methyl ketone formation is the release of free fatty acids by lipases produced by *Penicillium* spp. [26] and free fatty acids do not accumulate during methyl ketone formation. The fatty acids are then oxidized to β-

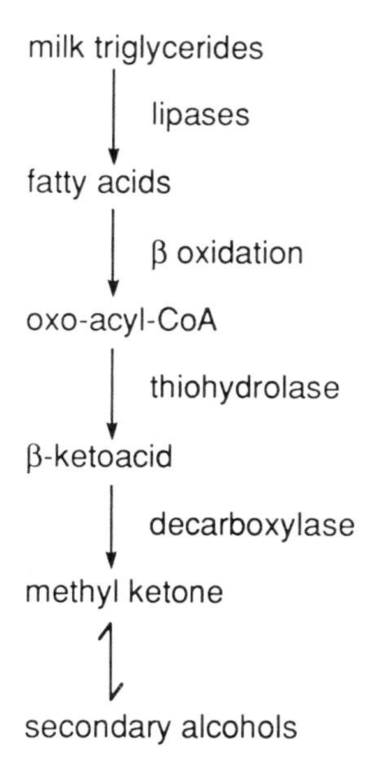

Figure 18.4 Metabolic pathway for the formation of fatty acids and methyl ketones by mycelium and spores of *Penicillium roqueforti*; *n* has a value of 6, 8, 10 or 12.

ketoacids, followed by decarboxylation to methyl ketone [27]. Although free fatty acids are present in a very large concentration, in good-quality cheeses they are not considered as character-impact compounds [4], but they contribute to the overall flavour. Using β-ketoacids as a substrate, the moulds are also able to form small concentrations of short-chain carbonyl compounds [28].

The type of ketones formed does not depend on the available precursor, since 2-heptanone always predominates in blue cheese, whether it is made from ewes' or cows' milk, while 2-nonanone predominates in soft cheese [29]. The concentrations of individual methyl ketones during the maturation of blue cheese fluctuates, indicating interconversion reactions. The methyl ketones can be metabolized into secondary alcohols by *Penicillium roqueforti*, which is a reversible reaction under aerobic conditions [30]. In Camembert and Blue cheeses made from refrigerated milk with high counts of psychotrophic bacteria with lipase activity, high concentrations of methyl ketones and 2-alcohols are found, including some unusual unsaturated methyl ketones such as 2-undecenone and tridicinone [31].

A limited amount of oxidation may explain the occurrence of oct-1-en-3-ol in Camembert, which gives a pleasant mushroom note to this cheese [32]. The mechanism of its formation is unknown but it could be formed from linoleic acid, as has been demonstrated in oxidized butter [33,34]. Several *Penicillium* spp. are able to produce oct-1-en-3-ol by oxidation of unsaturated fatty acids [35,36]; indeed, large concentrations formed by *Penicillium caseicolum* cause an off-flavour.

The γ- and δ-lactones form another group of compounds reported to contribute to cheese flavour [37], especially in Cheddar. The quality of Blue-cheese flavour was increased by increasing the concentrations of γ-tetradecalactone and δ-dodecalactone [38]. There are two possible mechanisms for the formation of δ-lactones. Possible traces of δ-hydroxy acids, perhaps released by lipases during cheese maturation may spontaneously undergo ring closure to form lactones, or they may be converted by enzymes [39]. The other possible mechanism reported to exist in yeasts and moulds, shown in Figure 18.5, involves the microbial reduction of the δ-ketoacids released by lipases, followed by spontaneous ring closure [39].

18.5 Proteins

18.5.1 Proteolytic enzymes

Proteins represent the only continuous solid phase of cheese and play an important role in determining the texture of cheese [16]. Besides the influence on texture, breakdown products of proteins influence the background flavour intensity and form an important pool of flavour

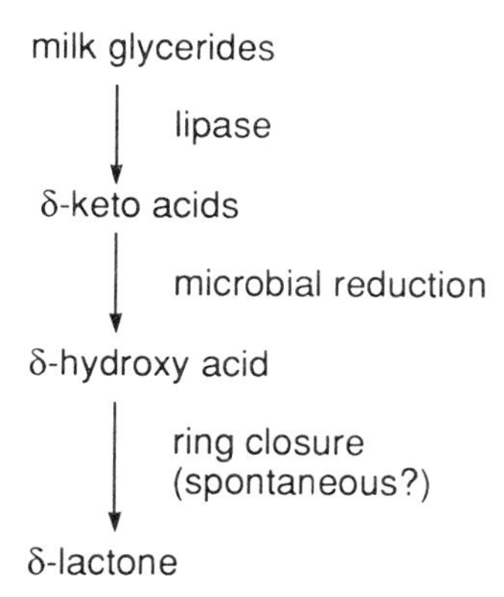

Figure 18.5 Metabolic pathway for the formation of δ-lactones in cheese; n has a value of 4 or 6.

precursors in all matured cheese varieties [16]. Therefore proteolysis has been intensively studied, although owing to its complexity the precise role in flavour development still needs to be elucidated.

Regarding texture, raising the concentration of protein by membrane ultrafiltration has been reported to influence the quality of the structure [40]. An increase in proteins leads to an increase in firmness and a harder texture of cheese [17]. In fact, in samples of 11 different cheeses, it has been determined that the protein level was more significant in explaining the observed differences in structure than pH value, water, sodium chloride and fat content [41].

There are several sources of proteolytic enzymes in milk, with different specificities towards the different caseins, which have been well described in a review [42]. Cows' milk itself contains contains several different proteinases, of which one appears to contribute to the breakdown of β-casein in surface-ripened cheeses [43].

The lower storage temperatures of milk allows the growth of psychotrophic bacteria, derived from soil, water and teat surfaces. These organisms produce heat-resistant proteinases and lipases, which may influence adversely the manufacture and maturation of cheese [42], such as the reduction in yield [44,45], spoilage by rancidity or the production of bitter peptides [43].

Although the addition of rennet to milk is made primarily to initiate the formation of the milk gel, a proportion of the enzyme activity is retained in the curds and plays a role during maturation of cheese – except in cheese manufactured with a high-temperature cooking stage, which denatures the residual enzyme activity [46]. Over a short period of time its main action would be specifically to breakdown the Phe-Met bond in κ-casein;

however, during the maturation period the enzyme induces the release of large molecular weight peptides but no free amino acids [16].

Researchers have long suspected that peptides produced from milk proteins by enzymes of lactic acid bacteria play an important role in determining both the quality and intensity of flavour in cheese.

Lactococci contain a complex proteolytic system that is nutritionally essential, providing the amino acid nutrients from high protein medium such as milk to enable rapid growth. Currently only some of the enzymes of the proteolytic system of the lactic acid bacteria have been isolated and characterized, thus preventing understanding of the complete process of total protein hydrolysis. However, rapid progress in laboratories world-wide (including that of the authors) is revealing the relationships between multiple-enzyme specificity and the yield of flavour-enhancing peptides such that the development of new starter culture constructs containing cloned genes for peptidases may not be too far in the future. Such cultures could combine the excellent acid-producing properties of selected strains with an enhanced flavour-producing capability.

The secondary microfloras such as *Brevibacterium linens* and *Penicillium* spp. growing in cheese also produce proteolytic enzymes, which contribute to the development of structure and flavour.

18.5.2 Flavour peptides

The best characterized peptides involved in flavour are those associated with bitterness [47]. Bitter peptides have now been isolated and characterized from cheeses such as Cheddar [48,49] and Gouda [50] and from protein hydrolysates formed by the action of chymosin [51] and trypsin [52] on caseins. The hydrophobic amino acid content of the peptides was demonstrated to be the major factor in bitterness. From synthetic studies [53] peptides with a high hydrophobic content were demonstrated to be more bitter than their constituent free amino acids. The actual sequence of the hydrophobic amino acids in the bitter peptides was considered not crucial, although blocking of the N-terminal amino and C-terminal carboxyl groups increased bitterness.

Other, more polar and hydrophilic peptides have also been reported to have more positive flavour properties [54]. Glutamyl oligopeptides such as Glu-Asp, Glu-Glu, Glu-Ser, and Glu-Gly-Ser, isolated from a proteinase-modified soy bean protein, were the first peptides demonstrated to have favourable taste properties. Similar acidic oligopeptides were found when fish protein concentrate was hydrolysed with pronase producing an acidic fraction of molecular weight less than 1000 (but not free aspartate and glutamate) that was reported to be brothy with favourable aftertaste [55]. Further analysis of this fraction revealed over 30 oligopeptides, from dipeptides to hexapeptides with a high glutamate and other hydrophilic amino acid content [56].

A peptide with the sequence Lys-Gly-Asp-Glu-Ser-Leu-Ala, has been isolated from the gravy of beef meat that is reported to have a delicious flavour [57]. Known as 'delicious' peptide, its flavour properties have been confirmed by chemical synthesis [58]. Studies on the taste-active sites revealed the peptide had an umami and a sour taste and that the overall taste was due to an interaction between the basic and the acidic components [59]. This work also showed that di- and tripeptide fragments of the delicious peptide, when combined in correct quantities, had the same taste threshold as the parent molecule. This result indicates that interactions between small peptides produced in protein hydrolysates may be a significant factor to be considered in future studies.

18.5.3 Peptides in cheese flavour

The evidence that proteolysis, and by inference peptides, are involved in cheese flavours largely stems from the accelerated cheese-ripening work where the use of proteinase combined with a peptidase-containing lactic acid bacteria extract increased the cheese-maturation rate. This system was proposed by Law and Wigmore in 1983 to achieve a balanced increase in the production of savoury flavour notes using a neutral proteinase from *Bacillus subtilis* and a *Streptococcus lactis* NCDO 712 intracellular extract to accelerate the ripening process [13]. A development of this enzyme system is now commercially available as a method of accelerating cheese-ripening.

The structure of the flavour components in Cheddar cheese and the enzymes involved in their generation are still unknown. The proposed mechanism for accelerated ripening is based on the initial finding that neutral proteinase supplements the starter culture proteinase activity acting on the caseins but causes the production of undesirable bitter cheeses when used in high concentrations [60,61]. This is presumed to be due to the production of hydrophobic bitter peptides. The addition of the intracellular extract of *Streptococcus lactis* NCDO 712 was proposed as a method of breaking down these peptides and allow development of the normal Cheddar cheese flavour. The development of a method for preparing and chromatographically analysing the peptide-containing water-soluble fraction of cheeses [62], which is considered to make the greatest contribution to the intensity of cheese flavour [63], has allowed studies that support this proposal. The use of the neutral proteinase and the intracellular peptidase extract in cheese-curd slurries showed a rapid development of both bitter and savoury flavours, which corresponded with the appearance and disappearance, respectively, of late-running, and

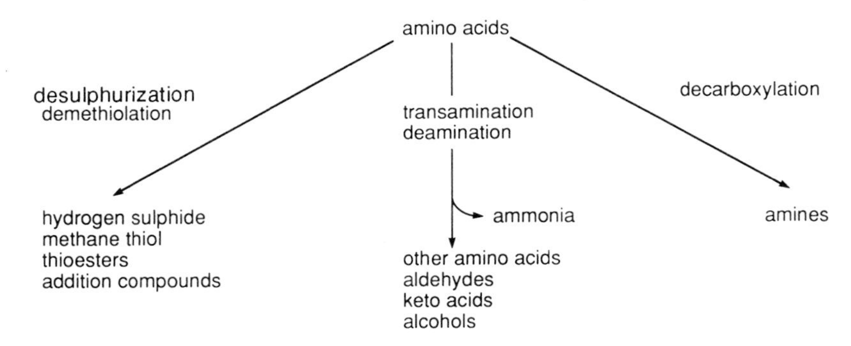

Figure 18.6 General pathways of the breakdown mechanisms for amino acids.

therefore probably hydrophobic components, on reverse-phase liquid chromatography [64].

18.5.4 Formation of volatiles

Amino acids, the end products of proteolysis, can be broken down further by decarboxylation, deamination, desulphurylation and demethiolation. In Figure 18.6 the general pathways for the breakdown mechanisms of amino acids in cheese are shown. These breakdown processes are important for the flavour development in cheese [5]. Therefore, the concentrations of amino acids present in cheese are dynamic. However, each type of cheese has its own characteristic pattern, in Camembert the percentage of tyrosine and lysine is lower than would be expected from the hydrolysis of casein, while alanine, leucine and phenylalanine are present in higher percentages [65].

Volatile sulphur compounds have been found in many cheese varieties and it is generally agreed that they are formed from sulphur-containing amino acids. These compounds may contribute to the typical flavour of Cheddar cheese. Selective removal of methanethiol from its headspace destroyed the typical aroma of Cheddar cheese; the intensity of the Cheddar aroma was found to correlate closely with the concentration [22,66]. Later publications did not confirm the direct involvement of methanethiol in Cheddar aroma [67,68]; however, there remains a strong possibility that it contributes to the sensory characteristics via further reactions and interactions [42].

The low redox potential in cheese is important for the formation of sulphur-containing flavour compounds such as hydrogen sulphide and methanethiol. In an artificially reduced cheese their nonenzymic formation has been observed [8]. It was proposed that hydrogen sulphide is produced by an unknown mechanism from cystine or cysteine; the sulphide then reacts with free methionine, produced by starter peptidases, or with casein and leads to the final release of methanethiol. Observations made during normal cheese-ripening confirmed that hydrogen sulphide is present, before methanethiol and hydrogen sulphide was found in fresh cheese curd [13]. The reducing conditions created by starter bacteria in normal cheese may be sufficient to initiate methanethiol production in the same way, and these conditions would allow the compounds to be maintained in its reduced state. This would explain why only cheeses produced with starters produce methanethiol, even though the starters themselves are unable to produce this compound from methionine [69].

Brevibacterium linens may be a source of methanethiol in surface-smear cheeses [70] and the enzyme responsible has been isolated. The presence of cheese-flavoured thioesters in some surface-ripened cheese may also be explained by the presence of smear bacteria, which appear to produce them from free fatty acids and methanethiol. *Penicillium camemberti* can produce hydrogen sulphide, dimethyl sulphide and methanethiol from methionine [71].

Volatile and nonvolatile amines can be formed by decarboxylation reactions, although the formation of secondary and tertiary amines is difficult to explain [72]. Tyramine is usually the most abundant amine, followed by histidine and tyrosine [73].

Brevibacterium linens can also accumulate a wide variety of deamination and transamination products [74]; branched-chain aldehydes, formed by additional spontaneous decarboxylation, are not normally found in cheese but are probably reduced to alcohols such as 3-methyl-1-butanol, phenylethanol, and 3-methylthiopropanol, derived from leucine, phenylalanine and methionine respectively, and contribute to the general flavour characteristics of surface-ripened cheese. Deamination followed by hydroxylation explains the formation of hydroxyphenylacetic acid in Camembert [75].

Acylation of amines could also explain the formation of *N*-isobutylacetamide, identified in Camembert [76], even though the mechanism of formation was not demonstrated in cheese.

Volatile fatty acids can be formed by oxidative deamination of amino acids. Glycine, alanine and serine give acetic acid, threonine gives propionic acid, valine gives isobutyric acid and isovaline gives isovaleric acid [77].

Aromatic amino acids are the source of interesting aroma compounds such as acetophenone, indole, cresol and phenol. Their occurrence is related to the presence of certain bacteria, but no mechanism for their formation has been determined [16].

Pyrazines have also been found in cheese, but there is no certainty about their formation [78]. In American processed cheese 2,5-dimethylpyrazine and 2,3,5-trimethylpyrazine were considered to impart a nutty background flavour [79]. In mature Cheddar-cheese distillates 2-acetylpyrazine and 2-methoxy-3-ethylpyrazine were found, along with several others [80]. Despite their low concentrations, they may have a beneficial effect on the flavour.

18.6 Conclusion

Many flavour compounds, which were not all discussed in this review, have been identified in cheese. However, their formation and contribution to cheese flavour is not always clear. Neither does there appear to be a common view regarding the important parameters that determine a good-quality flavour and structure of cheese. In spite of these limitations, the chemistry and biochemistry of cheese-flavour development has advanced sufficiently to support the use of enzymes to accelerate maturation.

References

1. Law, B.A., in *Fermented Foods, Economic Microbiology*, Vol.7, (ed. A.H. Rose), Academic Press, London, 1982, p. 147.
2. Chapman, H.R., and Law, B.A., in *Dairy Microbiology, The Microbiology of Milk*, Vol. 2, (ed. R.K. Robinson), Applied Science Publishers, London, 1981, p. 157.
3. Law, B.A. and Sharpe, M.E, in *Lactic Acid Bacteria in Beverages and Food*, (eds J.G. Carr, C.V. Cutting and G.C. Whiting), Academic Press, London 1975, p. 233.
4. Adda, J., Roger, S. and Dumont, J.P., in *Flavor of Foods and Beverages*, (ed. G. Charalambous), Academic Press, London, 1978, p. 65.
5. Adda, J., in *Milk – The Vital Force*, (edited by the organising committee of the Twenty-second International Dairy Congress), D. Reidel Publishing Company, 1987, p. 169.
6. Kandler, O., *Antonie van Leeuwenhoek*, 1983, **49**, 209.
7. Galestoot, T.E. and Kooy, J.S., *Neth. Milk Dairy J.* 1960, **14**, 1.
8. Manning, D.J. *J. Dairy Res.*, 1979, **46**, 531.
9. Lindsay, R.C., Day, E.A. and Sandine, W.E., *J. Dairy Sci.*, 1965, **48**, 863.
10. Bills, D.D. *et al.*, *J. Dairy Sci.*, 1965, **48**, 765.
11. Reiter, B. *et al.*, *J. Dairy Res.*, 1967, **34**, 257.
12. Law, B.A., Castanon, M.J. and Sharpe, M.E., *J. Dairy Res.*, 1976, **43**, 301.
13. Law, B.A. and Wigmore, A.S., *J. Dairy Res.*, 1983, **50**, 519.
14. Law, B.A., Castanon, M.J. and Sharpe, M.E., *J. Dairy Res.*, 1976, **43**, 117.
15. Noomen, A., *Neth. Milk Dairy J.*, 1977, **31**, 75.
16. Adda, J., Gripon, J.C. and Vassal, L., *Food Chem.*, 1982, **9**, 115.
17. Steffen, P., *Schweiz. Milchztg.*, 1975, **101**, 72.
18. Ohren, J.A. and Tuckey, S.L., *J. Dairy Sci.*, 1969, **31**, 598.
19. Aston, J.W. and Dulley, J.R., *Austr. J. Dairy Technol.*, 1982, **35**, 59.
20. Law, B.A., Sharpe, M.E. and Chapman, H.R., *J. Dairy Res.*, 1976, **43**, 459.
21. Stadhouders, J. and Veringer, H.A., *Neth. Milk Dairy J.*, 1973, **27**, 77.
22. Manning, D.J. and Price, J.C., *J. Dairy Res.*, 1977, **44**, 357.
23. Nakae, T. and Elliot, J.A., *J. Dairy Res.*, 1965, **48**, 287.
24. Dulley, J.R. and Grieve, P.A., *Austr. J. Dairy Technol*, 1974, **29**, 120.
25. Hettinga, D.H. and Reinbold, G.W., *J. Milk Food Technol.*, 1972, **35**, 358.
26. Kinsella, J.E. and Hwang, D., *Biotechnol. Bioengng.*, 1976, **18**, 927.
27. Hawke, J.C., *J. Dairy Res.*, 1966, **33**, 43.
28. Dartey, K.C. and Kinsella, J.E., *J. Agric. Food Chem.*, 1971, **19**, 771.

29. Anderson, D.F. and Day, E.A., *J. Agric. Food Chem.*, 1966, **14**, 241.
30. Fan, T.Y., Hwang, D.H. and Kinsella, J.E. *J. Agric. Food Chem.*, 1976, **24**, 443.
31. Dumont, J.P. *et al., Lait*, 1977, **57**, 619.
32. Stark, W. and Forss, D.A., *J. Dairy Res.*, 1964, **31**, 253.
33. Stark, W. and Forss, D.A., *Nature*, 1965, **208**, 190.
34. Stark, W. and Forss, D.A., *J. Dairy Res.*, 1966, **33**, 31.
35. Groux, M. and Moinas, M., *Lait,* 1974, **54**, 531.
36. Kaminski, E., Stawicki, S. and Wasowicz, E., *Appl. Microbiol.*, 1974, **27**, 1001.
37. Wong, N.P. *et al.*, *J. Dairy Sci.*, 1973, **56**, 636.
38. Jolly, R.C. and Kosikowski, F.V., *J. Agric. Food Chem.*, 1975, **23**, 1175.
39. Boldingh, J. and Taylor, R.J., *Nature*, 1962, **194**, 909.
40. Rousseaux, P., Maubois, J.L. and Mahaut, M., 1978, *Twentieth International Dairy Congress*, p. 805.
41. Chen, A.H. *et al.*, *J. Dairy Sci.*, 1979, **62**, 901.
42. Law, B.A., In *Chemistry, Physics and Microbiology*, Vol. 1, General aspects, (ed. P.F. Fox), Elsevier Applied Science, 1987, p. 365
43. Trieu-Cuot, P. and Gripon, J.C., *J. Dairy Res.*, 1982, **49**, 501.
44. Law, B.A., *J. Dairy Res.*, 1979, **46**, 559.
45. Cousin, M.A., *J. Food Protect.*, 1982, **45**, 172.
46. Matheson, A.R., *N. Z. J. Dairy Sci. Technol.*, 1981, **16**, 33.
47. Guigoz, Y. and Solms, J., *J. Chem. Senses Flavor*, 1976, **2**, 71.
48. Harwalkar, V.R. and Elliot, J.A., *J. Dairy Sci.*, 1971, **54**, 8.
49. Hamilton, J.S., Hill, R.D. and van Leeuwen, H., *Agric. Biol. Chem.*, 1974, **38**, 375.
50. Visser, S. *et al.*, *Neth. Milk Dairy J.*, 1983, **37**, 181.
51. Visser, S. *et al.*, *Neth. Milk Dairy J.*, 1983, **37**, 169.
52. Matoba, T. *et al.*, *Agric. Biol. Chem.*, 1970, **34**, 1235.
53. Matoba, T. and Hata, T., *Agricol. Biol. Chem.*, 1972, **36**, 1423.
54. Arai, S. *et al.*, *Agric. Biol. Chem.*, 1972, **36**, 1253.
55. Fujimaki, M. *et al.*, *Agric. Biol. Chem.*, 1973, **37**, 2891.
56. Noguchi, M. *et al.*, *J. Agric. Food Chem.*, 1975, **23**, 49.
57. Yamasaki, Y. and Meakawa, K., *Agric. Biol. Chem.*, 1978, **42**, 1761.
58. Yamasaki, Y. and Meakawa, K., *Agric. Biol. Chem.*, 1980, **44**, 93.
59. Tamura, M. *et al.*, *Agric. Biol. Chem.*, 1989, **53**, 1625.
60. Law, B.A. and Wigmore, A.S., *J. Dairy Res.*, 1982, **49**, 137.
61. Law, B.A. and Wigmore, A.S., *J. Soc. Dairy Technol.*, 1982, **49**, 75.
62. Cliffe, A.J. *et al.*, *Food Chem.*, 1989, **34**, 147.
63. McGugan, W.A., Emmons, D.B. and Larmond, E., *J. Dairy Sci.*, 1979, **62**, 398.
64. Cliffe, A.J. and Law, B.A., *Food Chem.*, 1990, **36**, 73.
65. DoNgoc, M., Lenoir, J. and Choisy. C., *Rev. Lait Fr.*, 1971, **288**, 447.
66. Manning, D.J., Chapman, H.R. and Hosking, Z.D., *J. Dairy Res.* 1976, **43**, 313.
67. Lamparsky, D. and Klimes, I., in *Flavour 81 Proceedings of the third Weurman Symposium*, Munchen 1981, Vol. 3, (ed. P. Schreier), Walter de Gruyter, Berlin, p. 557.
68. Aston, J.W. and Douglas, K., *Aust. J. Dairy Technol.*, 1983, **38**, 66.
69. Law, B.A. and Sharpe, M.E., *J. Dairy Res.*, 1978, **45**, 267.
70. Collin, J.C. and Law, B.A., *Sciences des Alimentes*, 1989, **9**, 805.
71. Tsugo, T. and Matsuoko, H., *Proceedings of the Sixteenth International Dairy Congress*, Copenhagen, 1962, Vol. B, p. 385.
72. Golovnya, R.V., Zhuravleva, I.L. and Kharatyan, A., *J. Chromatog.*, 1969, **44**, 262.
73. Smith, T.A., *Food Chem.*, 1981, **6**, 169.
74. Hemme, D. *et al.*, *Sciences Aliment.*, 1982, **2**, 113.
75. Siminart, P. and Mayaudon, J. *Neth. Milk Dairy J.*, 1956, **10**, 156.
76. Dumont, J.P. and Adda, J., in *Progress in Flavour Research*, Applied Science Publishers, London, 1978.
77. Nakae, T. and Elliott, J.A., *J. Dairy Sci.*, 1965, **48**, 293.
78. Morgan, M.E., *Biotechnol. Bioengng.*, 1976, **18**, 953.
79. Lin, S.S., *J. Agric. Food Chem.*, 1976, **24**, 1252.
80. McGugan, W.A., *J. Agric. Food Chem.*, 1975, **23**, 1047.

19 Savoury flavours – an overview

D.G. LAND

Abstract

Savoury flavour is defined as far as possible and discussed with examples of savoury foods. It is characterized as much by its complexity of taste, odour and trigeminally mediated attributes as by its combination of stimulation by mouth with lack of sweetness. The major sources of savoury flavour, originating from various food components are discussed in outline but not in terms of detailed chemistry. Factors that influence release of savoury flavour stimulating substances are then discussed, while attention is drawn to the lack of research resource in these areas, and the consequential empirical nature of savoury flavouring.

19.1 Introduction

The term 'savoury' covers a very wide range of perceived flavour experiences that must first be defined. The *Concise Oxford Dictionary* gives several uses of the word savoury, the most food/drink/oral (pharmaceutical context)-orientated of which is 'a flavour stimulating to taste, in contradistinction to sweet'. This all-embracing but somewhat negative definition could be interpreted as any flavour acceptable by mouth that is not sweet and is so wide that it has limited operational value.

An alternative approach is by induction from the components of flavour, defined by the International Standards Organisation as a:

> Complex combination of the olfactory, gustatory and trigeminal sensations perceived during tasting; it may be influenced by tactile, thermal, painful and/or kinaesthetic effect. (*ISO/DIS 5492, 1990*)

Note that this arises specifically from stimuli in the mouth. The odour element, stimulated retronasally from food within the closed mouth, is generally recognized as being predominant in most flavours, as demonstrated by 'food has no taste' when one is suffering from a common cold. This is almost entirely the result of nasopharyngeal congestion, which occludes nasal olfactory and trigeminal receptors but leaves taste and oral trigeminal receptors almost unaffected. The absence of these key notes

Table 19.1 Major classes of odour description

Linnaeus 1752	Zwaardemaker 1895	Crocker and Henderson 1927	Amoore 1962
Aromatic	Aromatic	–	–
–	Ethereal	–	Ethereal
Fragrant	Fragrant (including floral)	Fragrant	–
–		–	Floral
–	–	–	Minty
–	–	–	Camphor
Ambrosial	Ambrosial	–	Musky
–	Empyreumatic	Burnt	–
–	–	–	Pungent
Alliaceous	Alliaceous	–	–
Hircine	Hircine	Capryllic	–
–	–	Acid	–
Foul	Repulsive	–	Putrid
Nauseating	Foetid (Nauseous)	–	–

almost destroys our ability to recognize the complex perceptual 'patterns' of sensation that one has learnt to label as particular flavours.

19.1.1 Odour

However, few of the classical major elements of odour, for example, from the systems of Linnaeus and Zwaardemaker or more recently from Crocker and Henderson or Amoore [1] (Table 19.1) fall clearly into the acceptable by mouth but not sweet interpretation. The only classes that could be imagined to be acceptable by mouth but not sweet are alliacious (broadly onion-like), pungent, empyreumatic (broadly burnt, smoky), hircine/capryllic (goat-like) and acid.

19.1.2 Taste

Furthermore, of the four classical 'primary' tastes, only sweet can be excluded, since salt, sour and bitter are not sweet, and if the more recent claim [2] that umami is a primary taste is accepted, it certainly would be included in savoury flavour.

Thus of the two major sensory contributors to savoury flavour, there appears to be more emphasis on taste than on odour, although admittedly this is based on what are far from perfect classifications from many years of human experience of trying to relate adequately sensations in response to food and drink; they cannot be lightly dismissed.

19.1.3 Trigeminal sensations

It is also clear that many of the sensations that are not mediated via the olfactory or gustatory nerves but via the trigeminal nerves are also much more associated with savoury than with sweet sensations. These include hotness (e.g. of curry or chillies), cooling (e.g. of butter), astringency (e.g. of persimmons or many aperitifs) and pungency (e.g. of pickles). Trigeminally mediated (i.e. neither olfactory nor gustatory) sensations associated with sweet flavour are relatively rare and it is difficult to think of a clear example that is not associated with deteriorative changes. Again one is forced to conclude that savoury flavour is an almost global term that excludes only sweet tastes and a wider range of the mouth-compatible odours such as fragrant, fruity and minty.

19.1.4 Complexity

However, practical experience of usage suggests that numerically most different flavour sensations are not savoury. If this impression is correct, it leads to the further conclusion that a dimension other than that of component notes is involved. This dimension might be 'complexity', that whereas most sweet flavours are perceptually relatively simple, savoury flavours are perceptually more complex, involving especially more taste and trigeminally mediated notes. If this supposition is correct it implies that research to improve understanding of savoury flavours will be more difficult. It will take longer than research into sweet and other flavours because it must involve both volatile and non- or less-volatile chemical constituents and their sensory assessment, whereas most sweet flavours can probably be explained to a large extent by the composition and odour of the volatile fraction. For many years now the volatile fraction has been investigated readily by the very powerful separation and identification techniques of gas chromatography coupled with mass spectrometry, and more recently coupled with Fourier-transform infrared spectrometry.

These techniques have provided much data, which have improved understanding of volatile flavours. Nevertheless there are few sweet flavour extracts or flavourings for which the chemical composition is not known in great detail, at least by the major flavour houses; this is not the same as fully understanding what exactly is responsible for the sensory attributes when present in the intended medium. It represents, however, a greater level of knowledge than exists for savoury flavours, where analytical techniques for the separation, identification and quantitative analysis of the nonvolatile elements have not yet reached the same power as those for relatively volatile odour stimulating components. Many savoury flavours are probably still the result of empirical formulation and testing by trial and error.

19.2 Savoury foods

Although akin to stating the obvious, it is relevant to list briefly what is understood by the term savoury foods, if only to provide a basis for discussion, and to indicate what other chapters have already covered and what will be covered in this chapter.

The major savoury foods are those based on protein, whether of animal or plant origin. Of these, those based on flesh, fish and fowl are dominant in most parts of the world, although in the East and in many developing countries, foods based on legume (e.g. soya) protein and fermentation are at least equally important. Dairy products (e.g. butter and cheese) and eggs are also important savoury foods, although interestingly milk is not, probably because of the high content of lactose (milk sugar), which is not present in most dairy products. Soured milks and products such as yogurt are not generally considered as savoury in the West, although they are in Eastern Europe and the Middle East. Butter is not a protein food but mainly fat, and most fats are used in a context of savoury food, although not always; most are now of plant origin.

Salad and cooked vegetables are other low-protein savoury foods, as are the cereal-based foods, although the latter are also extensively used in the West as the base for sweet dishes. The other major seed-based food ingredients, nuts, can also be used in both contexts, and fruits are almost invariably regarded as sweet rather than savoury.

19.3 Sources of savoury flavour

Although some sources, for example added salt, may be seen as 'only' taste stimuli, most are very complex, involving taste, odour by mouth, and trigeminally mediated sensations. Most can occur during processing, and a selection is outlined below.

19.3.1 Sodium chloride

The outstanding contributor to savoury flavour, known since antiquity, is common salt. It is present to some level in all foods and drinks, and added more to many savoury foods for several reasons such as preservation at ambient temperature, as a flavour improver in cooking and on the plate. Its use is so widespread that it is largely taken for granted until the cook forgets to add it to the water in which potatoes are boiled, with disastrous results to the flavour and, interestingly, which is not really retrieved by subsequent addition of salt after cooking. This 'secondary effect' involved in flavour release is also very important in producing the complex of taste and odour sensations that are recognized as potato flavour; it is not merely the absence of the sensation of saltiness. Little is known about the mechanisms of such phenomena, and this is an area where it is very difficult and time-consuming to make progress. It is nevertheless a major challenge that awaits a champion.

19.3.2 Umami

This is the name given by the Japanese to the taste sensation produced by certain substances present in sea tangle, traditionally used to enhance the flavour of meats, fish, cheese, certain vegetables and mushrooms. It was shown to be produced by monosodium glutamate (MSG, the salt of a very common protein amino-acid) in 1909, and much more recently associated also with the presence of 5′ nucleosides (inosine monophosphate, IMP, and guanosine monophosphate, GMP). They are all now widely used as additives that enhance mainly savoury flavour, although MSG has no enhancing effect on any of the four basic tastes [3]. As intermediate metabolites in man, their physiological effects are very wide-ranging. These substances are also formed *post mortem* (in plant and animal) during deterioration and on fermentation, and are therefore innate components of most foods, particularly those above regarded as savoury.

19.3.3 Protein hydrolysates

These have been used for many years as meaty flavours, whether derived from meat (e.g. Bovril) or from yeast (e.g. Marmite), and all will contain some of the umami substances. Strictly speaking they are not just protein hydrolysates, for they are the result of thermal and/or enzymic breakdown of complex materials, not just of protein. They contain amino acids and peptides, the hydrolysis products of protein, many of which have pronounced tastes. Also, important for savoury flavours, they provide starting materials for Maillard reactions, which are major contributors to the odour and taste, flavour and colour – usually as a result of complex reactions with sugars.

19.3.4 Maillard reactions

These are a very complex series of reactions between amino acids or peptides and a range of sugars present, some only in trace amounts, in foods, and particularly the pentoses (which are formed by breakdown of nucleic acids, e.g. by chemical or enzymatic hydrolysis). Many of the numerous products can contribute to taste, odour and to appearance (e.g. in the browning that occurs on cooking) and many have now been identified. However, many complex reactions can occur in real foods, and there are many ill-defined variables. Despite the hundreds of studies carried out on model systems and on real foods, together with the enormous wealth of knowledge of the chemical reactions that can and do occur, the understanding of the kinetics of formation, the dynamic contribution to release and the interactions that influence the sensations that are produced on eating food containing the products of Maillard reactions is still a long way off. These include all heat-processed or cooked foods, which covers most savoury foods.

19.3.5 Fat reactions

As lipids are an integral part of all living systems, all foods contain some fats. The complex and polyunsaturated structural lipids of the flesh foods can be readily oxidized or altered when some structural integrity of biological tissue is lost on preparing it as food. These changes are often a combination of initial enzymic changes and subsequent chemical changes; they occur when the *in vivo* compartmentalized separation of enzymes and structural lipid breaks down during either *post-mortem* storage or processing damage. They can either contribute to the expected, essential normal flavour that develops on cooking or, in certain cases, can occur to an extent that is perceived as taint or off-flavour. Such changes are not uncommon in savoury foods.

19.4 Release of flavour

It is essential to appreciate that, whatever the content in food of any flavour substances determined analytically, what is perceived must be determined by the rate of release from the food and the consequent concentration reaching the various receptors in the mouth and nose. These rates of release will be greatly influenced by the rate at which new surfaces of food are exposed on chewing, and the composition of the fluid (e.g. juices from the food and saliva) in which the flavour-stimulating substances are dissolving and diffusing. In the case of volatile substances, there is also the rate of volatilization into the air in the mouth, dilution with air moving back and forth in the pharynx and rate at which it reaches the olfactory epithelia high up in the nose. All of these factors will influence the time course of perception as the stimulus concentration increases to a maximum and then decays.

It has been known for a long time that dissolved solutes, such as the ions and mucopolysaccharides in the saliva, ions (e.g. salt) and particularly proteins in the chewed mixture of food, can have a marked influence on the vapour concentration and therefore perceived intensity of any single savoury flavour stimulus [4–6]. Examples are shown of the influence of the physical composition and structure on the equilibrated vapour pressures of three common savoury odour-contributing substances in simple model systems (Figures 19.1–19.3).

Figure 19.1 shows that dimethyl sulphide obeys Henry's Law at perceived flavour concentrations in water, oil and in a two-phase water-oil emulsifier system [6]. However, when that system was emulsified, the vapour

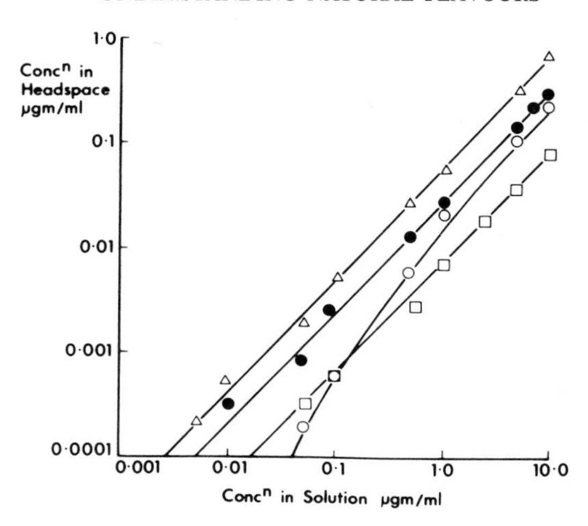

Figure 19.1 The influence of the liquid phase on the headspace concentration of dimethylsulphide: △ = water; □ = oil; ○ = emulsion; ● = water + oil; unemulsified. From [6].

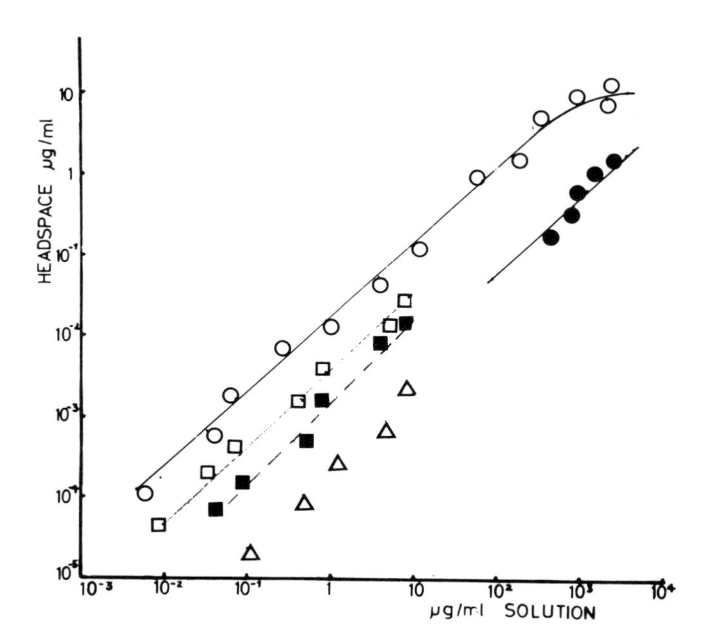

Figure 19.2 The influence of the liquid phase on the headspace concentration of allyl isothiocyanate: ○ = water; ● = mustard paste; □ = emulsion; ■ = unemulsified; △ = oil. From [6].

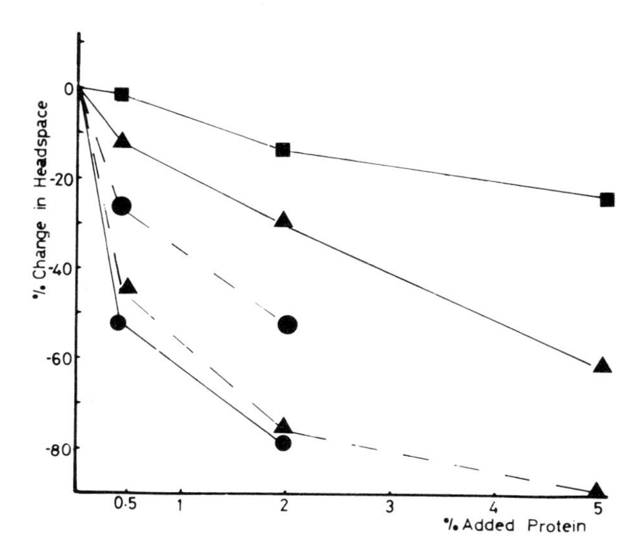

Figure 19.3 The influence of native proteins on the headspace concentrations of allyl isothiocyanate (solid lines) and diacetyl (dashed lines) in water: ▲ = egg white; ● = BSA; ■ = casein.

pressure dropped remarkably at the lower concentrations from a readily perceptible level to below threshold. Most foods are like emulsions, with a very large interfacial area. A similar but opposite effect occurs with allyl isothiocyanate [6], a substance important in mustard and horseradish sauce, while in mustard paste a much higher concentration is required to give the same vapour pressure and odour intensity, as in water (Figure 19.2).

Many nonvolatile solutes present in foods can markedly influence the vapour pressure. Figure 19.3 shows the large and differing effects when reducing the vapour pressure of allyl isothiocyanate and diacetyl caused by the presence of even 0.5% of certain proteins [7]. These ligand-binding effects are now well known and influence the maximum intensity reached and the persistence of the time–concentration curve [8,9], thus influencing the time pattern of what is perceived. Even studies of retronasal presentation of the odours produced from the mouth are very few in number. Such studies are often difficult to conduct at realistic levels, although techniques are available [10–12].

19.5 Conclusions

Savoury flavour appears to be characterized by a complexity of taste, odour and trigeminal stimuli, which is greater than for the nonsavoury flavours. This has resulted in a lower level of understanding of the underlying causes than for sweet flavours, although several identified areas for research are well known. These require much more research resource if understanding is to be increased. However, given that simple, single-flavouring substances are very rare in savoury flavours and that most foods are actually eaten as mixtures of different foods, it is not surprising that flavouring, particularly for savoury flavours is still largely an empirical, if very successful, art.

References

1. Harper, R., Bate Smith, E.C. and Land, D.G., *Odour Description and Odour Classification*, Churchill Livingstone, London, 1968, pp. 16–35.
2. Kurihara, K., Recent progress in the taste receptor mechanism, in *Umami: A Basic Taste*, (eds Y. Kawamura and M.R. Kare), Marcel Dekker, New York, 1986, pp. 35–37.
3. Yamaguchi, S. and Kimizuka, A., in *Glutamic acid: Advances in Biochemistry and Physiology*, (eds L.J. Filer, S. Garattini, M.R. Kare, W.A. Reynolds and R.J. Wurtman), Raven Press, New York, 1979, pp. 35–54.
4. Maier, H.G. Volatile flavouring substances in foodstuffs. *Angew. Chem. Internat. Edit.*, 1970, **9**, 917.
5. Solms, J., Osman-Ismail, F. and Beyeler, M. The interaction of volatiles with food components. *Canad. Inst. Food Sci. Technol. J.*, 1973, **6**, A10.
6. Land, D.G. Some factors influencing the perception of flavour-contributing substances in food, in *Progress in Flavour Research*, (eds D.G. Land and H.E. Nursten), Applied Science, London, 1979, p. 53–66.
7. Land, D.G. and Reynolds, J. The influence of food components on the volatility of diacetyl, in *Flavour '81*, (ed. P. Schreier), de Gruyter, Berlin, 1981, pp. 701–705.
8. Dumont, J.P., Flavour–protein interactions: A key to aroma persistance, in *Flavour Science and Technology*, (eds M. Martens, G.A. Dalen and H. Russwurm, Jr), John Wiley, Chichester, 1987, pp. 143–148.
9. Overbosch, P., Flavour release and perception, in *Flavour Science and Technology*, (eds M. Martens, G.A. Dalen and H. Russwurm, Jr), John Wiley, Chichester, 1987, pp. 291–300.
10. Lee, W.E. II, A suggested instrumental technique for studying dynamic flavor release from food products. *J. Food Sci.*, 1986, **51**, 249–250.
11. Marie, S., Land, D.G. and Booth, D.A. Comparison of flavour perception by sniff and by mouth, in *Flavour Science and Technology*, (eds M. Martens, G.A. Dalen and H. Russwurm, Jr), John Wiley, Chichester, 1987, pp. 301–8.
12. Voirol, E. and Daget, N. Nasal and retronasal perception of a meat aroma, in *Flavour Science and Technology*, (eds M. Martens, G.A. Dalen and H. Russwurm, Jr), John Wiley, Chichester, 1987, pp. 309–316.

Index